Modeling and Design Photonics by Examples Using MATLAB®

Series Editor

R Barry Johnson a Senior Research Professor at Alabama A&M University, has been involved for over 50 years in lens design, optical systems design, electro-optical systems engineering, and photonics. He has been a faculty member at three academic institutions engaged in optics education and research, employed by a number of companies, and provided consulting services.

Dr Johnson is an IOP Fellow, SPIE Fellow and Life Member, OSA Fellow, and was the 1987 President of SPIE. He serves on the editorial board of Infrared Physics & Technology and Advances in Optical Technologies. Dr Johnson has been awarded many patents, has published numerous papers and several books and book chapters, and was awarded the 2012 OSA/SPIE Joseph W Goodman Book Writing Award for Lens Design Fundamentals, Second Edition. He is a perennial co-chair of the annual SPIE Current Developments in Lens Design and Optical Engineering Conference.

Foreword

Until the 1960s, the field of optics was primarily concentrated in the classical areas of photography, cameras, binoculars, telescopes, spectrometers, colorimeters, radiometers, etc. In the late 1960s, optics began to blossom with the advent of new types of infrared detectors, liquid crystal displays (LCD), light emitting diodes (LED), charge coupled devices (CCD), lasers, holography, fiber optics, new optical materials, advances in optical and mechanical fabrication, new optical design programs, and many more technologies. With the development of the LED, LCD, CCD and other electo-optical devices, the term ‘photonics’ came into vogue in the 1980s to describe the science of using light in development of new technologies and the performance of a myriad of applications. Today, optics and photonics are truly pervasive throughout society and new technologies are continuing to emerge. The objective of this series is to provide students, researchers, and those who enjoy self-teaching with a wide-ranging collection of books that each focus on a relevant topic in technologies and application of optics and photonics. These books will provide knowledge to prepare the reader to be better able to participate in these exciting areas now and in the future. The title of this series is Emerging Technologies in Optics and Photonics where ‘emerging’ is taken to mean ‘coming into existence,’ ‘coming into maturity,’ and ‘coming into prominence.’ IOP Publishing and I hope that you find this Series of significant value to you and your career.

Modeling and Design Photonics by Examples Using MATLAB®

Dan T Nguyen
Corning Research and Development Corporation, Corning, NY 14831, USA

IOP Publishing, Bristol, UK

ISBN 978-0-7503-2272-0 (ebook)
ISBN 978-0-7503-2270-6 (print)
ISBN 978-0-7503-2273-7 (myPrint)
ISBN 978-0-7503-2271-3 (mobi)

DOI 10.1088/978-0-7503-2272-0

Version: 20210701

IOP ebooks

British Library Cataloguing-in-Publication Data: A catalogue record for this book is available from the British Library.

Published by IOP Publishing, wholly owned by The Institute of Physics, London

IOP Publishing, Temple Circus, Temple Way, Bristol, BS1 6HG, UK

US Office: IOP Publishing, Inc., 190 North Independence Mall West, Suite 601, Philadelphia, PA 19106, USA

This book is dedicated to my parents.

Contents

Preface

Over the past few decades, photonics has become a major field of science and technology. Now covering a broad range of physics from silicon waveguides and fiber lasers to photonics band-gap structures, and many other approaches of light manipulation, it is evident that this discipline is undergoing a period of extremely rapid progress.

As many physicists have entered this exciting field, they have driven the development of photonics systems through their unique skill sets. For example, theoretical physicists have played important roles not only for exploring new concepts and new photonics systems, but also for modeling and designing these new systems. These tasks require robust knowledge of fundamental theories of physics, ability to manipulate equations, development of new models for various systems, and the effective computation of these models. Nowadays, although most computational modeling relies on supercomputing and sophisticated software, a comprehensive understanding of physics remains critical for guiding research in novel, exploratory systems.

Back in 2000, when I worked as a theoretical physicist, I was offered the opportunity to model and design fiber amplifiers for applications in optical telecommunication. At first, I didn't think I could take the job; not only did I not have any experience in the field, but I also didn't think *any* theorist could effectively develop real-world devices. When Professor Peyghambarian, the head of the University of Arizona's photonics division in the Optical Sciences Center, invited several of us to a project meeting, I was surprised to see *actual* theoretical physicists present, such as Professor N Bloembergen, a Nobel laureate in physics, and Professor S Koch, a leading theorist in condensed matter physics, and others. The event changed my thinking completely about working as a modeler in photonics.

From 2001 to 2016, I have experienced both highs and lows in the field of photonics, both in the Optical Science Center and at NP Photonics, a research company based in Tucson, Arizona. However, as I delved deeper into my work, the more I enjoyed the process of modeling and design, especially in photonics. As a theorist it is incredible to see that many observed effects of light manipulation can be predicted precisely from solutions of Maxwell's equations established more than 160 years ago. Theoretical models can play a critical role in designing and optimizing a photonics system, and I truly appreciate the role of physical theories in the creation of groundbreaking technology and discovery of novel systems.

This book was a product of my experience of modeling and design of photonics after nearly 20 years working in academy and industry. It was initiated by Ms Ashley Gasque from IOP Publishing (Institute of Physics, UK), who had patiently convinced me to write this book. I know very well that there are a number of excellent books on modeling and computational photonics. Several also include computing programs that are very useful for students, scientists, and engineers. However, because photonics is a very broad area of science and technology, and it is relatively new to many, these books have mostly focused on the introductory and foundational parts of the field with some fundamental examples. Furthermore, most have been written mostly in an academic language, and may not adequately address many real-world problems that industrial scientists and engineers have to deal with in their daily work. This book is an effort to bridge the gap between the academic and industrial worlds of photonics. I have been quite lucky to have had opportunities to work in multiple research projects sponsored by leading agencies such as DAPRA, NSF, NASA, NIST, NRL, AFL, ONR etc and also in research and development of fiber lasers, amplifiers and other productions. This book aims to provide basics but robust explanations of the fundamental physics, mathematical models of several problems in the field of photonics, and the MATLAB® programs for studying examples. I do think errors and inaccuracies are unavoidable while writing the book, and would be happy to receive suggestions and critical comments.

In writing this book I would like to express my deep gratitude to Professors Nguyen Ba An, and Hoang Xuan Nguyen, my PhD advisors, Professor Eiichi Hanamura, my postdoctoral supervisor in Tokyo University who later offered me a chance to work with him in University of Arizona, Professors Truong Nguyen Tran, F Bassani, V M Agranovich, and finally Professor Peyghambarian for offered me opportunities in the journey.

Additionally, I would like to thank my various other colleagues: Dr Arturo Person-Chavez, the CTO of NP Photonics and a long-time colleague during both good and bad times, Dr Kam Ng for reading parts of the manuscript, Dr Bin Zhang for editing the book cover, Corning Corporation Fellows Daniel Nolan, Nick Borrelli for their support, and many others.

Finally, last but not least. This book is my present to my parents and my family; without their support this work would not be possible.

Corning, New York
February 2021

Author biography

Dan T Nguyen

Senior Optical Scientist, Corning Research and Development Corporation, Corning, New York (USA) and Adjunct Professor, College of Optical Sciences, The University of Arizona, Tucson, AZ, USA.

Dan T Nguyen received his PhD in 1993 from Institute of Physics, National Academy of Sciences, Hanoi, Vietnam, on theory of nonlinear optics in semiconductors. Since then, he had worked in several international academic institutions, and in the optics/photonics industry: Visiting Researcher of the International Center for Theoretical Physics (ICTP), Trieste, Italy (1992, 1994, 1996, 1997) where he was elected as a Regular Associate Member (five-year term from 1995 to 2000); a DAAD Researcher at Institut fur Festkorpertheorie und Theoretische Optik, Friedrich-Schiller-University, Jena, Germany (8/1994 to 4/1995); Nishina Memorial Postdoctoral at Department of Applied Physics, Tokyo University, Tokyo, Japan (4/1995–6/1996), Chair of Theoretical Physics, Hue University, Hue, Vietnam (from 7/1996 to 7/1998). From 7/1998 to 1/2017 he had been with College of Optical Sciences, the University of Arizona, Tucson, Arizona (USA) from a Research Associate to Associate Research Professor. During that period of time, he had also worked as Principal Modeling Scientist at NP Photonics Inc., Tucson, Arizona (USA). He worked in multiple research projects sponsored by leading agencies such as DAPRA, NSF, NASA, NIST, Naval Research Laboratories (NRL), Air Force Lab (AFL) and Office of Naval Research (ONR), and The Exploratory Research for Advanced Technology (ERATO) by Japan Science and Technology (JST) Agency. He, himself, was a Principal Investigator for a number of research projects sponsored by NRL, AFL and ONR.

Dr Nguyen joined Corning Research and Development Corporation, Corning, NY in 2/2017 as Senior Optical Scientist. Since then, he has also been Adjunct Professor of College of Optical Sciences, the University of Arizona. Dr Nguyen is a member of SPIE. He has authored and co-authored more than 75 research publications, two book chapters and five US patents and several pending patents. His research interests include photonics computation, modeling and design of optics and photonics, fiber designs, quantum communication, quantum integrated photonics, optical computing, fiber lasers and fiber amplifiers, quantum computing, VCSELs, GAWBS noise, laser cooling, micro-cavity bio-sensing, optical limiting, nonlinear optics, plasmonics etc.

IOP Publishing

Modeling and Design Photonics by Examples Using MATLAB®

Dan T Nguyen

Chapter 1

Introduction

1.1 Overview of the book

The term photonics was coined in the late 1960s, first as a research field that performs similar functions of electronics by photons (light), such as telecommunications, information processing, and later extended to describe a much broader area of science and technology involving generation, manipulation, detection of light and related applications.

After the invention of the laser in 1960, photonics as a field started with researches and applications of semiconductor lasers, and then optical fibers for transmitting information, and the rare-earth doped fiber amplifiers and fiber lasers. Because of that, nowadays photonics covers a huge range of science and technology applications, including fiber amplifiers and lasers, photonics chips, sensing, medical diagnostics and therapy, optical communication, quantum communication and quantum information etc.

Although photonics plays very important roles in enabling numerous technologies, it is still a relative new area of science with which many scientists and engineers are not familiar. The main reason for that is the term had been first used and then became popular during the research and development of semiconductor lasers, optical fiber technology and Er-doped fiber amplifiers, all of which later became the foundations of the telecommunication revolution. Therefore, the term 'photonics' is very popular for scientists and engineers who work in those fields but it may not be known or even heard by others, although later the term 'photonics' covered a much broader area of science and technology.

It is worthwhile to note that there are so many excellent textbooks [1–5], a huge number of research and review papers, and several journals are solely devoted to many different fields belonging to photonics. Because of the broadness and complexity of photonics sciences and technologies, it is our hope that this book provides readers, especially newcomers just those examples in modeling and designing of photonics systems that they could find useful in their first learning and working projects. To make it easier for beginners to understand the problems and to follow the MATLAB® programs, fundamental theories of physics and mathematical models are also provided for each set of problems covered in this book. However, it is important to stress that the theoretical presentations are not detailed and deep enough to convey many important features of the problems. That is because it would take a large volume to present such detailed theories which are already available in many excellent textbooks. It is my own experience that without a solid background of physics and mathematics from deeply understanding the theory, we may not be able to understand a model for a problem, much less the needs of building new models. However, sometimes we can easily get lost after studying a theory with so many mathematical details and complexity. Therefore, there is a real need for many to have good balancing between understanding the theory and the use of this knowledge to deal with the real problems that we face in our daily working tasks. Therefore, readers are always encouraged to find the comprehensive presentations of these topics in other sources, or at least in the above references. On the other hand, the theoretical presentations in this book can be considered as a bridge between mathematically rigid presentation of the theory and one that ends up with models of problems in practices. Following the theoretical presentations in each chapter is the key to understanding the physics and mathematical models of the problems in followed examples. Once one has understood the models, the MATLAB® programs for simulation become easy to understand.

Chapter 2 is devoted to modeling and simulation of one-dimensional (1D) periodic and quasi-periodic photonics crystal (PhC) structures [6]. The direct applications from the theory will be presented in several examples with MATLAB® programs. First, multi-layers of periodic dielectric structures are presented to illustrate the 1D PhC systems. Moreover, the general concept of the photonics band gap (PBG), which is a very important concept in physics is introduced theoretically. Second, the simplest model of mathematical equations for calculating optical properties of the PhCs based on transfer matrix method (TMM) is derived. Linear TMM formalism will be presented as an effective method for calculating optical spectra of several real systems. For instance, PBG of periodic dielectric layers (1D PhC) can be realized for reflection spectrum of mirrors, as shown in figure 1.1(a). Meanwhile, a laser cavity or an optical resonator can be built with the two mirrors, as shown in figure 1.1(b).

It can be seen that although the theoretical models of 1D PhC structures are quite simple, they can be used to simulate and design important applications such as Bragg reflectors, fiber gratings, optical resonators, laser cavities including distributed Bragg reflectors (DBRs) and distributed feedback (DFB) fiber lasers, mirrors with quasiperiodic multiple reflection windows.

It should be noted that both models and computing programs for simulation presented in this part can be easily found in literature. What can be considered benefits for readers are to show even very simple models can play important roles in practical applications; they are especially very effective for dealing with several important problems such as DBR and DFB lasers which are quite difficult to simulate.

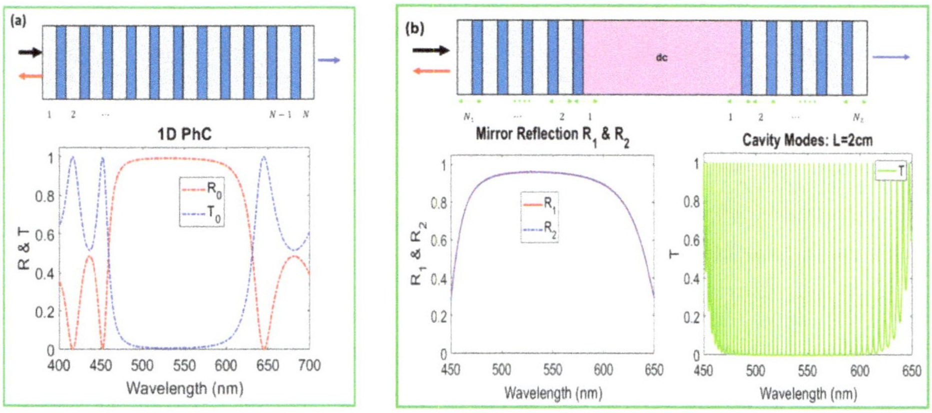

Figure 1.1. Upper: diagram of 1D PhC of N pairs of layers of two dielectric layers (a) and scheme of an optical cavity (b). Lower: calculated reflectivity (R) and transmissivity (T) of 1D PhC structures (a), and reflection of two mirrors (left) and resonant modes of the cavity (right).

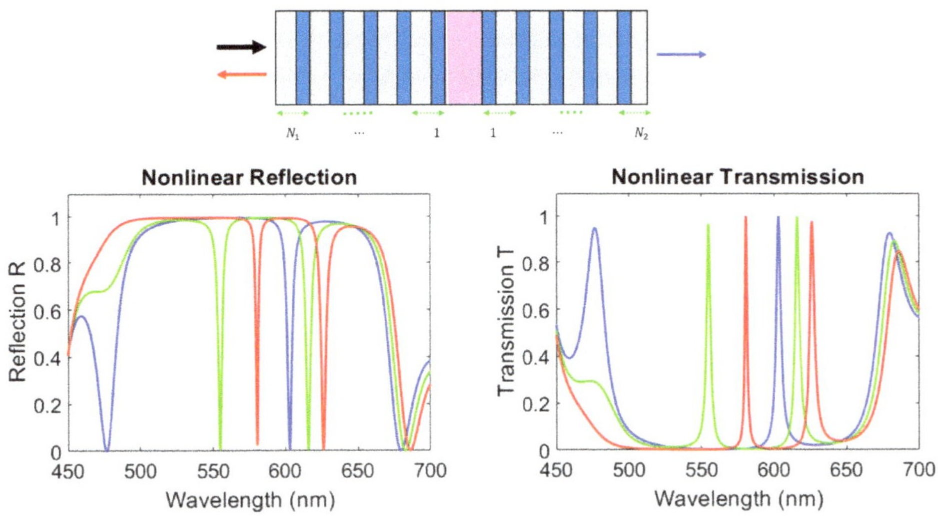

Figure 1.2. Nonlinear reflection R (left) and nonlinear transmission T (right) of a nonlinear defected 1D PhC (see example in chapter 2).

The last part of this chapter will be devoted to a nonlinear TMM formalism for calculations of nonlinear problems such as nonlinear spectra of defected cavities as shown in figure 1.2, all-optical switching based on optical bistability etc. Note that both modeling, simulation and experimental results of these problems are well known [7]. However, detailed modeling and simulation of the systems are still not available in literature. Therefore, MATLAB® programs that are presented in this part for numerically studying and analyzing several examples would be of real benefit for many.

It is worth noting that although the dielectric multilayer structures can be considered as simplest 1D PhC structures, they have been used for many important applications. After studying examples presented in chapter 2, readers can apply the linear and nonlinear TMM for different problems.

Chapter 3 is devoted to a new approach based on beam propagation method (BPM) for modeling fiber amplifiers, especially the scheme cladding-pumped fibers. Rare-earth-doped fiber lasers and fiber amplifiers have become key components of numerous technologies in communication, medical equipment, materials processing, and military applications etc. We want to stress again, that there are many good references on these topics, for example, references [4, 5] are just two of the excellent textbooks for readers who want to have comprehensive knowledge of the topics. It is well known that fiber lasers and fiber amplifiers can be divided into two categories based on the optical pumping approach: single-mode (SM) or multimode (MM) pump schemes. In the SM pump scheme (SMP), SM laser sources are usually used to couple the pump power to the fiber core, which is typically in the range of a few μm to few tens of μm for single-mode fibers (SMF), as shown in figure 1.3(a). As a result, there are limitations in pump power, and a lot of effort for coupling in an SMP scheme. To overcome these problems, a different pumping method has been developed in which pump power from high-power MM laser sources is coupled to a large cladding of double-clad fiber (DCF) surrounding rare-earth-doped core. The pump power is gradually coupled to the core and is absorbed by rare-earth doped materials. The method therefore is called an MM cladding-pumped or MMP scheme since MM laser sources are used, as shown in figure 1.3(b).

Modeling SMP fiber amplifiers are presented in detail in literature, especially in [4, 5], and commercial software is also available. The problem though is not simple but in general all difficulties are resolved numerically without any challenges. The main reason for that is the physics of the problem is very clear and straightforward. A single-mode pumping beam propagates in a single-mode fiber core, where the processes of pump absorption and signal amplification occur, can be very well modeled by combining rate equations and propagation of pump and signal beam powers. Both pump and signal beams are well confined in the core, therefore they can be described by propagation equations of SM beams along the fiber length. Most often, the fiber core is nearly perfectly circular and the 3D problem can be simplified as a 2D one with one transverse to and one parallel with propagation direction. Further simplifications can be made if the core is small, which is typical for SM fibers, resulting in power distribution inside the core being nearly uniform.

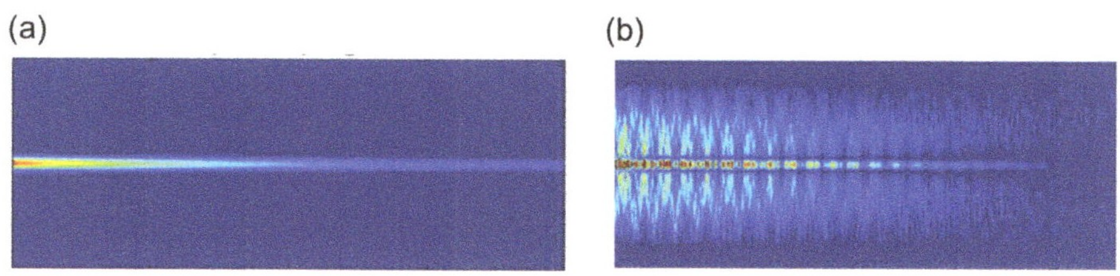

Figure 1.3. Pump beam propagates in SMP fiber amplifier (a) and in MM cladding pump fiber amplifier (b).

In such conditions, the 2D problem can be considered approximately as 1D propagation and the beam confinements can be characterized by an overlap factor for each beam. Meanwhile, modeling MMP rare-earth doped fibers has been very challenging as we can imagine from the physics of the problem. Instead of propagating in the core as in the case of SMP, now an MM pumping beam is launched in large cladding, propagates in the clad and gradually couples to a small rare-earth doped core. It is clear that the effects of cladding shapes and fiber structures can strongly impact the pump absorption and the amplifier performance. The common simplifications of 1D propagation used in SMP scheme is not acceptable since power distribution of pump is very non-uniform. The problem of propagation of MM beam in a cladding that supports hundreds or thousands of modes is very challenging. Because of that, until now the problem of modeling cladding-pumped fiber amplifiers is still not very well described in literature, except for a few publications including our works [8–12]. In chapter 3, we will try to bridge the gap providing details of modeling MMP scheme fiber amplifiers, especially using examples with MATLAB® programs.

The modeling method presented in this chapter can be applied for complicated structures like multicore fiber amplifiers, as shown in figure 1.4. This fiber has 19 cores doped with Yb, and is used for making an optical Ising machine based on fiber lasers [11, 12].

Chapter 4 presents the problem of modeling and simulation of ultrafast mode-locking fiber lasers. The modeling examples with MATLAB® programs in this chapter aim to provide an effective method of simulation of mode-locked (ML) lasers, especially ML fiber lasers. It is worth noting that mode-locked lasers have been investigated thoroughly and comprehensively for more than 30 years, and excellent reviews of the field can be found in [13, 14]. However, details of theoretical modeling and simulation of all the mode-locked laser systems are still lacking, at least in literature. To the best of our knowledge there are few publications on details of modeling works of mode-locked lasers, although all components and configurations of the lasers can be characterized very well experimentally and theoretically. This does not mean that no one knows how to simulate the lasers. It just means anyone who wants to learn and work with modeling and simulation of ML lasers would face challenges that they can avoid when studying other topics. This chapter aims to help them, especially for readers who want to work with lasers, and even want to design the lasers. The materials in this book could be helpful for anyone who has little or even no prior knowledge of the topics who can learn very quickly both the physics and mathematics, especially they can start to design a system based on what they just learned. However, if they really want to understand better, they must study the topics in more comprehensive review or textbook, not least from the above references.

Figure 1.5 shows a general ML ring fiber laser and the evolution of the laser in the cavity as correlated with the changes of saturable absorber.

In order to provide a model of the mode-locked fiber laser in examples, short but quite detailed descriptions for the key elements of the fiber lasers including nonlinearity, dispersion, saturable absorption, gains and losses in the laser cavity

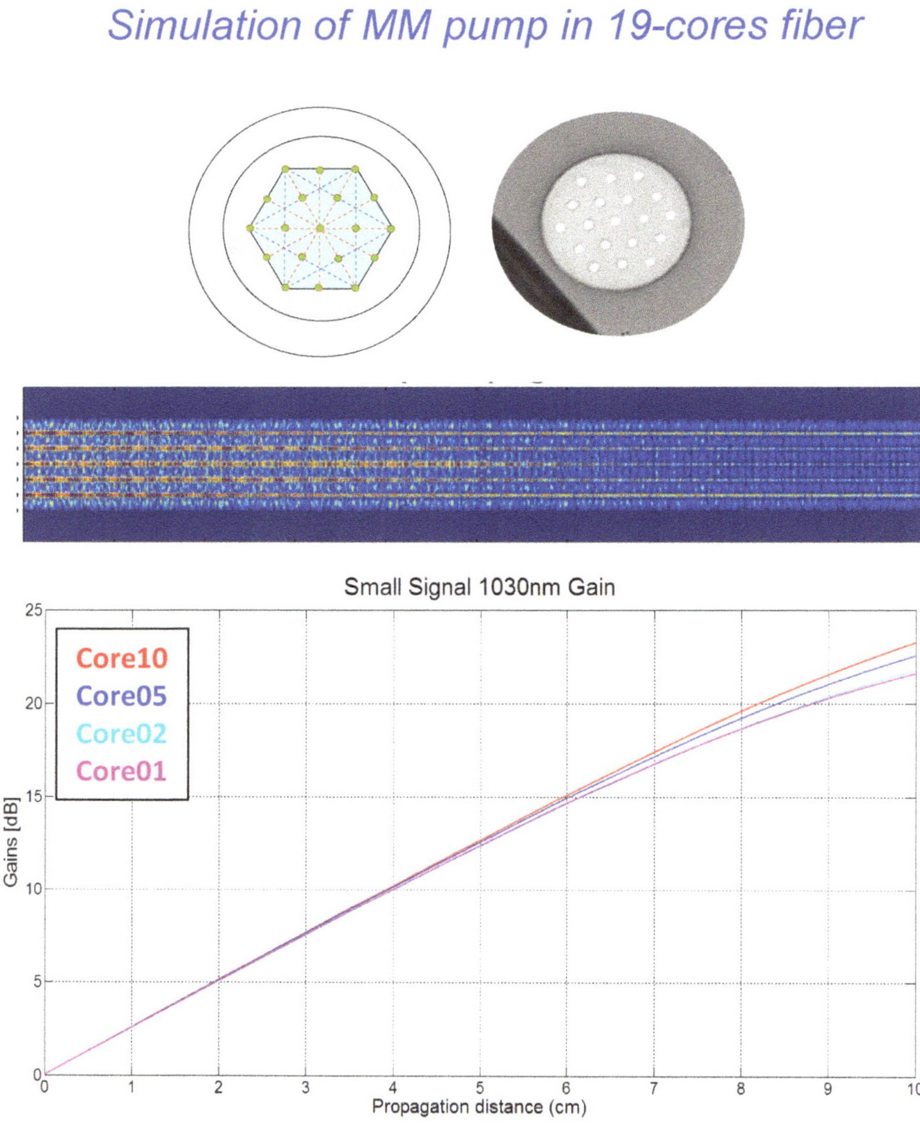

Figure 1.4. Simulation of 19 cores fiber amplifiers. Top: design and image of the fiber with 19 cores doped with Yb. Middle: Pumping MM beam propagate along the fiber. Bottom: gain of 1030 nm-signal in different cores. Reprinted by permission from Macmillan Publishers Ltd: [12], copyright (2019) (CCBY 4.0).

are provided. One of the most important parts for the task of modeling the ML lasers is how to solve the propagation of the laser pulses in the ML laser cavity. The equation, in this case is the nonlinear Schrödinger equation (NLSE) and generalized NLSE (GNLSE) [15]. There are different numerical methods for solving NLSE and GNLSE, however, the RK4IP has been proved to be the best in numerical errors and efficient algorithm as compared to other methods [16], and that becomes very important for numerical solving light that circulates multiple rounds with gain in

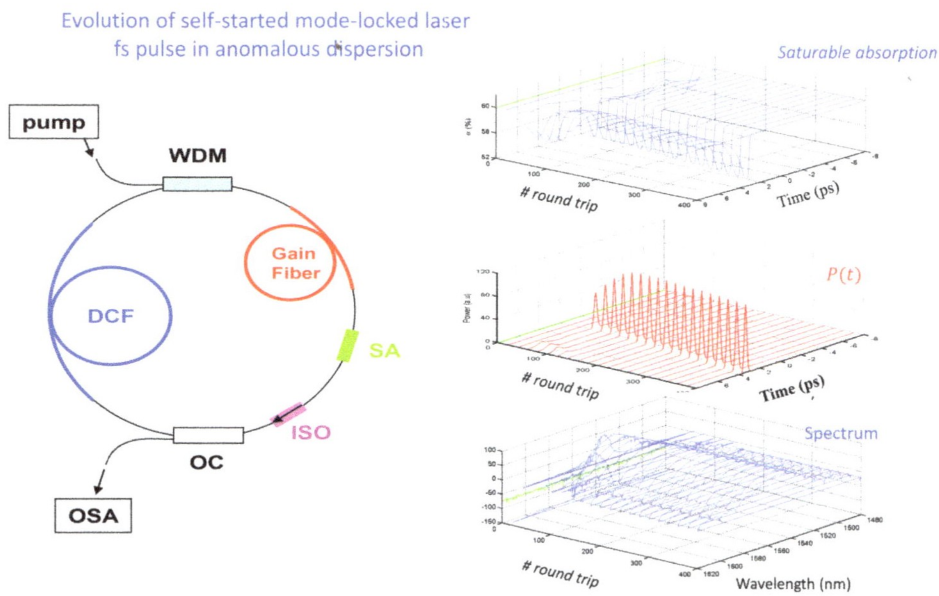

Figure 1.5. A schematic of an ML fiber ring laser (left), and the evolution of laser with different number of round trips inside the cavity (right).

highly complicated structures including nonlinear absorption by SA, nonlinear index, dispersion etc. Therefore, we will describe the modeling method based on RK4IP for solving NLSE of laser pulse propagation in the cavity. After understanding all these components and their mathematical descriptions as well as the numerical method of solving the NLSE we then study some examples using MATLAB® programs.

Chapter 5 is devoted to modeling and simulating the system of chirped-pulse amplification (CPA), which is a technique for generating high energy, ultrashort pulse lasers. High power, ultrashort pulse lasers have been widely used in numerous applications from studying fundamental physics principles to important applications including cancer treatments, laser eye surgery, laser materials processing etc [17–19]. CPA systems have been developed for decades using different configurations, but a typical CPA system consists of four main stages: (i) short pulse laser source (usually in picosecond or femtosecond regimes), (ii) pulse stretcher, (iii) power amplifier and (iv) pulse compressor. It has been demonstrated that the CPA systems that are built on bulk crystals can deliver the highest energy pulses. However, these systems are usually complex, bulky, and difficult for alignment and maintenance. As a result, it is usually difficult and expensive to use these systems in many practical applications. Therefore, developments of CPA systems that are compact and highly integrated have attracted a lot of effort. Figure 1.6 is a schematic of a fiber-based CPA system.

Although the CPA technology has been developed for decades and is widely used in many applications, detailed descriptions of modeling and simulation of the CPA systems are still not available, at least in literature. Again, it is a similar situation to that in modeling and simulation of the ML laser presented in the previous chapter. By providing basic theory and simple models of several examples with MATLAB®

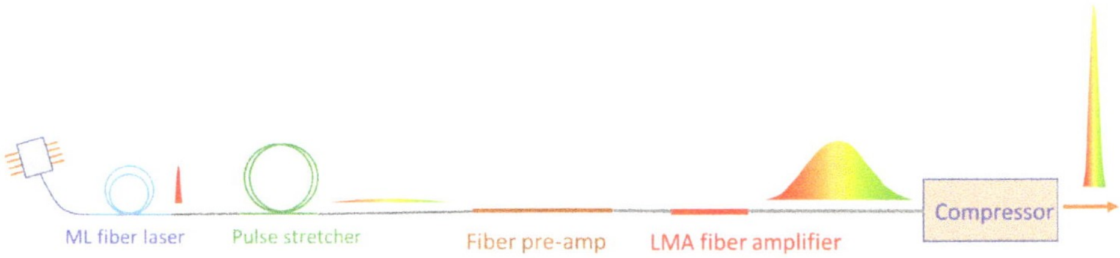

Figure 1.6. Schematic of a fiber-based CPA system.

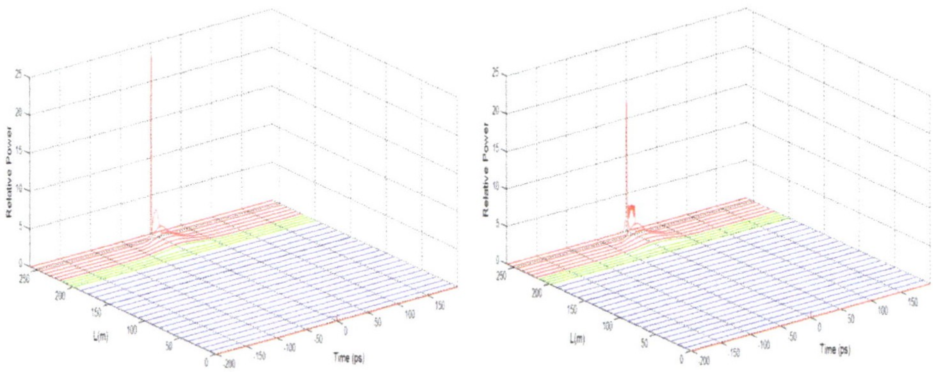

Figure 1.7. Illustration of the pulse evolution in all-fiber CPA systems. A system is with perfect matching dispersion between compressor and stretching fibers (left), and the other is with differences between high order of dispersion (right). For each plot, the input (red) from the left propagates in fiber stretcher (blue), fiber amplifier (green) and fiber compressor (red).

programs, it is our hope to help readers not only to understand the problem but to use the knowledge for designing the CPA in their work.

Figure 1.7 illustrates the pulse evolution in two all-fiber CPA systems, one is with perfect matching dispersion between compressor and stretching fibers (left), and the other is with differences between high order of dispersion (right). The pulse evolution in different fiber parts is plotted with different colors. The input on the right: red, stretching in stretcher fiber: blue, amplifier (green) compressing fiber: red.

The simulations are clearly useful not only for understanding the pulse behavior in the CPA system, but also for optimization of the system design. However, in order to reach the point where the whole CPA system can be modeled, it is necessary to understand both physics and mathematical models of all parts of the system. Therefore, chapter 5 also provides some fundamental theory required to understand the models of the problems and also the MATLAB® programs.

In the following section a short introduction to MATLAB® will be given, aiming to help readers who have very little or even no prior experience with MATLAB®, to understand the examples in the book. However, these are the first steps for them; as usual they are strongly encouraged to further study the computing language to master it.

1.2 An instruction using MATLAB® programs

MATLAB® is a language of technical computing (an abbreviation of 'matrix laboratory') that was first developed in the late 1970s by Cleve Moler. Over time, MATLAB® has become the computing language most used in academia, industries, economics etc, with more than 4 million users worldwide in 2020 [20]. Quite likely, a beginner can learn MATLAB® for computing relatively quickly, utilizing a very large set of functions, properties of graphics objects and operators of MATLAB®. One of the best features of MATLAB® is that one does not need to master the language before using it in real computing works. Instead, we can study and use MATLAB® at the same time, and the more you use it in practice the better and more effectively you can learn the language. It is our hope that readers can learn the physics, the mathematical models of the problems and MATLAB® programming at the same time through examples in this book.

For each example in this book, there is a MATLAB® program for simulation of the problem. The whole program is listed in the book so that the most important points, computational and mathematical of the program can be explained. Furthermore, the programs are intentionally presented in a way that closely follows the theoretical models examined early.

It is assumed that a reader who wants to run a MATLAB® program, already has MATLAB® installed in his/her computer. Click the MATLAB® icon to start it, we have several windows as below

We will learn all functionalities of the windows and commands in figure 1.8 while studying the programs in the book. The next step is to create a folder for each chapter —as can be seen for folders CH01, CH02… CH05 under Current Folder in figure 1.8.

The next step is to open the New Script icon with the red circle in the top panel of figure 1.8, we can see the Editor window opened as in figure 1.9. Copy the whole program, for example the MATLAB® Program 2.1 with the head below and paste it into the Editor window (indicated by a red sentence). Save it with a name PhC_1D.m into folder CH02.

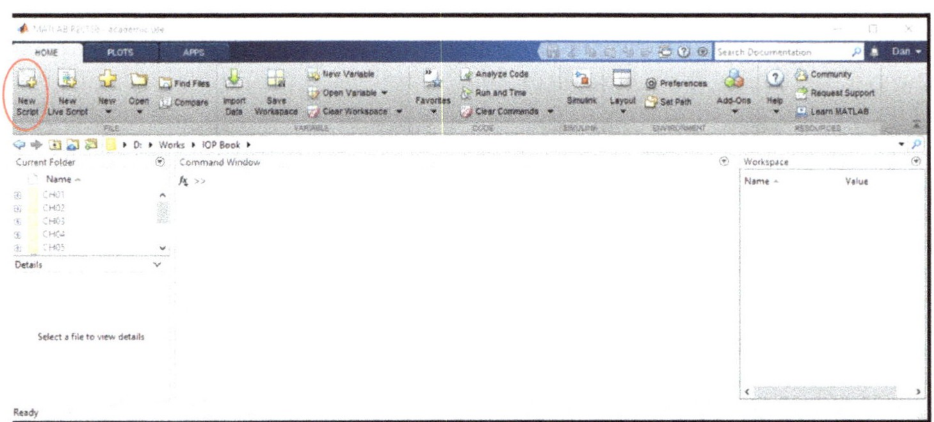

Figure 1.8. MATLAB® windows when started.

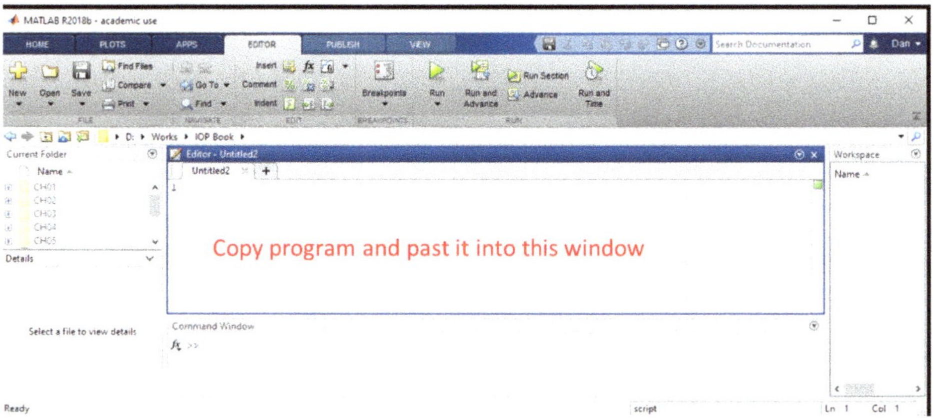

Figure 1.9. Copy the whole program in Editor window.

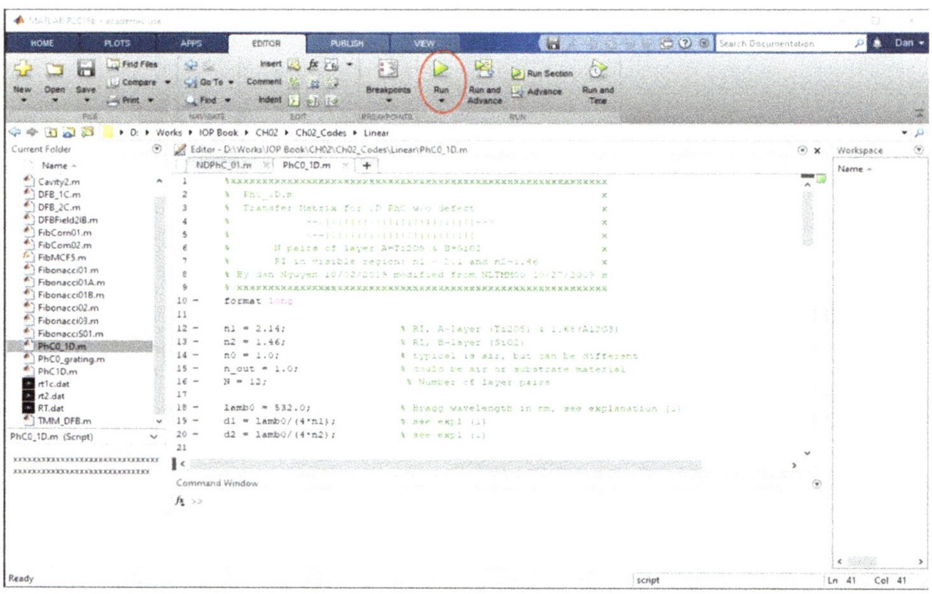

Figure 1.10. Copy the whole program in Editor window.

```
%xxxxxxxxxxxxxxxxxxxxxxxxxxxxxxxxxxxxxxxxxxxxxxxxxxxxxxxxxxxxxxxx
%   PhC_1D.m                                                      x
%   Transfer Matrix for 1D PhC w/o defect                         x
%               -->||||||||||||||||||||||||-->                    x
%          N pairs of layer A=Ti2O5 & B=SiO2                      x
%          RI in visible region: n1 ~ 2.1 and n2~1.46             x
%xxxxxxxxxxxxxxxxxxxxxxxxxxxxxxxxxxxxxxxxxxxxxxxxxxxxxxxxxxxxxxxx
```

Now, a window with the program like figure 1.10 appears.

Click the Run button (with red circle) in the top panel in figure 1.10. After a few seconds, the results of calculations from this program appear, as in figure 1.11 below.

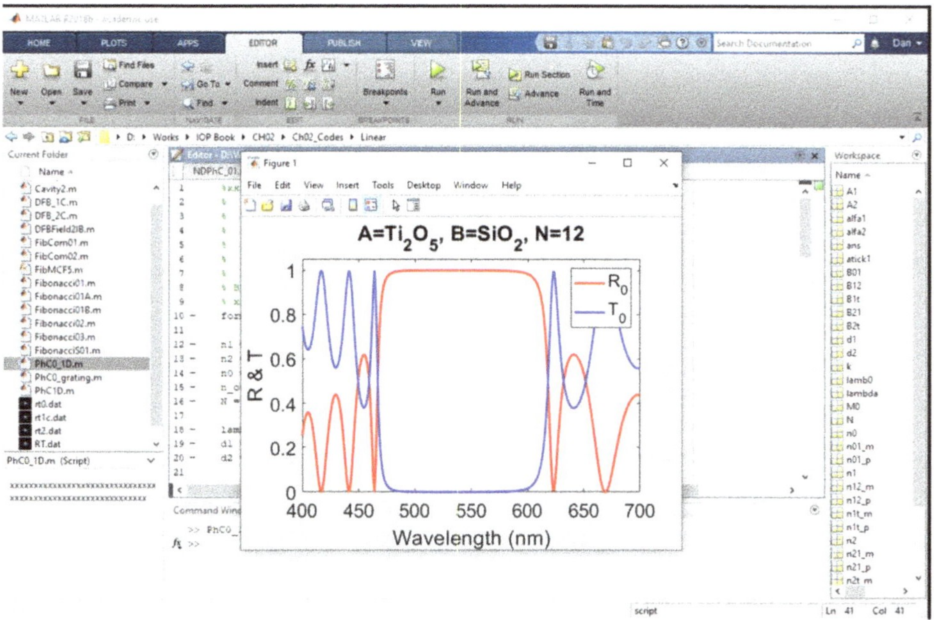

Figure 1.11. Results of calculations from Program PhC_1D.m. Reprinted by permission from Macmillan Publishers Ltd: [12], copyright (2019) (CCBY 4.0).

The best way to study the examples in this book is to follow ordering steps: first, it is important to understand the general theory, and then the models of problems of corresponding examples. After understanding the model, try to understand the MATLAB® programs for simulating the problems. Note that the programs are written closely following the models so that it is easy to understand the programs once knowing the models. It is our intention to present the programs not in optimization of programming since it could make it harder to learn the problems through the examples with MATLAB® programs, especially for readers without prior experience with MATLAB®. Furthermore, there are explanations before, after and also inside the programs using the comment % extensively.

One of the best features of MATLAB® is its enthusiastic community that exchanges many good codes, experiences solving numerous problems from very simple to extremely difficult https://www.mathworks.com/matlabcentral/. Therefore, though the programs are presented in the simplest way in this book, it is expected that readers can find many answers for questions that they still have.

Finally, MATLAB® itself changes very rapidly—every year it has a new version that sometimes does not work for too old programs. All programs in this book have been tested and run well in MATLAB® 2016 version.

References

[1] Saleh B E A and Teich M C 1991 *Fundamentals of Photonics* (Wiley Series in Pure and Applied Optics) (New York: Wiley)

[2] Joannopoulos J D, Johnson S G, Winn J N and Meade R D (ed) 2008 *Photonic Crystals: Molding the Flow of Light* 2nd edn (Princeton, NJ: Princeton University Press)
[3] Siegman A E 1986 *Lasers* (Mill Valey, CA: Lasers, University Science Books)
[4] Desurvire E 2002 *Erbium-Doped Fiber Amplifiers, Principles and Applications* (New York: Wiley)
[5] Digonnet M J F (ed) 1993 *Rare Earth Doped Fiber Lasers and Amplifiers* (New York: Marcel Dekker, Inc.)
[6] Yeh P 1988 *Optical Waves in Layered Media* (Wiley Series in Pure and Applied Optics) (New York: Wiley)
[7] Danckaert J *et al* 1991 Dispersive optical bistability in stratified structures *Phys. Rev.* B **44** 8214
[8] Valley G C 2001 Modeling cladding-pumped ER/YB fiber amplifiers *Opt. Fiber Tech.* **7** 21–44
[9] Kouznetsov D, Moloney J V and Wright E M 2001 Efficiency of pump absorption in double-clad fiber amplifiers. II. Fiber with circular symmetry *J. Opt. Soc. Am.* B **18** 743–9
[10] Nguyen D T *et al* 2007 A novel approach of modelling cladding-pump highly doped Er/Yb fiber amplifiers *IEEE J. Quantum Electron.* **43** 1018–27
[11] Babaeian M, Nguyen D T, Demir V, Akbulut M, Blanche P-A, Kaneda K, Neifeld M and Peyghambarian N 2019 A single shot coherent Ising machine based on a network of injection-locked multicore fiber lasers *Nat. Commun.* **10** 3516
[12] Demir V, Akbulut M, Nguyen D T, Kaneda K, Neifeld M and Peyghambarian N 2019 Injectionlocked, single frequency, multi-core Yb-doped phosphate fiber laser *Sci. Rep.* **9** 356
[13] Keller U 2010 Ultrafast solid-state laser oscillators: a success story for the last 20 years with no end in sight *Appl. Phys.* **B100** 15–28
[14] Haus H A 1975 Theory of mode locking with a fast saturable absorber *J. Appl. Phys.* **46** 3049
[15] Agrawal G P 2007 *Nonlinear Fiber Optics* 4th edn (San Diego, CA: Academic)
[16] Johan H 2007 A fourth-order Runge–Kutta in the interaction picture method for simulating supercontinuum generation in optical fibers *J. Lightwave Tech.* **25** 3770–75
[17] Strickland D and Mourou G 1985 Compression of amplified chirped optical pulses *Opt. Commun.* **56** 219–21
[18] Keppler S *et al* 2016 The generation of amplified spontaneous emission in high-power CPA laser systems *Laser Photonics Rev.* **10** 264
[19] Fermann M E, Galvanauskas A and Sucha G 2003 *Ultrafast Lasers: Technology and Applications* (New York: Marcel Dekker)
[20] The MathWorks 2020 Company overview

Dan T Nguyen

Chapter 2

One-dimensional periodic and quasi-periodic photonics crystal structures

In this chapter we will first present the general concept of one-dimensional (1D) photonics crystal (PhC) structures, and then derive the simplest mathematical model for calculating optical properties of the PhCs based on transfer matrix method (TMM). Linear TMM formalism will be presented as an effective method for calculating optical spectra of several real applications such as Bragg reflectors, fiber gratings, optical resonators, laser cavities including distributed Bragg reflectors (DBR) and distributed feedback (DFB) fiber lasers, mirrors with quasiperiodic multiple reflection windows. The last part will be devoted to a nonlinear TMM formalism for calculations of nonlinear problems such as non-linear spectra of defected cavities, all-optical switching based on optical bistability etc. MATLAB® codes will be presented for numerically studying and analyzing several examples.

It is worth noting that although 1D PhCs can be considered as the simplest ones among photonics structures, they have been used for many important applications. The examples mentioned above are just a few of the well-known devices. In this chapter we will learn how to model and calculate optical properties of these systems by studying several examples with MATLAB® codes. We want to stress that the codes are presented in a way that follows closely to the physics and mathematical models of the problems, not in the optimized ways of programming. By doing it this way it is our hope that readers, especially beginners, can easily understand the physics and mathematics models that are calculated by these computing programs. The optimized codes are left for readers who want to make their programs run best after understanding the physics of the problems.

doi:10.1088/978-0-7503-2272-0ch2

2.1 1D photonics crystals and mathematics models

2.1.1 One-dimensional photonics crystals

For many of us, from the beginning 'photonics crystal' sounds curiously interesting as 'photon' and 'crystal' are both very well-known terms, but 'photonics crystal' is completely strange, at least for the first time. It turns out the term itself bears a similarity to the well-established concept in solid state physics. The concept of crystals is of periodic structures of atoms or molecules in which electrons as quantum particles propagate in periodic potentials leading to energy bands. Electrons with energies that fall into allowable energy bands can propagate in the medium 'freely'. However, there are band gaps in between the allowable energy bands, and electrons are forbidden to propagate if their energies are in those band gaps. The same concept can be applied to photons—the quantum particles of light, in an optically periodic medium. PhCs therefore can be considered the periodic and quasi-periodic structures where the scales of the lattices are comparable with wavelengths of the light or electromagnetic waves. In such PhC structures, if frequencies or wavelengths of light fall into some regions, the photonics band gap, the light can be completely reflected. The photonics band gap is the frequency (or wavelength) band in which light is not allowed to propagate through the structures but is highly reflected or even completely reflected. Readers can find a very comprehensive theory of photonics crystals in several excellent textbooks [1–3].

Figure 2.1 illustrates two simple cases of so-called 1D PhCs without and with a defect, and their reflectivity (R) and transmissivity (T) spectra. Those look quite simple structures, but they have several important applications that we will study later in this chapter. The upper diagrams in figure 2.1(a) and (b) show a periodic structure of N pairs of two (or more than two, in general) different material layers, and the defected PhC with a defect layer in the middle of the structure, respectively. These pairs of material layers are the fundamental blocks of the 1D PhC, and the

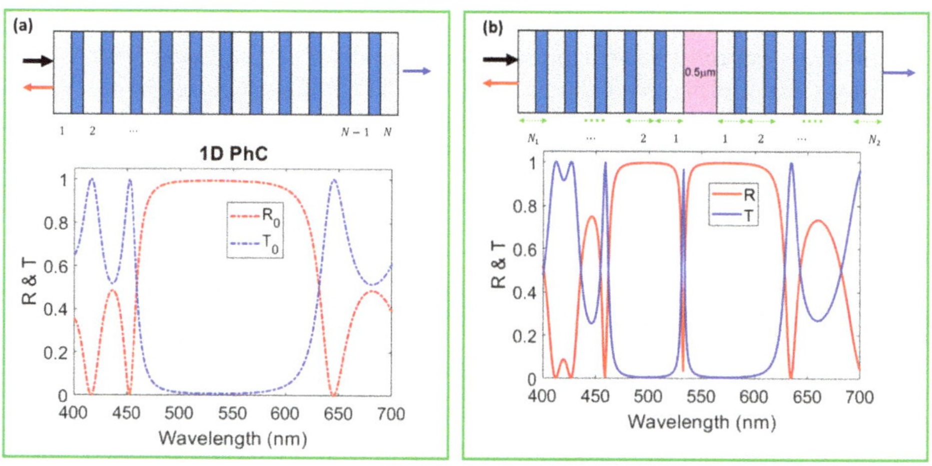

Figure 2.1. Upper: diagram of 1D PhC of N pairs of layers of two materials without defect (a) and with a defect (b). Lower: calculated reflectivity (R) and transmissivity (T) of corresponding structures.

defect is a thin layer whose thickness and/or material can be different with these two fundamental layers. In general, the position of the defect can be different inside the structure. The refractive indexes of the materials of layers A (gray) and B (blue) are n_A and n_B, respectively, and C (pink) is n_C. The reflectivity and transmissivity are calculated with the assumption that an incident light (black arrow) hits the leftmost interface, the reflection (red arrow) and transmission (blue arrow) are shown in the upper diagrams in figure 2.1(a) and (b).

We will study these 1D PhC structures in detail, in the following sections of this chapter, but let's discuss briefly the calculated spectra in figure 2.1. First, the most important feature of reflectivity spectra R_0 (red, dotted) and transmissivity T_0 (blue, dotted) of the PhC without defect (figure 2.1(a)) is the concept of photonics *band gap* (BG)—a wavelength region in which the light is reflected completely, or light transmission is not allowed in this region. Meanwhile, it can be seen clearly from figure 2.1(b) that a narrow window is created inside BG or inter-band (IB), in which the light is transmitted completely in a defected PhC. Those special features, the photonics BG and IB can be utilized for different applications, as will be presented later in this chapter.

As mentioned earlier, the concept of BGs in photonics crystals is similar to the energy band gap in crystals—a consequence of the wave propagation in periodic structures [1–3]. In crystals, electrons as quantum particles with dual wave–particle nature that propagate in periodic lattices of atoms and molecules with crystal lattices are in the scale of wavelength of quantum particles. Similarly, the light can strongly sense the photonics lattices if the period lattices are in scales of wavelength of the light. That's the main reason why photonics crystals have been investigated extensively only since about 30 years ago, when technologies were advanced enough for making such small-scale period structures for observing the special properties of photonics crystals. Meanwhile, the periodic lattices of atoms and molecules in natural crystals are in the scales of wavelengths of quantum-particle electrons, therefore both experimental and theoretical investigations of electronic band gap of the crystals were carried out much earlier.

In the next sections, we will present the formalism of the linear TMM and its applications for multilayered media including 1D PhC structures [4–8]. We then apply the TMM formalism to study optical properties of some structures that are directly applied in many devices. The MATLAB® codes are presented to calculate and analyze the properties for designing some real devices. For example, mirrors, resonant cavities, DBR and DFB fiber laser cavities are presented in detail. The TMM can then be applied to the quasiperiodic multilayered structures or quasi-periodic PhC. The effect of localization of light can be understood by numerical results. As an example, the quasi-periodic Fibonacci mirrors with multiple reflection windows will also be presented emphasizing their applications. We want to stress again that to make it as simple as possible for beginners to understand the physics and mathematics models, the MATLAB® codes are intended to be presented in non-optimized ways so that the physics behind the formulae can be explained. The last part of this chapter will be devoted to more complicated situations when the nonlinear refractive indexes of the materials are considered in the multilayered structures. In such situations, nonlinear TMM formalism is needed. We will follow the nonlinear TMM formalism that has been developed by [9] and then apply the method for investigating nonlinear defected PhC structures for nonlinear switching applications. It should be stressed that the TMM formalism presented in this chapter is for the steady state regime. The TMM formalism in transient regime has been proven powerful for nonlinear dynamic systems, and can be found in [10, 11].

Note that TMM is a well established method and its comprehensive theory has been described in many textbooks [4, 5] and references [6–8]. Readers can learn those fundamentals from the excellent references mentioned above. Here, in order to make it easier to understand the fundamental concepts, especially for the beginners of photonics design, a simple but still rigorous formalism will be presented. Once the formalism is deeply understood, modelers will be later able to extend or modify the learned problems to other complicated ones. At the same time, we would like to avoid going deeply into long and tedious derivations that are quite often counter-productive, especially for the beginners. In the examples under consideration, it is assumed that readers know the very fundamentals of MATLAB®. Explanations will be provided to explain each point if necessary in the programs. We believe even readers without any MATLAB® experience can also easily understand and effectively use the codes for their study.

2.1.2 Transfer matrix method

In this section, the TMM formalism is derived in detail to study 1D PhC structures, and several examples will be numerically studied by MATLAB® programs. The examples are chosen to be very close to real problems, but the models are presented as simply as possible, otherwise we may get lost in too many equations. For more comprehensive theory of TMM formalism, readers are referred to several textbooks mentioned earlier [4, 5], and other references [6–8].

Let's first consider a system consisting of two layers of two different materials, as shown in figure 2.2. Once we have derived the formula for this simple structure we can easily extend it to multilayered periodic and quasi-periodic systems that contain any number of different material layers. This two-layer structure serves as a fundamental block for multilayered structures that are used in many photonics devices. Next, we assume that these two materials have refractive indexes (RI) n_1 and n_2 and thicknesses d_1 and d_2, respectively. In general, the refractive index of the incident and output media (before and after the structure) can be different and are

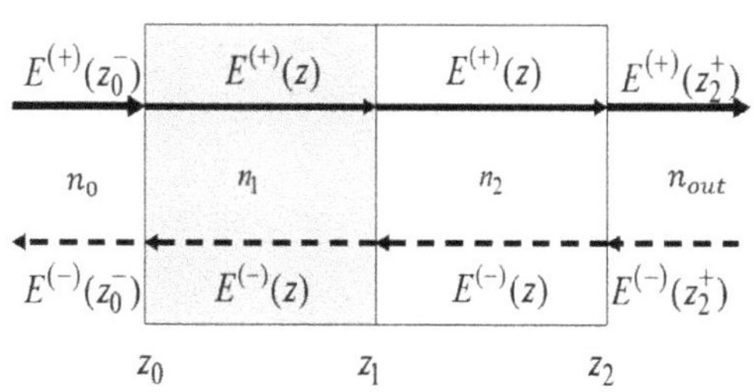

Figure 2.2. Calculation scheme in TMM.

denoted as n_0 and n_{out}, respectively. We also assume the system is spatially homogeneous in the transverse directions and that *the light travels normally to the layers* from left to right (in the z-direction). In general, the electric field at any given point is a superposition of the forward-moving and backwards-moving electromagnetic waves. Physically, this decomposition of forward and backward light fields is very meaningful because the light gets reflected and transmitted at any boundary interface between two different media. Mathematically, both forward and backward fields are solutions of wave equations, e.g., the Maxwell equations, and therefore, the total field can be expressed as a linear superposition of these two fields. We now denote $E^{(\pm)}(z)$ the frequency-domain electric field amplitudes of the forward (+) and backward (−) traveling plane waves along the z-direction, and $z_m^{(\pm)}$ the right (+) and left (−) boundaries of the coordinates z_m, respectively, ($m = 1, 2$ in two-layer structures in figure 2.2).

The general formulae between the electric fields at z_0 and z_2 in TMM can be written as

$$\begin{pmatrix} E^{(+)}(z_0^-) \\ E^{(-)}(z_0^-) \end{pmatrix} = \begin{pmatrix} M_{11} & M_{12} \\ M_{21} & M_{22} \end{pmatrix} \begin{pmatrix} E^{(+)}(z_2^+) \\ E^{(-)}(z_2^+) \end{pmatrix}. \tag{2.1}$$

where $z_m^{\pm} = \lim_{\varepsilon \to 0}(z_m \pm \varepsilon)$ and M_{ij} are the elements of the transfer matrix M.

For simplicity, let us start with the problem when a light beam incident E_{in} propagates from the left to the right of the first boundary z_0^- and there is no incident light from the right $E^{(-)}(z_2^+) = 0$. From the diagram in figure 2.2, we can denote the amplitude of the incident field as $E_{in} = E^{(+)}(z_0^-)$, the reflected field $E_{ref} = E^{(-)}(z_0^-)$ and the transmitted field $E_{tr} = E^{(+)}(z_2^+)$. From matrix equation (2.1), we can derive the complex amplitude of reflected and transmitted fields r and t, respectively, as follows:

$$\left.\begin{aligned} E^{(+)}(z_0^-) &= M_{11}E^{+}(z_2^+) \\ E^{(-)}(z_0^-) &= M_{21}E^{+}(z_2^+) \end{aligned}\right\} \rightarrow r = \frac{E^{(-)}(z_0^-)}{E^{(+)}(z_0^-)} = \frac{E_{ref}}{E_{in}} = \frac{M_{21}}{M_{11}}, \tag{2.2}$$

and

$$E^{(+)}(z_0^-) = M_{11}E^{+}(z_2^+) \rightarrow t = \frac{E^{(+)}(z_2^+)}{E^{(+)}(z_0^-)} = \frac{E_{tr}}{E_{in}} = \frac{1}{M_{11}}. \tag{2.3}$$

The complex amplitudes of reflected and transmitted field can be re-written as

$$r = \left.\frac{E^{(-)}(z_0^-)}{E^{(+)}(z_0^-)}\right|_{E^{(-)}(z_2^+)=0} = \frac{M_{21}}{M_{11}}, \qquad t = \left.\frac{E^{(+)}(z_2^+)}{E^{(+)}(z_0^+)}\right|_{E^{(-)}(z_2^+)=0} = \frac{1}{M_{11}}. \tag{2.4}$$

Note that what we are going to calculate are the spectra of reflectivity R and transmissivity T of optical systems which are observable and measurable, and are defined as

$$R = \frac{I_{ref}}{I_{in}} = \frac{n_0|E_{ref}|^2}{n_0|E_{in}|^2} = \frac{|E^{(-)}(z_0^-)|^2}{|E^{(+)}(z_0^-)|^2} = |r|^2,\ T = \frac{I_{tr}}{I_{in}} = \frac{n_{out}|E_{tr}|^2}{n_{in}|E_{in}|^2} = \frac{n_{out}}{n_0}|t|^2. \tag{2.5}$$

Similarly, if the incident field propagates from the right to the right z_2^+, (no incident light from the right $E^{(+)}(z_0^-) = 0$). In this case the amplitude of the incident field is denoted as $E_{in} = E^{(-)}(z_2^+)$, the reflected field $E_{ref} = E^{(+)}(z_2^+)$ and the transmitted field $E_{tr} = E^{(-)}(z_0^-)$. If we define the amplitudes of reflected and transmitted field in this case as r' and t', respectively, then we have

$$0 = M_{11}E^{+}(z_2^+) + M_{12}E^{-}(z_2^+) \rightarrow r' = \frac{E^{(-)}(z_0^-)}{E^{(+)}(z_2^+)} = \frac{E_{ref}}{E_{in}} = -\frac{M_{12}}{M_{11}}, \tag{2.6}$$

and

$$\left.\begin{aligned} 0 &= M_{11}E^{+}(z_2^+) + M_{12}E^{-}(z_2^+) \\ E^{-}(z_0^-) &= M_{21}E^{+}(z_2^+) + M_{22}E^{-}(z_2^+) \end{aligned}\right\} \rightarrow t' = \frac{E^{(-)}(z_0^-)}{E^{(-)}(z_2^+)} = \frac{E_{tr}}{E_{in}} = -\frac{M_{11}M_{22} - M_{12}M_{21}}{M_{11}}. \tag{2.7}$$

Equations (2.6) and (2.7) can be re-written similarly to (2.4) as

$$r' = \left.\frac{E^{(+)}(z_2^+)}{E^{(-)}(z_2^+)}\right|_{E^{(+)}(z_0^-)=0} = -\frac{M_{22}}{M_{21}}, \qquad t' = \left.\frac{E^{(-)}(z_0^-)}{E^{(-)}(z_2^+)}\right|_{E^{(+)}(z_0^-)=0} = \frac{M_{11}M_{22} - M_{12}M_{21}}{M_{11}}. \tag{2.8}$$

These above equations are very useful for us to write explicitly the transfer matrix elements for any multilayered structures.

In the next section, we will develop models based on TMM formalism for some specific systems whose parts or the whole structure have common features of periodic and quasiperiodic multilayered structures. We will see TMM formalism is very effective for calculating and analyzing the properties of those periodic and quasi-periodic 1D PhCs.

2.2 Linear transfer matrix method and modeling examples

2.2.1 Linear transfer matrix method formalism

In this subsection we will derive the explicit matrix formulae of the linear TMM for light propagation in multilayered structures or 1D PhCs. The term 'linear' is used here to stress that this formalism is valid only in the linear regime in which optical properties of the materials are independent of light-intensity. In contrast, if optical properties of materials are intensity-dependent, a nonlinear TMM formalism will be needed and is presented in section 2.3. We will start from a basic block depicted in figure 2.2 above. For convenience the TMM formalism will be developed step by step, layer by layer from the left to the right in the propagation direction. It is important to note that although we consider here only the simplest situation with the zero-incident angle, the derivation is very useful for readers to understand TMM formalism for 1D PhC structures and the corresponding MATLAB® codes. Once we understand the TMM models for 1D PhCs we can easily modify or extend the method to other complicated problems, and that is the main goal of this book. As stated earlier, we also want to avoid too lengthy and tedious calculations that may cause readers get lost at this earlier stage of study. Therefore, we start from a well-known formulae for reflected and transmitted amplitudes r and t in a general case, in which light propagates from the left to right of a boundary interface between two materials with refractive indexes of the incident and transmitted materials that are n_l (left side of interface) and n_r (right side of interface), respectively [4–7].

$$r_{\parallel} = \frac{n_l \cos(\theta_l) - n_r \cos(\theta_r)}{n_l \cos(\theta_l) + n_r \cos(\theta_r)}, \quad \text{and} \quad t_{\parallel} = \frac{2n_l \cos(\theta_l)}{n_l \cos(\theta_l) + n_r \cos(\theta_r)} \tag{2.9}$$

where $\theta_{l,r}$ are the incident and transmitted angles, and are denoted as the angles in the left (θ_l) and in the right (θ_r) medium for convenient and consistent notations in the description. As stated earlier, in this chapter we will restrict ourselves to consider the simplest case, where $\theta_l = \theta_2 = 0$. However, after understanding the TMM readers can easily expand the formalism to the general cases with any incident angle using equation (2.9). From the boundary conditions, the electric fields at the boundary coordinate z_m between two materials, one can write the relationships as (see, problem 2.1 at the end of the chapter):

$$\begin{aligned} E^{(+)}(z_m^-) &= \frac{n_l + n_r}{2n_l} E^{(+)}(z_m^+) + \frac{n_l - n_r}{2n_l} E^{(-)}(z_m^+) \\ E^{(-)}(z_m^-) &= \frac{n_l - n_r}{2n_l} E^{(+)}(z_m^+) + \frac{n_l + n_r}{2n_l} E^{(-)}(z_m^+) \end{aligned}. \tag{2.10}$$

Equation (2.10) can be re-written in the matrix form as

$$\begin{pmatrix} E^{(+)}(z_m^-) \\ E^{(-)}(z_m^-) \end{pmatrix} = \frac{1}{2n_l} \begin{pmatrix} n_l + n_r & n_l - n_r \\ n_l - n_r & n_l + n_r \end{pmatrix} \begin{pmatrix} E^{(+)}(z_m^+) \\ E^{(-)}(z_m^+) \end{pmatrix} = \begin{pmatrix} B_{11} & B_{12} \\ B_{21} & B_{22} \end{pmatrix} \begin{pmatrix} E^{(+)}(z_m^+) \\ E^{(-)}(z_m^+) \end{pmatrix}. \tag{2.11}$$

Through this chapter we will use consistently the notation $B(n_l|n_r)$ or B_{lr} to represent a transfer matrix for an interface between the left- and right-materials with refractive indexes n_l and n_r, respectively, and with the assumption that the light travels from the left to the right. We can now apply equation (2.11) for the first interface at coordinate z_0 in figure 2.2, where the refractive indexes of the incident and transmitted materials are n_0 and n_1, respectively. The matrix equation at z_0 can be written as

$$\begin{pmatrix} E^{(+)}(z_0^-) \\ E^{(-)}(z_0^-) \end{pmatrix} = \frac{1}{2n_0} \begin{pmatrix} n_0 + n_1 & n_0 - n_1 \\ n_0 - n_1 & n_0 + n_1 \end{pmatrix} \begin{pmatrix} E^{(+)}(z_0^+) \\ E^{(-)}(z_0^+) \end{pmatrix} = B(n_0|n_1) \begin{pmatrix} E^{(+)}(z_0^+) \\ E^{(-)}(z_0^+) \end{pmatrix}. \tag{2.12}$$

Let us now consider the light propagation in the first layer of our structure. As indicated in the diagram, refractive index (RI) and thickness of the first layer are denoted as n_1 and $d_1 = z_1^- - z_0^+$, respectively. The propagations of the forward

$E^{(+)}(z)$ field from z_0^+ to z_1^- and the backward $E^{(-)}(z)$ field from z_1^- to z_0^+ in the homogenous material with RI n_1 can be expressed in simple equations

$$\left.\begin{matrix} E^{(+)}(z_1^-) = e^{-ik_0n_1d_1}E^{(+)}(z_0^+) \\ E^{(-)}(z_1^-) = e^{ik_0n_1d_1}E^{(-)}(z_0^+) \end{matrix}\right\} \rightarrow \begin{pmatrix} E^{(+)}(z_1^-) \\ E^{(-)}(z_1^-) \end{pmatrix} = \begin{pmatrix} e^{-ik_0n_1d_1} & 0 \\ 0 & e^{ik_0n_1d_1} \end{pmatrix} \begin{pmatrix} E^{(+)}(z_0^+) \\ E^{(-)}(z_0^+) \end{pmatrix}, \tag{2.13}$$

where $k_0 = \frac{2\pi}{\lambda}$, $d_m = z_m - z_{m-1}$ and λ is wavelength of the light wave.

Equation (2.13) can be rewritten as

$$\begin{pmatrix} E^{(+)}(z_0^+) \\ E^{(-)}(z_0^+) \end{pmatrix} = \begin{pmatrix} e^{ik_0n_1d_1} & 0 \\ 0 & e^{-ik_0n_1d_1} \end{pmatrix} \begin{pmatrix} E^{(+)}(z_1^-) \\ E^{(-)}(z_1^-) \end{pmatrix} = A(n_1, d_1) \begin{pmatrix} E^{(+)}(z_1^-) \\ E^{(-)}(z_1^-) \end{pmatrix}. \tag{2.14}$$

From equations (2.12) and (2.14), the matrix equation for propagation from z_0^- to z_1^- can be written as

$$\begin{pmatrix} E^{(+)}(z_0^-) \\ E^{(-)}(z_0^-) \end{pmatrix} = B(n_0|n_1) \begin{pmatrix} E^{(+)}(z_0^+) \\ E^{(-)}(z_0^+) \end{pmatrix} = B(n_0|n_1)A(n_1, d_1) \begin{pmatrix} E^{(+)}(z_1^-) \\ E^{(-)}(z_1^-) \end{pmatrix}. \tag{2.15}$$

Repeating the above steps, we have transfer matrix (TM) equation from z_0 to z_2 as follows:

$$\begin{pmatrix} E^{(+)}(z_0^-) \\ E^{(-)}(z_0^-) \end{pmatrix} = B(n_0|n_1)A(n_1, d_1)B(n_1|n_2)A(n_2, d_2)B(n_2|n_{out}) \begin{pmatrix} E^{(+)}(z_2^+) \\ E^{(-)}(z_2^+) \end{pmatrix}. \tag{2.16}$$

From equation (2.16), the TM for the whole system described in figure 2.2 can be written in explicit form

$$\begin{aligned} M = {} & \frac{1}{2n_0} \begin{pmatrix} n_0 + n_1 & n_0 - n_1 \\ n_0 - n_1 & n_0 + n_1 \end{pmatrix} \begin{pmatrix} e^{-ik_0n_1d_1} & 0 \\ 0 & e^{ik_0n_1d_1} \end{pmatrix} \frac{1}{2n_1} \begin{pmatrix} n_1 + n_2 & n_1 - n_2 \\ n_1 - n_2 & n_1 + n_2 \end{pmatrix} \\ & \times \begin{pmatrix} e^{-ik_0n_2d_2} & 0 \\ 0 & e^{ik_0n_2d_2} \end{pmatrix} \frac{1}{2n_2} \begin{pmatrix} n_2 + n_{out} & n_2 - n_{out} \\ n_2 - n_{out} & n_2 + n_{out} \end{pmatrix}. \end{aligned} \tag{2.17}$$

It is easy to see how cumbersome it is to write down explicitly each element of the transfer matrix M in equation (2.17) and use them to calculate the reflection and transmission coefficients using equations (2.4)–(2.5) and (2.6)–(2.7) above, especially in typical structures for real applications when tens, hundreds or even thousands of layers are used. As will be shown in studying examples in the next sections, MATLAB® programming becomes so useful and powerful for dealing with the matrix equations. Equation (2.17) can be rewritten in a simpler and compact form as

$$M = B(n_0|n_1)A(n_1, d_1)B(n_1|n_2)A(n_2, d_2)B(n_2|n_{out}). \tag{2.18}$$

Equation (2.18) is not only compact but also very convenient for extending to different layered structures, periodic or un-periodic and different materials. All we need is to define two different types of TMs. First, the interface matrixes (IMs) for all boundary interfaces between two different materials characterized by refractive indexes n_l and n_l of the left and right materials of the boundary, respectively:

$$B(n_l|n_r) = \begin{pmatrix} B_{11} & B_{12} \\ B_{21} & B_{22} \end{pmatrix} = \frac{1}{2n_l}\begin{pmatrix} n_l + n_r & n_l - n_r \\ n_l - n_r & n_l + n_r \end{pmatrix}, \tag{2.19}$$

and the propagation matrixes (PMs) for any homogenous layer characterized by physical thickness d and refractive index n:

$$A(n|d) = \begin{pmatrix} A_{11} & 0 \\ 0 & A_{22} \end{pmatrix} = \begin{pmatrix} e^{ik_0 nd} & 0 \\ 0 & e^{-ik_0 nd} \end{pmatrix}. \tag{2.20}$$

For convenience, we will keep using these notations $A(n, d)$ and $B(n_l|n_r)$ through out this chapter. Readers will see how useful they are when studying our MATLAB® programs in the following subsections.

Note that in the above equations (2.19) and (2.20), for the case of the absorbing materials, the refractive index can be expressed in complex number as $n = n' + in''$, where n' and n'' are the real and imaginary parts of the complex index. The imaginary part of index is related to the absorption coefficient α_m of mth-layer as $\alpha_m = 2\omega\, n''_m/c$, where ω is frequency of the light wave.

In the following we will develop TMM models for several examples, and present MATLAB® programs for calculating and analyzing optical properties of the structures. The studied examples are aiming at the design tasks for applications.

2.2.2 Modeling example: 1D photonics band gap structures

Let us now consider a problem of modeling light propagation in a periodic multilayer dielectric structure depicted in figure 2.3. We already mentioned some special features of these structures earlier, such as the photonics band gap—a region of wavelength (or frequency) in which light is completely reflected. Clearly, this property can be used for many applications such as mirrors, optical cavities including laser cavities. This example will help us not only to understand but also how to manipulate the reflection and transmission of light by varying just a few

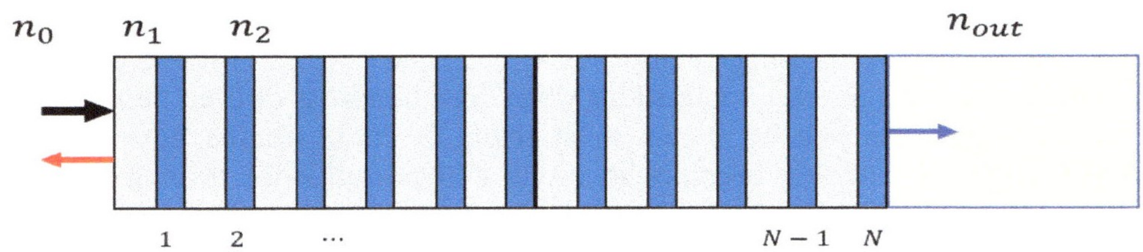

Figure 2.3. Schematic of periodic N pairs dielectric material layers whose thicknesses and refractive indexes (d,n) of first layer (gray color) and second layer (dark blue) are $(d_1,n_1)$and (d_2,n_2), respectively. An incident beam of light propagates from the left (black arrow), reflected to the incident medium of RI n_0 (red arrow), transmitted field (blue arrow) enters the output medium of RI n_{out}.

parameters such as RI of materials, thicknesses of the layers, and also the number of layers. We will also see the RI of incident and output media can affect to the optical properties of the structure.

Figure 2.3 shows a typical 1D PhC, which is a periodic multilayered structure. In the following example, we consider a structure consists of N pairs of material layers A = Ti_2O_5 (n ~ 2.14 in visible region) and B = SiO_2 (n ~ 1.46) which are widely used in many applications for making special mirrors by coating technique. We want to calculate the optical properties such as spectra of reflectivity and transmissivity of light in the visible region (400–700 nm) of this structure. As will be seen, the high index contrast between the two layers (~0.68) plays an important role for generating a very broad band of high reflection mirrors with relatively small number of layers. In general, the model can deal with any number of layers, and it does not need to be a number of layer pairs. However, for simplicity of the programming, in the following example we will consider the system with a number of layer pairs. It is easy to modify this model for a case with any number of layers. We assume the media before the structure (incident medium) is air ($n_0 = 1$), and n_{out} of output medium can be air or substrate material.

Now, let us present the MATLAB® program to do calculation of the optical transmission and reflection based on TMM formalism we have developed earlier. To understand the program, we assume readers have a minimum knowledge of programming. For the first few programs, we will list most parts of the codes, but in many cases only the most important commands with explanations will be presented.

***MATLAB® program* 2.1**

Before presenting the MATLAB® code for calculation of refection and transmission spectra of the 1D PhC structure that is depicted in figure 2.3, let's do some preparations for writing the program. As mentioned earlier, we will need to define two different types of TMs. The first are IMs $B(n_l|n_r)$ for every boundary between two different material layers characterized by refractive indexes n_l and n_r. The other type of matrix is propagation matrixes PMs $A(n, d)$ in any homogenous material layer characterized by physical thickness d and RI n. From the diagram in figure 2.3, it can be seen that there are only four different IMs for all interfaces in the structure. First, the IM $B(n_0|n_1)$ for the first interface between incident medium and the first material layer having refractive indexes n_0 and n_1, respectively:

$$B(n_0|n_1) = \frac{1}{2n_0}\begin{pmatrix} n_0 + n_1 & n_0 - n_1 \\ n_0 - n_1 & n_0 + n_1 \end{pmatrix} = \begin{pmatrix} n_{01,p} & n_{01,m} \\ n_{01,m} & n_{01,p} \end{pmatrix}, \tag{2.21}$$

Here, we have used a short notation $n_{01,p}$ and $n_{01,m}$ (the index p stands for plus, m: minus)

$$n_{01,p} = \frac{1}{2n_0}(n_0 + n_1) \text{ and } n_{01,m} = \frac{1}{2n_0}(n_0 - n_1). \tag{2.22}$$

Inside the structure, there are only two different IMs, $B(n_1|n_2)$ and $B(n_2|n_1)$ for $n_1|n_2$ and $n_2|n_1$ interfaces, respectively. These two IMs can be written using $n_{12,p}(n_{12,m})$ and $n_{21,p}(n_{21,m})$ similar to the notations in equation (2.22). In general, the number of layers can be odd or even. The last layer of the PhC can be n_1 or n_2, corresponding to IM $B(n_1|n_{out})$ and $B(n_2|n_{out})$ in the program, but only one of them will be needed for each situation.

For the propagation matrixes, there are only two different PMs to describe propagation in two layers of materials 1 and 2 with thicknesses and RI (d_1, n_1) and (d_2, n_2).

$$A(n_1|d_1) = \begin{pmatrix} e^{ik_0 n_1 d_1} & 0 \\ 0 & e^{-ik_0 n_1 d_1} \end{pmatrix} \text{ and } A(n_2|d_2) = \begin{pmatrix} e^{ik_0 n_2 d_2} & 0 \\ 0 & e^{-ik_0 n_2 d_2} \end{pmatrix}. \tag{2.23}$$

Once all TMs are defined and understood, then the codes are very easy to understand and also easy to extend to more complicated structures.

In the following code listed below, some explanations will be given for beginners who are not familiar with MATLAB®, even at a very basic level. For example, in MATLAB® anything after '%' is just comments or explanations that will be ignored during calculations. It is good practice to use this comment feature for making the codes easy to understand for some important details of the models such as calculation methods, algorithms etc. Sometimes, we use this feature '%' to explain the model of problem, even the mathematic equations that we may forget, especially if we haven't work on the problem in a long period of time. It is also good practice to have a short description of the problem at the beginning of the program, as shown below for example.

```
%xxxxxxxxxxxxxxxxxxxxxxxxxxxxxxxxxxxxxxxxxxxxxxxxxxxxxxxxxxx
%   PhC_1D.m                                               x
%   Transfer Matrix for 1D PhC w/o defect                  x
%             -->|||||||||||||||||||||||||-->              x
%        N pairs of layer A=Ti2O5 & B=SiO2                 x
%        RI in visible region: n1 ~ 2.1 and n2~1.46        x
%xxxxxxxxxxxxxxxxxxxxxxxxxxxxxxxxxxxxxxxxxxxxxxxxxxxxxxxxxxx

format long
n1 = 2.14;                   % RI, A-layer (Ti2O5)
n2 = 1.46;                   % RI, B-layer (SiO2)
n0 = 1.0;                    % typical is air, but can be different
n_out = 1.0;                 % could be air or substrate material
N = 8;                       % Number of layer pairs
lamb0 = 532.0;               % Bragg wavelength in nm, see explanation (1)
d1 = lamb0/(4*n1);           % see Explanation (1)
d2 = lamb0/(4*n2);           % see Explanation (1)

% Input ITM for interface n0 -|- n1
n01_p = (n0 + n1)/(2*n0); n01_m = (n0 - n1)/(2*n0);
B01   = [n01_p, n01_m; n01_m, n01_p];

% ITM for interface n1-|-n2
n12_p = (n1 + n2)/(2*n1); n12_m = (n1 - n2)/(2*n1);
B12   = [n12_p, n12_m; n12_m, n12_p];

% ITM interface n2-|-n1
n21_p = (n2 + n1)/(2*n2); n21_m = (n2 - n1)/(2*n2);
B21   = [n21_p, n21_m; n21_m, n21_p];

% Output ITM for interface n1-|-n_out
n1t_p = (n1 + n_out)/(2*n1); n1t_m = (n1 - n_out)/(2*n1);
B1t   = [n1t_p, n1t_m; n1t_m, n1t_p];
```

```
% Output ITM for interface n2-|-n_out
n2t_p = (n2 + n_out)/(2*n2); n2t_m = (n2 - n_out)/(2*n2);
B2t   = [n2t_p, n2t_m; n2t_m, n2t_p];

rt = fopen('rt0.dat','w+');                % for saving data in rt0.dat file
wavelength = 400:1:700;                    % wavelength range (nm) in calculation
R0 = zeros(1,length(wavelength));          % Initiate arrays for storing value of
T0 = zeros(1,length(wavelength));          % reflectivity & transmissivity

for k = 1:length(wavelength)
    lambda = wavelength(k);

% Propagation Matrix PTM in material-1 layer
    alfa1 = exp(1j*2*pi*n1*d1/lambda);
       A1 = [alfa1, 0; 0, 1/(alfa1)];

% Propagation Matrix PTM in material-1 layer
    alfa2 = exp(1j*2*pi*n2*d2/lambda);
       A2 = [alfa2, 0; 0, 1/(alfa2)];


    %=============================================================
    % Mirror M0 (see explanation 2)                             x
    %   n0-|-n1-|-|-n2-|...|-n1-|-|-n2-|-|-n1-|-|-n2-|-n_out    x
    %   B01-(A1-B12-A2-B21)...(A1-B12-A2-B21)-A1-B12-A2-B2t     x
    %   B01-(A1-B12-A2-B21)^(N-1)-A1-B12-A2-B2t                 x
    %=============================================================

    % TMM for the 1D PhC
    M0 = B01*(A1*B12*A2*B21)^(N-1)*A1*B12*A2*B2t;

    % reflection R0 & transmission T0 of PhC
    R0(k) = abs(M0(2,1)/M0(1,1))^2;
    T0(k) = n_out/n0*abs(1/M0(1,1))^2;    % see expl. (3)

    % save data in three column for lambda, R0, T0
    fprintf(rt,'%f %12.8f %12.8f\n',lambda,R0(k),T0(k));
end

% see expl. (4)
tick1 = 400:50:700;
figure(1)
  plot(wavelength,R0,'r', wavelength,T0,'b','linewidth', 2)
  legend('R_0', 'T_0');  grid minor
  set(gca,'FontSize',18);
  axis([400 700 0 1.05]);
  set(gca,'XTick',tick1);
  set(gca,'YTick',0:0.2:1);
  xlabel('Wavelength (nm)');
  ylabel('R & T');
  title( 'A=Ti_2O_5, B=SiO_2, N=6');

fclose(rt);                                  % for saving data
%=====================================================================
```

Let's see what we obtain when running the code by clicking on a green triangle  'Run' button on the top panel. Note that, for MATLAB® earlier 2016, we have to type the name of the program to run it. The explanations of the code will

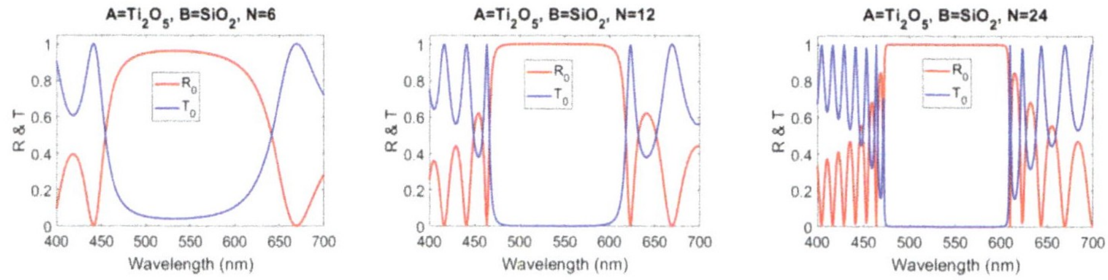

Figure 2.4. Calculated reflectivity (R) and transmissivity (T) of 1D PhC consists of $N = 6$, 12 and 24 pairs of layers Ti_2O_5 and SiO_2.

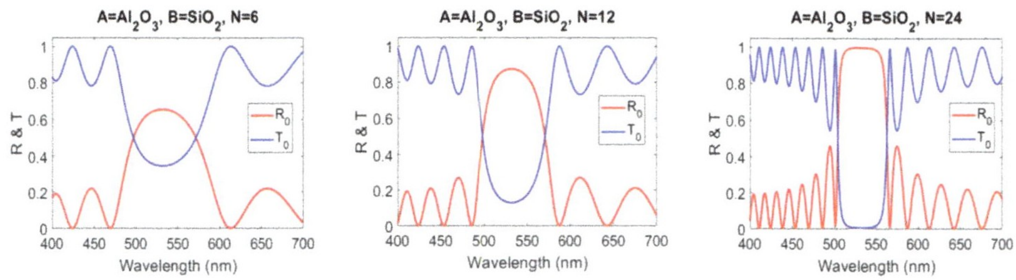

Figure 2.5. Calculated reflectivity (R) and transmissivity (T) of 1D PhC consists of $N = 6$, 12 and 24 pairs of layers Al_2O_3 and SiO_2.

be given later. Figure 2.4 below shows reflectivity R and transmissivity T spectra of 1D PhC with material layer A = Ti_2O_5, B = SiO_2, $N = 6$, 12 and 24.

At first glance, one can see clearly that the photonics band-gap is built up by increasing the number of layers. Note that the index contrast between the two materials in this example is very high $\Delta n = n_1 - n_2 = 0.68$. In such situations, usually a small number of layers is needed for fully establishing a band gap in which light is completely reflected. For example, figure 2.4 shows reflection reaches ~100% at and around λ_0 ($R \sim 1$) with $N = 6$, but a broad band of nearly 100 nm with $N = 12$.

Figure 2.5 shows the results of optical spectra of a PhC with different materials A = Al_2O_3 ($n \sim 1.68$) B = SiO_2 ($n \sim 1.46$) which are also widely used in coating applications with number of layer pairs $N = 6$, 12 and 24. In this case, the index contrast between the two materials $\Delta n = 0.22$ is smaller than the previous case with A = Ti_2O_5, B = SiO_2 with $\Delta n = 0.68$.

It is clear from the results in figures 2.4 and 2.5, that by changing the refractive indexes or the refractive index contrast between the two materials of the 1D PhC we can manipulate the refraction and transmission of the whole structure. For the PhC structures with smaller index contrasts the larger numbers of layers are required for completely establishing the band gap. Another important feature is that the band gap is also narrower in structures with smaller index contrast.

***Explanations of MATLAB® Program* 2.1**

Explanation-1

In the MATLAB® code 2.1 above for 1D PhC structure, the thicknesses of the two layers `d1` and `d2` are expressed as a function of Bragg wavelength `lamb0` and refractive index `n1, n2`:

```
lamb0 = 532.0;           % Bragg wavelength in nm, see Explanation (1)
d1 = lamb0/(4*n1);       % see Explanation-1
d2 = lamb0/(4*n2);       % see Explanation-1
```

First of all, assume that a reader does not know anything about the physics of Bragg's diffraction and Bragg's law [12, 13], and try to run program 2.1 with different thicknesses. After spending a lot of time trying with different parameters and in a very broad range of wavelength, he or she may come to realize that the photonics band gap will appear around the wavelength `lamb0` that relates to the thicknesses as defined in these expressions from Bragg's law, which describes the conditions for the constructive interference of diffraction to be at its strongest [13]

$$\lambda_0 = \frac{2d}{m}\sin(\theta), \tag{2.24}$$

where m is a positive integer (diffraction order), θ is glancing angle. Originally, the theory was developed for x-ray diffraction in crystalline solid. The waves are scattered from lattice planes separated by the interplanar distance d, and λ_0 is the resonant or Bragg wavelength of the crystal structure (see more details in [13]). The diffraction theory is outside the scope of our study, however, applying the similarity of scattering in periodic crystal and photonics crystal one could find great help in understanding PhCs. Without going into the detail of the diffraction theory, for the 1D PhC problem under consideration in which a light beam incident with electrics field is perpendicular to the layer plan of 1D PhC, Bragg's law can be written as

$$\lambda_0 = 2d, m = 1, \ \theta = \pi/2. \tag{2.25}$$

Note that in our structure the optical lattice $d = n_1d_1 + n_2d_2$ where d_1 and d_2 are the physical thickness of the two material layers (n_jd_j is optical thickness or optical path of j-layer). A simple design with $n_1d_1 = n_2d_2 = d/2$ that satisfies Bragg's law (2.25), then we can choose

$$\lambda_0 = 2(2n_1d_1) = 4n_1d_1 = 4n_2d_1 \rightarrow d_{1(2)} = \frac{\lambda_0}{4n_{1(2)}} \tag{2.26}$$

Equation (2.26) is the condition for strongest scattering or strongest reflection in photonics crystals, and in this case the photonics band gap occurs in the system. For that reason, the structures are usually called Bragg reflectors. This expression is used in our program so that band gap is obtained in the first run without spending a lot of time trying.

Explanation-2
This explanation is for the comments in the box below

```
%==========================================================
% Mirror M0 (see explanation 2)                            x
%  n_in-|-n1-|-|-n2-|...|-n1-|-|-n2-|-|-n1-|-|-n2-|-n_out  x
%    B01-(A1-B12-A2-B21)...(A1-B12-A2-B21)-A1-B12-A2-B2t   x
%    B01-(A1-B12-A2-B21)^(N-1)-A1-B12-A2-B2t               x
%==========================================================
```

The whole part is a comment. This is a simple but very effective way for us to remember important details of the problem, Sometimes, we want to use old programs for problems or fields that we haven't worked on for a long time, and we don't remember every detail of the models. Therefore, it is a good custom to make comments of some details of the problem that we are working on.

In this example, a diagram of a multilayered structure is characterized by the refractive indexes of each layer, including the input and output media, and below are the IMs for each interface and PMs for each material layer.

```
%  n_in-|-n1-|-|-n2-|...|-n1-|-|-n2-|-|-n1-|-|-n2-|-n_out  x
%    B01-(A1-B12-A2-B21)...(A1-B12-A2-B21)-A1-B12-A2-B2t   x
```

Taking into account the periodic property of the structure, the total matrix of the system can be rewritten as

```
%    B01-(A1-B12-A2-B21)^(N-1)-A1-B12-A2-B2t               x
```

The expression becomes very compact and useful for utilizing matrix utilities of MATLAB® to calculate a very complicated formula by a simple line:

```
% TMM for the 1D PhC
M0 = B01*(A1*B12*A2*B21)^(N-1)*A1*B12*A2*B2t;
```

As can be seen, the comment '`%`' is very useful for making clear some points in the program. It can be used to describe some details of the model. In this case, the whole TMM model of 1D PhC can be described just in few comment lines.

***Explanation*-3**
The expression of $R = |r|^2$, and $T = \frac{n_0}{n_{out}}|t|^2$ are the reflectivity and transmittivity that are calculated from amplitude of reflected and transmitted field from equation (2.4) above

```
% reflection R0 & transmission T0 of PhC
R0(k) = abs(M0(2,1)/M0(1,1))^2;
T0(k) = n_out/n0*abs(1/M0(1,1))^2;     % see expl. (3)
```

***Explanation*-4**
This explanation is for readers who have no prior or little experience with MATLAB®, especially the function `plot` in MATLAB®. It is worth noting that MATLAB® has a very powerful function of plotting figures. However, the commands for control of the features of the figures such as color and axis are not easy to remember, therefore, to master the experience, readers should try to understand all commands in this example

of the figure plot. Once we understand the main functionalities of those commands, we can re-use them later in future codes very conveniently.

A plot can be started by command `figure(n)`, `n` is the order number of the figure, follow by `plot(x,y,'c1', x,z,'c2' ...)`. The colors of curves in figures are controlled by `'c1', 'c2' for example 'r'` for red, `'b'` for blue etc.

```
plot(wavelength,R0,'r', wavelength,T0,'b','linewidth', 2)
```

2.2.3 Modeling example: distributed bragg reflector (DBR) fiber laser—cavity design

In the previous section, the TMM formalism has been used for calculating optical spectra of multilayered structures, especially the photonics band gaps and Bragg reflectors etc. In this section, we will show the formalism is very effective for calculating and analyzing resonant characterizations of optical resonators and laser cavities etc. Note that, the laser theory is complicated and can be found in many textbooks [14, 15], and is not the scope of our study in this book. Here, in this section we only want to show how to calculate and understand the optical properties of a very typical DBR laser cavity, especially single mode (SM), single frequency DBR fiber lasers. We can see that the TMM formalism is a very effective tool for designing the laser cavities, although it is quite simple.

First, let's study a simple resonant cavity which is the plane-parallel or Fabry–Pérot (FP) cavity, consisting of two opposing flat mirrors M_1 and M_2, as shown in figure 2.6. This cavity is also called linear cavity to distinguish it from a ring cavity, which is not a subject of our study. Typically, an FP laser includes a linear cavity, and a pump (electrical or optical field) used to excite ions of an active medium inside cavity, and amplify the stimulated emission light in the medium.

As the general and detailed theory of resonators or resonant cavity can be found in almost every textbook of optics and lasers, we will not repeat it here. The theory is so beautiful and is also a powerful example of using mathematic models for analyzing complex problems. The model provides very good characterizations of resonant mode of the cavity. Note that, for simplicity the mode characterizations that are widely presented in most of the textbooks are usually calculated by an *analytical model* with assumption that both mirrors M_1 and M_2 have the same reflectivity R. However, a *real* laser cavity typically has one mirror with very high refection (at the laser wavelength) $R \sim 100\%$ and the other with lower reflection for

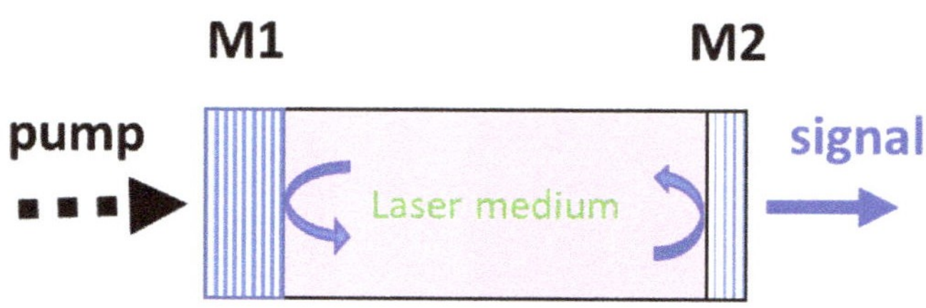

Figure 2.6. Schematic of a linear laser cavity with mirrors M1 and M2.

the laser output. At the same time, the mirror at the pumping side (if an optical pump is used) should be designed to have very high transmission at the pumping wavelength. Optimizing mirrors, especially reflection at the output side, play an important role for efficiency and total output power of the laser. We will see in the following the TMM formalism can be used for a real laser cavity with different mirrors. Interestingly, the TMM calculations though quite simple, can reveal some new properties of the cavity modes as compared to the results from the analytical model with equal reflections.

Inside the cavity, the amplified stimulated emission light is bounded by the mirrors or reflectors, and circulates multiple times. Under pumping excitation, the stimulated emission light gets amplified, and accumulates its power if the gain of the active medium (or laser medium) is higher than the total loss for the laser signal. At the same time, the counter-propagations of light in the cavity between the two mirrors produce standing waves for certain resonance frequencies or resonant modes of the cavity [14, 15]. Understanding the laser cavity including resonant modes and making use of it for designing and optimizing laser operation is very important in practice. It can be seen in the following example that the TMM formalism can be very effective for understanding and optimizing laser operation, especially for SM, and single frequency fiber lasers.

Note that a mirror that is constructed as a multilayered dielectric structure with Bragg resonance (see Explanation-1 for Program 2.1) can be called a Bragg reflector. Therefore, a laser cavity with two Bragg reflectors can be called a distributed Bragg reflector (DBR) laser cavity. Let us now calculating the mode characterization of the DBR laser cavity. First, from the structure in figure 2.6, we can present the main cavity as a structure consisting of an active medium embedded in between two mirrors M_1 and M_2 which are two periodic gratings, as shown in figure 2.7. In general, the refractive index of the active medium of the cavity (pink layer) n_c can be different from the n_a (gray) and n_b (blue) of the two layers of the gratings. Therefore, we can easily extend the TMM model for 1D PhC to study the resonant cavity.

For simplicity to understand the resonant modes of a cavity, let us start with a simple structure with small numbers of layers of materials. Therefore, we will examine a cavity with Bragg reflectors of grating with high index contrast, such as A = Ti_2O_5 and B = SiO_2 as in the previous example of 1D PhC.

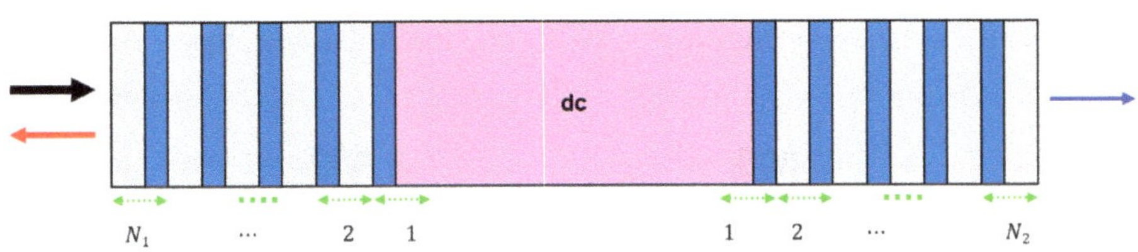

Figure 2.7. Schematic of a defected PhC having an active medium (pink layer) with thickness d and refractive index n embedded by two periodic gratings with N_1 and N_2 pairs of dielectric material layers whose thicknesses and refractive indexes of the first layer (gray color) and second layer (dark blue) are (d_1,n_1)and (d_2,n_2), respectively. An incident beam of light propagates from the left (black arrow), reflected to the incident medium of RI n_0 (red arrow), transmitted field (blue arrow) enters the output medium of RI n_{out}.

In terms of the TMM model, it is clear from figures 2.7 and 2.3 that the DBR laser cavity has few differences in comparison with the 1D PhC. In addition to all matrixes in 1D PhC, we would need two more interface matrixes IMs for two interfaces between the first and second gratings with the active medium layer, and one more PM for the active medium layer. Using the notations that are defined earlier, $B(n_1|n_c)$ and $B(n_c|n_1)$ can be written as

$$B(n_1|n_c) = \begin{pmatrix} n_{1c,p} & n_{1c,m} \\ n_{1c,m} & n_{1c,p} \end{pmatrix},\ B(n_c|n_1) = \begin{pmatrix} n_{c1,p} & n_{c1,m} \\ n_{c1,m} & n_{c1,p} \end{pmatrix}, \tag{2.27}$$

Here, as before, these notations $n_{ij,p}$ and $n_{ij,m}$ are defined

$$n_{ij,p} = \frac{1}{2n_i}(n_i + n_j),\ n_{ij,m} = \frac{1}{2n_i}(n_i - n_j), \tag{2.28}$$

and the PM for propagation inside the defect layer

$$A(n_c|d_c) = \begin{pmatrix} e^{ik_0 n_c d_c} & 0 \\ 0 & e^{-ik_0 n_c d_c} \end{pmatrix}. \tag{2.29}$$

Below is the MATLAB® code for calculating reflection and transmission of a resonant cavity that is built as an extension from Program 2.1.

MATLAB® Program 2.2.

```
%xxxxxxxxxxxxxxxxxxxxxxxxxxxxxxxxxxxxxxxxxxxxxxxxxxxxxxxxxxxxxxxxxx
%  LinearCavity.m                                                 x
%  Transfer Matrix for Defected PhC                               x
%                                                                 x
%          --->-|A|B|A|B|...|A|B|====|B|A|B|A|...|B|A|-->-        x
%            N1 pairs (A,B)       C      N2 pairs (A,B)           x
% xxxxxxxxxxxxxxxxxxxxxxxxxxxxxxxxxxxxxxxxxxxxxxxxxxxxxxxxxxxxxxxxx
format long

n1 = 2.14;                  % Ref. index, A-layer (Ti2O5)
n2 = 1.46;                  % Ref. index, B-layer (SiO2)
nc = 1.60;                  % Ref. index, C-layer (defect)
n_in = 1.0;
n_out = 1.0;

lamb0 = 532.0;              % Bragg wavelength in nm
d1 = lamb0/(4*n1);
d2 = lamb0/(4*n2);
dc = 20000;

N1 = 8;
N2 = 6;

% Input interface n0 -|- n1
n01_p = (n_in + n1)/(2*n_in); n01_m = (n_in - n1)/(2*n_in);
B01   = [n01_p, n01_m; n01_m, n01_p];

% Interface n1-|-n2
n12_p = (n1 + n2)/(2*n1); n12_m = (n1 - n2)/(2*n1);
B12   = [n12_p, n12_m; n12_m, n12_p];
```

```
% Interface n2-|-n1
n21_p = (n2 + n1)/(2*n2); n21_m = (n2 - n1)/(2*n2);
B21   = [n21_p, n21_m; n21_m, n21_p];

% Interface n2-|-nc
n2c_p = (n2 + nc)/(2*n2); n2c_m = (n2 - nc)/(2*n2);
B2c   = [n2c_p, n2c_m; n2c_m, n2c_p];

% Interface nc-|-n2
nc2_p = (nc + n2)/(2*nc); nc2_m = (nc - n2)/(2*nc);
Bc2   = [nc2_p, nc2_m; nc2_m, nc2_p];

% Output Interface n1-|-nL
n1t_p = (n1 + n_out)/(2*n1); n1t_m = (n1 - n_out)/(2*n1);
B1t   = [n1t_p, n1t_m; n1t_m, n1t_p];

% Output Interface n2-|-nL
n2t_p = (n2 + n_out)/(2*n2); n2L_m = (n2 - n_out)/(2*n2);
B2t   = [n2t_p, n2L_m; n2L_m, n2t_p];

%=====================================================================
% Model                                                              x
%                    Grating M1              Grating M2              x
%           --->-|A|B|A|B|...|A|B|====|B|A|B|A|...|B|A|-->-          x
%                    N1 pairs       dc        N2 pairs               x
%=====================================================================

wavelength = 450:0.002:650;
R01 = zeros(1,length(wavelength));
R02 = zeros(1,length(wavelength));
R = zeros(1,length(wavelength));
T = zeros(1,length(wavelength));

for k = 1:length(wavelength)
    lambda = wavelength(k);

% Propagation Matrix
    alfa1 = exp(1j*2*pi*n1*d1/lambda);
       A1 = [alfa1, 0; 0, 1/(alfa1)];

    alfa2 = exp(1j*2*pi*n2*d2/lambda);
       A2 = [alfa2, 0; 0, 1/(alfa2)];

    alfaC = exp(1j*2*pi*nc*dc/lambda);
       Ac = [alfaC, 0; 0, 1/(alfaC)];

    %=============================================================
    % Mirror M1                                                  x
    %   n0-|-n1-|-|-n2-|...|-n1-|-|-n2-|-|-n1-|-|-n2-|=nc          x
    %    B01-A1-B12-A2-B21...-A1-B12-A2-B21-A1-B12-A2-B2c         x
    %   B01-(A1-B12-A2-B21)^(N1-1)-A1-B12-A2-B2c                  x
    %=============================================================
    M1 = B01*(A1*B12*A2*B21)^(N1-1)*A1*B12*A2*B2c;
    % TMM for the single mirror M1
    M01 = B01*(A1*B12*A2*B21)^(N1-1)*A1*B12*A2*B2t;

    % TMM for the single mirror M2
    M02 = B01*(A1*B12*A2*B21)^(N2-1)*A1*B12*A2*B2t;
```

```
    %=============================================================
    % Mirror M2                                                  x
    %    nc=|-n2-|-|-n1-|...|-n2-|-|-n1-|-|-n2-|-|-n1-|-n_out    x
    %     Bc2-A2-B21-A1-B12...A2-B21-A1-B12-A2-B21-A1-B2t        x
    %    Bc2-(A2-B21-A1-B12)...(A2-B21-A1-B12)-A2-B21-A1-B2t     x
    %    Bc2-(A2-B21-A1-B12)^(N2-1)-A2-B21-A1-B2t                x
    %=============================================================
    M2 = Bc2*(A2*B21*A1*B12)^(N2-1)*A2*B21*A1*B1t;

    % TMM for cavity with active medium embedded between M1, M2
    Md = M1*Ac*M2;

    % Reflection & transmission of M1
    R01(k) = abs(M01(2,1)/M01(1,1))^2;
    R02(k) = abs(M02(2,1)/M02(1,1))^2;

    % Reflection & transmission of M2
    R(k) = abs(Md(2,1)/Md(1,1))^2;
    T(k) = n_out/n_in*abs(1/Md(1,1))^2;

end

atick1 = 450:50:650;
figure(1)
  plot(wavelength,R01,'r', wavelength,R02,'-.b','linewidth', 2)
  legend('R_1', 'R_2');
  set(gca,'FontSize',18);
  axis([450 650 0 1.05]);
  set(gca,'XTick',atick1);
  set(gca,'YTick',0:0.2:1);
  grid minor
  xlabel('Wavelength (nm)');
  ylabel('R & T');
  title( 'Reflection R_1 and R_2');

figure(2)
  plot(wavelength,T,'g','linewidth', 2)
  legend('T');
  set(gca,'FontSize',18);
  axis([450 650 0 1.05]);
  set(gca,'XTick',atick1);
  set(gca,'YTick',0:0.2:1);
  grid minor
  xlabel('Wavelength (nm)');
  ylabel('T');
  title( 'Cavity Modes: L=2cm');
%%================end==================================
```

We show in figure 2.8 resonant modes of a cavity with reflectivity $R_1 = R_2 \sim 0.9$ or 90% of reflection (left panel), and another cavity with $R_1 \sim 1$ (~100% of reflection) and $R_2 \sim 0.9$ (right panel). These results are obtained from running Program 2.2 with $N_1 = N_2 = 6$ pairs of A = Ti_2O_5 and B = SiO_2 (left panel), and with $N_1 = 8$, and $N_2 = 6$ (right panel). The cavity lengths in both cases are $L = 2$ cm.

It is very clear that transmissions of the resonant modes are quite different in the two cases. The cavity with equal reflections at both sides (left panel) provides uniform transmissivity of resonant modes. The result is well-known from the *analytical model* that has been often presented in many textbooks on optical resonators in which both mirrors have the same reflections. Meanwhile, the modes

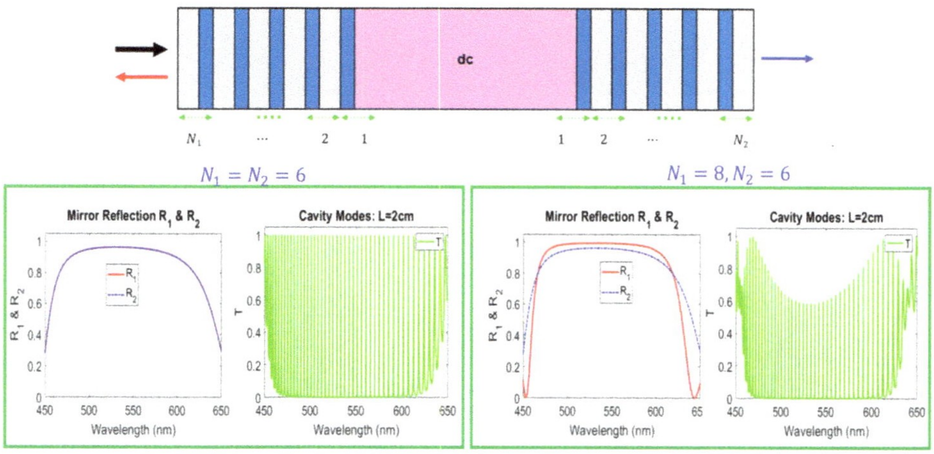

Figure 2.8. Calculated reflection R_1 and R_2 of two mirrors M_1 and M_2, and transmission of the whole resonant cavity T. Left panel: cavity with two mirrors have the same number of gratings $N_1 = N_2 = 6$. Right panel: cavity of different mirrors $N_1 = 8$, and $N_2 = 6$. Upper: schematic of resonant cavity with two mirrors M_1 and M_2.

have quite different transmissivities in the resonant cavities having two different mirrors which are typical structures of DBR laser cavity, or the modes experience different cavity losses. Cavity loss and gain for the signal are the key factors for laser operation. Therefore, although the TMM model for resonant cavity is quite simple, it is very effective for analyzing resonant modes of the real laser cavities with unequal mirror reflections.

It is worth pointing out that the frequency spacing between two adjacent resonant modes of a cavity is well-known from the analytical model as [14–16]

$$\nu_q = q\frac{c}{2d} = q\frac{c_0}{2nd}, \quad \Delta\nu_q = \frac{c_0}{2nd} \quad (q = 1, 2\cdots), \tag{2.30}$$

where c_0 is speed of light in vacuum, d is thickness of cavity material of RI n, q is the order number of resonant modes. In other words, the longer the length d, the closer the mode spacing. This property is clearly shown from TMM calculation in figure 2.9, in which mode spacing of cavity length $L = 10$ cm is about five times that of cavity $L = 2$ cm. However, the analytical model provides uniform resonant modes that may not be able to explain the different intensity distributions in lasers with multi-longitudinal modes.

We would like to stress here that the modes described above are longitudinal modes that are associated with the length of the cavity, in contrast with transverse modes which are the guiding modes in the waveguide that is used to form the cavity. The waveguide theory is out of the scope in this chapter, but it can be found in many textbooks [17, 18]. In chapter 3, we will have a brief discussion about the waveguide modes in the fiber amplifiers. In general, a laser cavity can have multiple longitudinal modes and multiple transverse modes. Typically, different longitudinal modes have different wavelengths. In general, the transverse modes (except for

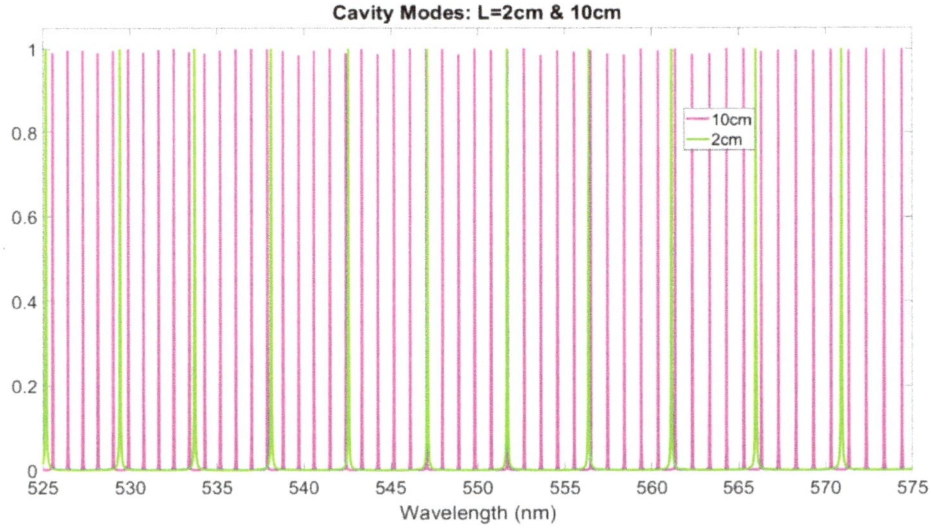

Figure 2.9. Calculated transmission T of the whole resonant cavity with different cavity lengths: $L = 2$ cm (green) and $L = 10$ cm (pink). Two mirrors have the same reflections as in the left panel of figure 2.8.

degenerated ones) of a waveguide have different effective indexes, and therefore propagate with different propagation constants. As a result, waveguides with multiple transverse modes would have a waveguide dispersion problem that is undesirable. Lasers that operate with SM transverse mode and SM longitudinal mode are usually called SM, single-frequency lasers and are very much desired for many applications. In the following, we will apply the concept that we have just learned about the laser cavity and resonant modes, and our TMM formalism to study another important application, the design of DBR fiber lasers.

Distributed Bragg reflector (DBR) fiber lasers: A DBR fiber laser usually has a linear laser cavity formed by an active (rare-earth-doped) fiber embedded between two fiber Bragg gratings (FGs). Optical fibers and fiber lasers are two big areas of modern technologies, and we are not trying to cover even a small part of these technologies. What we want to do is to help readers who already known fiber optics and fiber lasers but still find it difficult to understand the laser cavities and their behavior. The following example is very effective for designing the DBR fiber lasers, especially for SM, single frequency fiber lasers.

Let us consider the case in which the laser cavity is formed by SM transverse fiber, or usually called single mode fiber (SMF). An SMF is the fiber that allows only one transverse mode to propagate in the fiber core. The waveguide theory is too complicated to present it here. From the well-established theory [17–19], the condition for single mode fiber is quite simple $V = 2\pi \cdot a\sqrt{n_{core}^2 - n_{clad}^2}/\lambda \leqslant 2.405$, where a is radius of the core, n_{core} and n_{clad} are refractive indexes of the core and clad, respectively. The active fiber has a core doped with active materials to provide amplification for the signal under pumping conditions. Note that, the cores of SMFs typically have small diameters, typically less than 10 μm but they can be larger than

~50 μm in some special designs. Therefore, the active fibers of laser cavities are usually long, typically in the meter scale, sometimes much longer, to absorb enough pump power and to provide enough gain for the laser signal. From the results in figure 2.9 above, we can see that it is difficult to make truly single longitudinal mode fiber lasers when the cavity is made with fiber length of just 2 cm, which is extremely short for a fiber cavity. If the mirrors have a very broad reflection window the cavity would have many cavity modes spanning in a broad wavelength region. Those modes will get amplification and would be lasing if the gain is higher than cavity loss. As a result, the laser operates with multiple longitudinal modes. In the following, we show how TMM formalism can be very effective for designing SM, single frequency DBR fiber lasers.

First, let us describe a typical DBR fiber laser cavity. Usually (but not always) the cavity is formed by fusion splicing an active fiber with two FGs. The FGs are the fibers in which the refractive index of one section of fiber core is changed with a periodic modulation (grating). For some reason, the real profile of index modifications can be somehow different with a 1D PhC that we described in figure 2.1. However, the main function of the FG is to play a role of a Bragg reflector, and therefore it can be modeled very effectively by TMM model as presented earlier. The most important feature of an FG is that the index contrast between two layers of grating is very small, typically in the range from ~10^{-4} to 10^{-3}. That is because the index modification in the FG usually is made by UV light-induced refractive index change in glass. Readers who are interested in the technology can find details of it in [18, 19]. Depending on UV source conditions (wavelength, intensity and exposing time), the index modification δn in glass fiber can be different but is usually much smaller than the materials that are used in many mirror coating applications in previous examples.

Figure 2.10 shows a schematic of the fiber laser cavity, in which two FGs with different reflections are fusion spliced with a section of Er-doped glass fiber with fiber length L. It is worth noting that the fusion spliced technology makes the forming fiber cavities very simple. Typically, an experienced technician can create a simple cavity in a few minutes (however, it could take a much longer time in special cases with different glass fibers). in figure 2.10, the left figure shows calculated reflection spectra of two FGs, one has reflection ~100% at 1535 nm with $N = 4100$, and the other has reflection ~95% at 1535 nm with $N = 3000$. The right figure shows resonant modes of the cavity with length $L = 2$ and 10 cm. The spectra are calculated from the code above with small changes in indexes: $n_1 = 1.46$, $\delta n = 10^{-4}$, $n_2 = n_1 + \delta n$, and $N = 4100$ and 3000.

The results of cavity modes in figure 2.10 show that there are few modes in a very narrow band less than 0.1 nm (from 1535.2 to 1535.3 nm). The cavity with $L = 10$ cm has about 10 modes falling within the transmission band, while the cavity with $L = 2$ cm has two modes. Note that, the modes at the edges of the band gap are not actually resonant modes; we will examine these modes in more detail later in this chapter. Therefore, even with fiber cavity with only 2 cm length, which is extremely short for a fiber laser, it is not guaranteed that SM operation can be achieved.

There are several ways to have SM operation, one of these is to discriminate several modes in the short cavity $L = 2$ cm above by turning the reflections of two

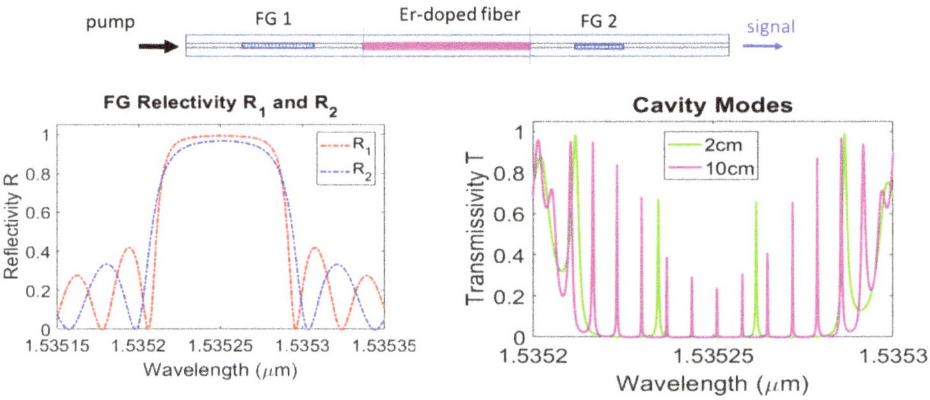

Figure 2.10. Calculated transmission and reflection of fiber grating with $\delta n = 10^{-3}$. N = 4100 (left), and N = 3000 (right). Top: schematic of DBR fiber laser cavity with two FGs fusion spliced with Er-doped fiber.

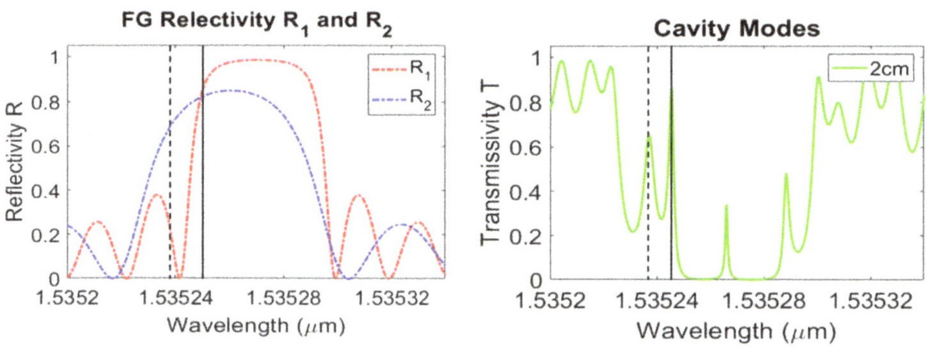

Figure 2.11. Turning reflections of the two FGs (left) and the only one strong resonant mode (at the solid line) will be lasing.

FGs slightly, as shown in figure 2.11 below. In that situation, by optimizing loss and gain we may achieve the laser with SM single frequency operation with the strongest resonant mode (solid line in the right figure). Other modes on the right of the lasing mode are discriminated by high cavity losses. Another mode (dashed line) is outside of the reflection window of one mirror, will therefore suffer huge loss and can be discriminated.

It is worth noting that in DRB fiber lasers that are SM, single frequency operation as described in this part with the requirement of very short fiber cavity is usually associated with some technical challenges. For example, usually FGs are written into photosensitive silicate fiber sections, but active dopants like Er, Yb etc for fiber laser at 1.5 and 1 μm cannot be doped with high concentration in silicate glass fiber. As a result, silicate glass Er-doped fibers should be long enough to have high gain for lasing conditions, typically in the scale of meters or even longer. Therefore, in an innovative way the active fibers are made from highly doped phosphate glass while the fiber gratings are written into photosensitive silicate fiber sections that are fusion spliced to the phosphate fibers [20]. In these devices both optical losses and mechanical instabilities at the splicing joints present inherent challenges owing to large differences in thermal properties, such as melting temperature and thermal expansion coefficient, between the different glasses. In practice, due to several problems such as thermal effect, the mode discrimination may not work perfectly, and there is hopping between the modes causing instability that requires special design to deal with.

Note that the scheme that is presented above is one among different ways. Another way is to further shorten fiber length. In that case, the active fiber must have very high gain and that is not easy because the concentration of active atoms that can be doped in glass is limited. Finally, it should be mentioned here that laser wavelength in the range of 1530–1550 nm is very popular for telecommunication and fiber lasers in that region that are made from Er-doped glass fibers. For a comprehensive theory of Er-doped and rare-earth-doped fiber amplifiers and fiber lasers, readers are referred to excellent books [21, 22].

2.2.4 Modeling example: distributed feedback fiber laser—cavity design

In this part, we will present another important example of designing a distributed-feedback (DFB) fiber laser using the TMM formalism. Again, we want to stress that understanding the whole laser system including laser medium, pumping schemes such as core pumped scheme or multimode cladding pumped ones would require advanced knowledge of laser technology, and this is beyond the scope of this book. Here, in this part only one aspect of the DFB fiber laser—the cavity design for an SM extremely narrow linewidth fiber laser can be understood and analyzed by the earlier experience with TMM. In other words, important characterizations of the DFB laser cavity can be calculated using the TMM model described earlier for 1D PhC.

As can be seen from the above example of DBR fiber laser, it is difficult to have an SM single frequency DBR fiber laser even with as short as 2 cm fiber cavity. In fact, there are very few commercial fiber lasers that are available [20–23]. Alternatively, a compact and very high beam quality laser can be made by a different approach, the distributed-feedback fiber lasers that we are going to study in the following. A distributed-feedback laser is a laser where the whole laser cavity consists of a periodic structure or grating with a defect on the gain medium. The grating with defect (see figure 2.12) acts as a distributed reflector in the wavelength range of laser action. This cavity structure essentially is a defected Bragg grating on the gain medium. As will be seen later in this part, the DFB fiber can be made for truly single mode both longitudinal and transverse operation. The mode hopping therefore can be avoided in a DFB laser. The fabrication involving the writing of a grating structure with ultraviolet (UV) light into an rare-earth doped fiber [24]. In addition, SM pump light can be sent into the fiber core that contains the active ions inside the laser cavity, leading to an alignment free resonator with optimum overlap of pump and signal light. A multimode cladding pump can also be used to achieve a higher output laser. The description and discussion of core pump and cladding pump scheme are presented in chapter 3. Readers who are interested in DFB fiber lasers

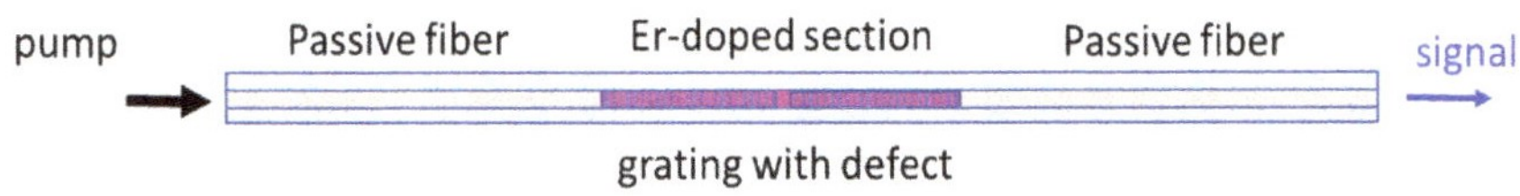

Figure 2.12. Schematic of DFB fiber laser in which the whole laser-medium resonator consists of periodic grating with a defect. A DFB fiber laser that is compact with SM operation can be achieved.

can find good references of DFB fiber lasers in [24, 25], in which details of the real Er^{+3}–Yb^{+3} co-doped fiber DFB lasers are described. Briefly, the DFB grating structure is written directly into the Er-doped doped SM fiber by exposure to 193 nm UV light through a phase mask. The grating consists of a 2 cm long section close to the pump side and a 1.5 cm long section at the single-mode DFB emission side that are separated by a 50 μm wide gap that creates the defect state inside the grating's reflection band. The overall ~3.5 cm long grating structure is located approximately at the center of the active fiber piece of 5 cm, and the asymmetric DFB grating design results in unidirectional DFB laser emission. The above description of the DFB fiber lasers seems quite simple, but it is difficult to model it analytically. In the following, we use TMM formalism to calculate and analyze the mode characterization of the DFB laser cavity, which is the most important for designing the cavity.

Figure 2.12 is a simplified diagram of a DFB fiber laser operating around telecom wavelength 1.55 μm, in which a fiber grating with a defect is written by UV light on a gain medium section of Er or Er-Yb co-doped glass fiber (pink section). The DFB fiber cavity—the section of grating on gain medium fiber—is then fusion sliced with two passive fibers for guiding the pump beam into the laser cavity and guiding out the lasing signal out of the cavity, as shown in figure 2.12. Note that the laser can be pumped by diode laser with 980 nm-wavelength for Er-Yb co-doped fiber, and at 1480 or 980 nm wavelength for Er-doped fiber [26, 27]. Pumping schemes with 1480 and 980 nm wavelength diode lasers will also be discussed in more detail in chapter 3. From the mode properties of the DFB cavity, we can see how SM, single frequency DFB lasers can be achieved. We want to stress again that the comprehensive theory of the DFB fiber laser is quite complicated. It includes several aspects that are not within the scope of our study. However, the calculations presented in the following will help readers to more easily understand the mode characterization of the DFB fiber cavity which is the most important property of the laser.

Before studying the resonant characterization of a DFB fiber cavity, let us consider a simpler model that can be regarded as a 1D PhC with a defect or defected PhC, as shown in the right panel of figure 2.13. Physically, the main difference between the defected PhC in figure 2.13, and the FP cavity in figure 2.6 is that the thickness of the defect is very small as to compared to the cavity length in FR cavity. Mathematically, they both can be described by the TMM model that we presented earlier for the DBR cavity. Using MATLAB® Program 2.2, we can obtain the reflection and transmission spectra of a defected PhC. Figure 2.13 shows spectra of PhC without defect (left) and with a defect (right), both structures have a total of 12 pairs of A = Ti_2O_5, B = SiO_2. Note that the resulting defected PhC's defect layer is 500 nm, which is very different

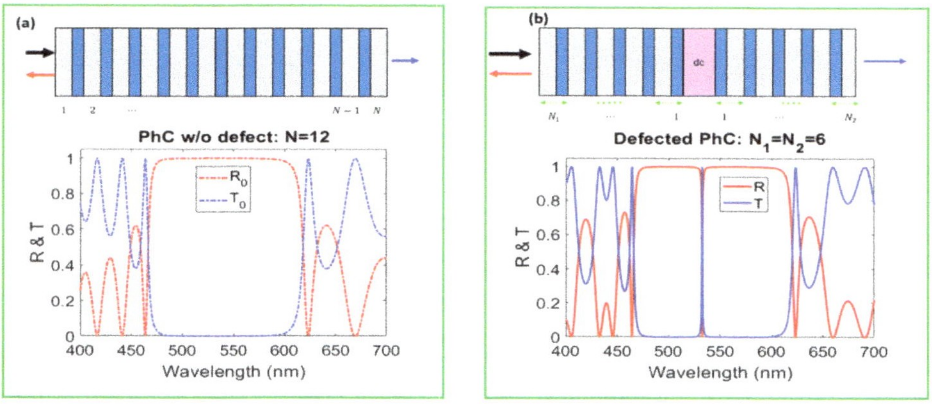

Figure 2.13. Upper: diagram of 1D PhC of N pairs of layers of two materials without defect (a) and with defect (b). Lower: calculated reflectivity (R) and transmissivity (T) with $N = 12$ for PhD without defect, and $N_1 = N_2 = 6$ for defected PhC. Defect has thickness dc = 500 nm and refractive index = 1.6.

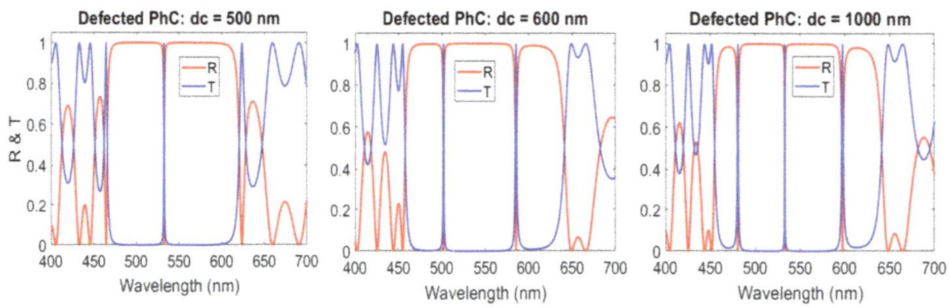

Figure 2.14. Calculated reflectivity (R) and transmissivity (T) of PhC with defects of 500, 600 and 1000 nm, and $N_1 = N_2 = 6$.

with the thickness of both layers $d_1 = \lambda_0/4n_1 = 530$ nm/(4 × 2.14) ~ 62 nm and $d_2 = \lambda_0/4n_2 = 530$ nm/(4 × 1.46)~ 91 nm, respectively.

Next, running the same Program 2.2 with different defect thicknesses of 600 and 1000 nm, we obtain the spectra shown in figure 2.14. The results show that depending on the defect thickness, there are different numbers of transmission modes inside the band gap. It appears that the number of these modes increases with thickness of the defect layer. We will show later that the light with wavelength at these modes has strongly resonant enhancement inside the structures. For that reason, these modes are called resonant modes of these structures.

Let us now apply the TMM formalism to calculate the resonant mode of a DFB fiber laser. A DFB fiber laser cavity consists of a defect grating that is written on a section of active fiber. Physically, the structure of a DFB fiber laser cavity is the same as in figure 2.13(b), but the defected grating is replaced by a defected FG written on a section of active fiber.

Figure 2.15(a) show the spectra of two FGs at two sides of the defect, and figure 2.15(b) is the spectra of the whole grating with a defect layer of 50 μm—the DFB fiber cavity.

Figure 2.16 shows experimental results of SM, single frequency DFB fiber laser with the DFB fiber cavity of about 3.5 cm, and total output power is limited by 160 mW, see more in [24].

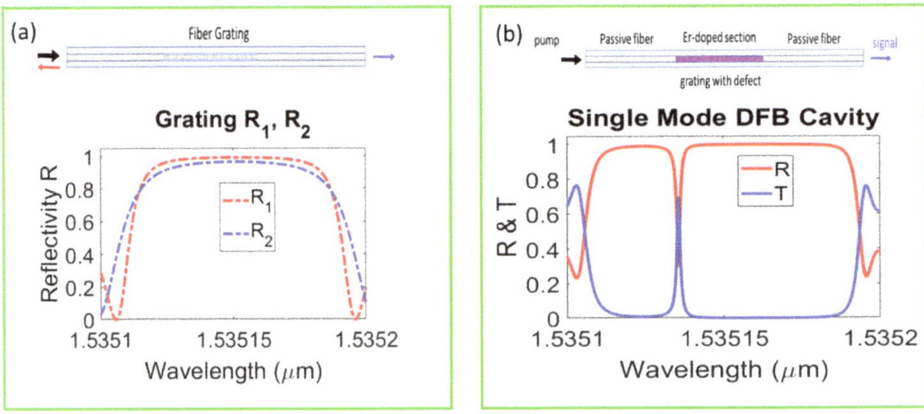

Figure 2.15. Upper: diagram of (a) FG, and (b) configuration of fiber cavity for an SM DFB fiber laser. Lower: calculated reflectivity (R) and transmissivity (T) of (a) fiber grating and (b) fiber DFB cavity.

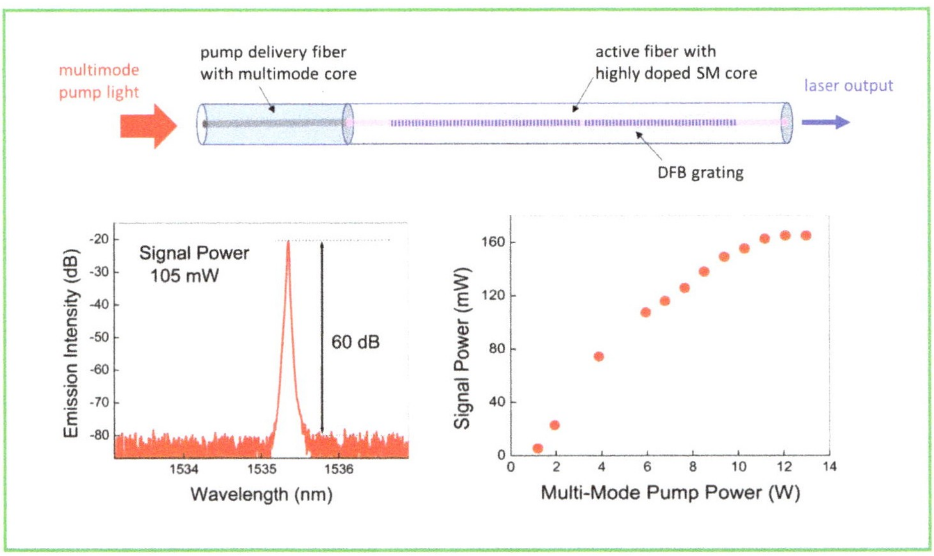

Figure 2.16. Upper: diagram of DFB fiber laser. Lower: experimental results of DFB single mode single frequency DFB laser (left) and output power of DFB laser versus pump power [24] (Permission from OSA).

The model calculations in this example have been used to design the real DFB fiber laser in [24]. Note that the length of active fiber is the whole cavity, which is only ~5 cm, but it is long enough for absorbing pumping power and providing a reasonable high output power [24, 25]. It is also important to stress here that the reflection band of the FG is very narrow ~0.05 nm, as shown in figure 2.15(a). Meanwhile, resonant mode in DFB cavity is much narrower ~0.005 nm. The very narrow resonant mode is important for a DFB fiber laser operating as an SM single frequency fiber laser. In practice, controlling index contrast δn of the grating is not perfect, and there are always fabrication imperfections, therefore, the linewidth of the laser is broader. However, in principle a DFB fiber laser can be a truly SM single frequency fiber laser with very narrow linewidth. It can also be very stable because it can avoid the mode hopping in the case of DBR fiber lasers. The details of the design will be presented in MATLAB® Program 2.3 below.

MATLAB® Program 2.3

```
%xxxxxxxxxxxxxxxxxxxxxxxxxxxxxxxxxxxxxxxxxxxxxxxxxxxxxxxxxxxxx
%   DFB_1C.m   Model of DFB cavity                            x
%   Transfer Matrix for DFB with 2 Grating + 1 Cavity         x
%   n = 1.5634, Delta_n = 1e-4                                x
%   L1 = 20 mm (N1~41000); L2 ~ 15mm (N2~30000)               x
%   Lc1 = 50micron                                            x
%   By dan Nguyen, 11/10/2019                                 x
% xxxxxxxxxxxxxxxxxxxxxxxxxxxxxxxxxxxxxxxxxxxxxxxxxxxxxxxxxxxxx
format long

N1 = 41000;                    % number of period
N2 = 30000;                    %
neff = 1.5634;                 % effective refractive index (Exp. 2-3.1)
d1 = 1.5352/(4*neff);
d2 = 1.5352/(4*neff);
dB = d1+d2;                    % Length of single period in micron
lambdaB = 2*neff*dB;           % Bragg wavelength in micron
Lc = 50.0;                     % Cavity length in micron

% Index
n1 = neff;                     % effective index of the propagation mode
dn = 1*1e-4;                   % Delta n
n2 = n1 - dn;
nc = n1;                       % neff in cavity = n1

n0 = 1.0;
n_out = 1.0;


% Input interface (1st from the left air-glass): n0-|n1
n01_p = (n0 + n1)/(2*n0); n01_m = (n0 - n1)/(2*n0);
B01   = [n01_p, n01_m; n01_m, n01_p];

% In grating: n1-|n2
n12_p = (n1 + n2)/(2*n1); n12_m = (n1 - n2)/(2*n1);
B12   = [n12_p, n12_m; n12_m, n12_p];
```

```
% In grating: n2-|n1
n21_p = (n2 + n1)/(2*n2); n21_m = (n2 - n1)/(2*n2);
B21   = [n21_p, n21_m; n21_m, n21_p];

% Output interface (rightest glass-air/substrate): n1-|n_out
n1t_p = (n1 + n_out)/(2*n1); n1t_m = (n1 - n_out)/(2*n1);
B1t   = [n1t_p, n1t_m; n1t_m, n1t_p];

% Unit Matrix
B11 =[1, 0; 0, 1];

%-----------------------------------------------------------------------
%
%     --->-||||||||||||||||||===|||||||||||||||||-->-
%                  M1          c1         M2
%
%-----------------------------------------------------------------------

dfb2 = fopen('rt1c.dat','w+');            % saving data

wavelength = 1.53505:0.0000001:1.53520;
R01 = zeros(1,length(wavelength));
R02 = zeros(1,length(wavelength));
R = zeros(1,length(wavelength));
T = zeros(1,length(wavelength));

for k = 1:length(wavelength)
    lambda = wavelength(k);

% Gratings
    alfa1 = exp(1i*2*pi*n1*d1/lambda);
    alfa2 = exp(1i*2*pi*n2*d2/lambda);
    A1 = [alfa1, 0; 0, 1/(alfa1)];
    A2 = [alfa2, 0; 0, 1/(alfa2)];

    %===============================================================
    % First Grating M1                                            x
    % Air --|-n1-|-(|-n2-|-|-n1-|)...-(|-n2-|--|-n1-|)-n1         x
    %     B01-A1-(B12-A2-B21-A1)-...-(B12-A2-B21-A1)-B11          x
    %     B01-A1-(B12-A2-B21-A1)^Na-B11                           x
    %===============================================================
      M1 = B01*A1*(B12*A2*B21*A1)^N1*B11;

    %===============================================================
    % Second Grating M2                                           x
    % n1--|n1|---(|n2|---|n1|)-...-(|n2|---|n1|--n_out            x
    %    B11-A1-(B12-A2-B21-A1)-...-(B12-A2-B21-A1)-B1t           x
    %    B11-A1-(B12-A2-B21-A1)^N2-B1t                            x
    %===============================================================
      M2 = B11*A1*(B12*A2*B21*A1)^N2*B11;

    % Propagation Matrix in Cavity of Lc
    alfaC = exp(1i*2*pi*n1*Lc/lambda);
    AC = [alfaC, 0; 0, 1/(alfaC)];

    % Transfer Matrix for FG1 & FG2
    M01 = B01*A1*(B12*A2*B21*A1)^N1*B1t;
    R01(k)  = abs(M01(2,1)/M01(1,1))^2;                  % FG1's R
```

```
    M02 = B01*A1*(B12*A2*B21*A1)^N2*B1t;
    R02(k) = abs(M02(2,1)/M02(1,1))^2;                    % FG2's R

    % Transfer Matrix for whole cavity
    M = M1*AC*M2*B1t;
    R(k) = abs(M(2,1)/M(1,1))^2;
    T(k) = n_out/n0*abs(1/M(1,1))^2;


 fprintf(dfb2,'%f %12.8f %12.8f %12.8f
 %12.8f\n',lambda,R01(k),R02(k),R(k),T(k));

 end

atick1 = 1.5351:0.00005:1.53520;
figure(1)
  plot(wavelength,R01,'-.r', wavelength,R02,'-.b','linewidth', 3)
  legend('R_1', 'R_2');
  set(gca,'FontSize',24);
  axis([1.5351 1.53520 0 1.05]);
  set(gca,'XTick',atick1);
  set(gca,'YTick',0:0.2:1);
  grid minor
  xlabel('Wavelength (\mum)');
  ylabel('Reflectivity R');
  title( 'Grating R_1, R_2');

figure(2)
  plot(wavelength,R,'r', wavelength,T,'b','linewidth', 3)
  legend('R', 'T');
  set(gca,'FontSize',24);
  axis([1.5351 1.53520 0 1.05]);
  set(gca,'XTick',atick1);
  set(gca,'YTick',0:0.2:1);
  grid minor
  xlabel('Wavelength (\mum)');
  ylabel('R & T');
  title( 'Single Mode DFB Cavity');

fclose(dfb2);

%%=====================end=======================================
```

The DFB fiber lasers that are designed as in the above example are truly SM and single frequency operation. The laser beam quality is therefore very good and in general the lasers are stable. However, due to very short active fiber of the cavity the output power of the laser is limited, see figure 2.16 and more details in [24, 25]. If the length of active fiber with written grating on it is increased to be longer, the grating itself prevents the laser signal from penetrating further into both sides of the grating to get more gain from the active medium under pump excitation in those parts of the cavity—remember the concept of photonics band gap. As a result, the laser signal cannot get more amplification in much longer DFB cavities as well as the output power laser. See more on that in Exercise 2.

A multiple frequencies, higher power laser with very stable operation would be important for some special applications. For example, dual-wavelength Er-doped fiber lasers have been used for optical fiber sensors [28, 29], microwave generation [30] etc, and various techniques have been developed to achieve stable dual-wavelength lasers. In the following, we show how a DFB fiber laser with a simple change can be a very effective approach for generating a dual-wavelength laser with reasonably high powers. Let us us first increase the thickness of the defect in DFB as we know from the previous example of defected PhC that multiple resonant modes would appear in that situation. However, there are some technical problems with that approach: (i) thickness of the defect layer must be well controlled to control the number of modes, and (ii) the total output power would not increase with multiple wavelengths as total length of active fiber is limited, as discussed earlier. However, if we design the DFB cavity as coupled cavities with three grating sections as shown in figure 2.17, we can increase the total active fiber length and output power, at the time dual-wavelength lasers can be achieved. Multiple wavelength with higher power can be obtained by the same concept using multiple grating sections. Figure 2.17 shows the design of a multiple section gratings DFB fiber laser, and the two resonant modes of the cavity. Readers can obtain different results themselves by running the program with varying index contrast of grating, thicknesses of the defects and lengths of each grating section etc.

This design concept has been applied for constructing a real DFB fiber laser with dual-wavelength [25]. The DFB fiber cavity has three grating sections with total length of active fiber of 10 cm as compared with ~5 cm in the previous case with only two grating sections. Figure 2.18 shows experimental results of a single (transverse) mode, dual-wavelength DFB fiber laser with the total output power of more than 1 W versus ~160 mW of single frequency DFB fiber laser, see more in [25]. The reason for higher out power is that in multiple grating sections the signal can penetrate in longer length of active fibers and therefore getting more amplification from pump-excited active medium.

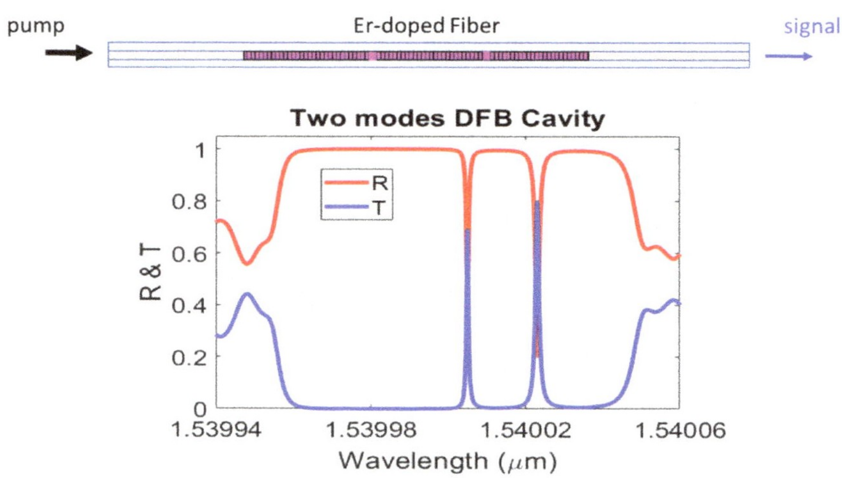

Figure 2.17. Upper: configuration of fiber cavity for two modes DFB fiber laser. Lower: calculated reflectivity (R) and transmissivity (T) fiber DFB cavity.

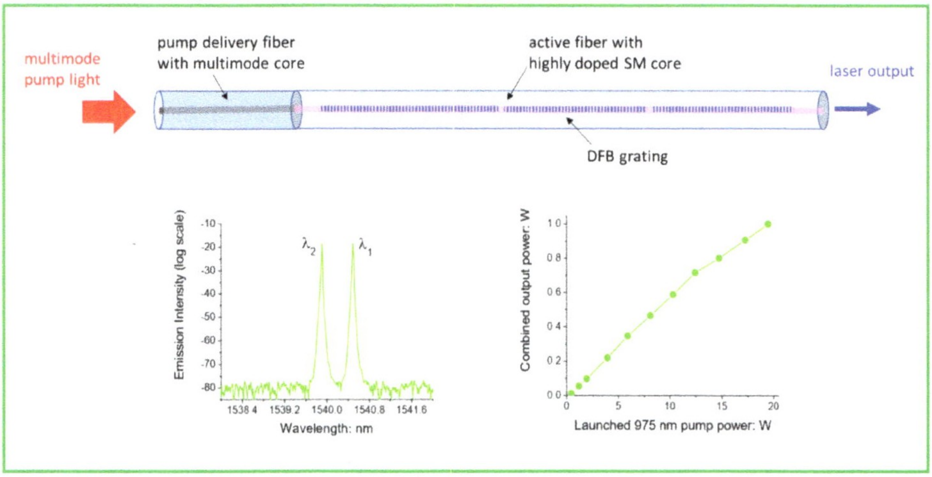

Figure 2.18. Upper: diagram of DFB dual-wavelength fiber laser. Lower: experiment results of DFB single (transverse) mode, dual-wavelength DFB laser (left) and output power of DFB laser versus pump power. Reprinted with permission from [25] copyright (2008) IEEE.

The model for this structure can be described in MATLAB® as:

```
%-----------------------------------------------------------------------
%     --->-||||||||||||||||||||===||||||===||||||||||||-->-
%                       M1          c1  M2    c2       M3
%-----------------------------------------------------------------------
```

Here, C1 and C2 are the two defects of the DFB cavity. We can extend Program 2.3 to describe the second defect with adding another PM $A(d_{C2}|n_{eff})$, and another fiber grating M3, which is similar to M2 in a real device [23], therefore, we do not need to write it here. The transfer matrix for the whole cavity can be expressed as:

```
N1 = 40000;                   % number of period
N2 = 20000;                   %
N3 = 30000;                   %
neff = 1.5634;                % effective index of the guided mode
d1 = 1.54005/(4*neff);
d2 = 1.54005/(4*neff);
Lc = 70.0;                    % cavity length in micron
```

The reflectivity and transmissivity of the cavity can be calculated as:

```
% Transfer Matrix for grating only
    M = M1*AC*M2*AC*M3;
    R(k) = abs(M(2,1)/M(1,1))^2;
    T(k) = nout/n0*abs(1/M(1,1))^2;
```

And running the program we obtain the results as shown in figure 2.16. The model calculations in this example have been used to design the real DFB fiber laser in [25].

2.2.5 Modeling example: quasi-periodic Fibonacci mirrors

Up to now, we have studied the systems that are periodic structures without or with defects, as presented in earlier sections, although the 1D PhC as periodic multiple layers structures have many interesting properties that can be applied for broad applications. It is well known that deviations from periodicity may result in higher complexity and give rise to a number of surprising effects. One such deviation can be found in the field of optics in the realization of photonic quasi crystals, a class of structures made from building blocks that are arranged using well-designed patterns but lack translational symmetry. It has been recognized that quasi-periodic systems could also lead to localization in optics [31–33]. A quasi-periodic system is neither a periodic nor a random one, so it could be considered as an intermediate between the two. Examples of such systems constructed with Fibonacci sequences include 1D quasi-crystalline Fibonacci dielectric multilayers (FDMLs) [34] and semiconductor quantum-wells [35] etc. In this example, we will use the very simple TMM formalism that we have learned earlier to calculate optical properties of complicated and interesting structures—the quasi-periodic Fibonacci dielectric multilayers [34].

Dielectric mirrors based on alternating stacks of dielectric layers have very broad applications in photonics and optoelectronic technology. Mirrors can be generated by sequentially depositing layers of two dielectric materials with significantly different refractive indices such as SiO_2 and Ta_2O_5, among other potential pairs. Conventionally, the alternating layers of dielectric materials are constructed periodically with some modifications; a typical periodicity is a quarter-wavelength. By changing the number of layers, layer thicknesses and periodicity, the value of the reflectivity and the reflection wavelength band can be accurately and precisely controlled. In this way, mirrors with reflectivity $0 \leqslant R \leqslant 100$ and reflection windows ranging from a few nanometers to hundreds of nanometers in width can be achieved [36]. However, it is very difficult to make mirrors with multiple reflection windows having similar values of reflectivity over a broad wavelength region, for example covering the entire visible region from 400 to 700 nm. Below we will study a new, elegant concept for mirror design in which mirrors can have tens or even hundreds of spectral windows with the same approximate reflectivity over a very broad spectral region. The new mirror concept is based on unique properties of 1D Fibonacci chains of dielectric layers that have been demonstrated experimentally for the first time. Multiple reflection window mirrors (MRWMs) can be used in many different photonics and optoelectronics applications. For example, MRWMs can be used in nonlinear cavities to enhance optical limiting effects for laser eye and sensor protection. It has been theoretically predicted and experimentally demonstrated that the nonlinear absorption in a micro-cavity filled with a nonlinear absorber is enhanced significantly [36]. This enhancement results in reduced transmission with increasing intensity of the incident radiation, e.g. nonlinear transmission (NT). Therefore, the new mirror design is likely to have many applications in broad areas of photonics and optoelectronics.

First, in mathematics the Fibonacci numbers are the integer numbers following a Fibonacci sequence, characterized by the fact that every number after the first two is the sum of the two preceding ones [31–35]:

$$S_j = S_{j-2} + S_{j-1}, \quad j \geqslant 3 \quad \text{with} \quad S_1 = 1, \; S_2 = 1. \tag{2.31}$$

From equation (2.31) we can easily write down the first numbers of the Fibonacci sequence as: 1, 1, 2, 3, 5, 8, 13, 21.... An important characterization of the Fibonacci sequence is the golden ratio which has been widely used in architecture, the arts, and also discovered in Nature.

At the beginning, quasi-periodic structures were mainly considered as suitable theoretical models to describe the conceptual transition from randomness to periodic order. Later, it was realized that the structures may offer interesting possibilities for technological applications as well. A particularly interesting 1D quasi-periodic model was proposed in [33] by Kohmoto *et al.* This model is based on the Fibonacci sequence which is constructed recursively with multiple layers of two materials [33–38].

$$S_j = S_{j-2}S_{j-1}, \quad j \geqslant 3 \text{ with } S_1 = A, \quad S_2 = B, \tag{2.32}$$

where $A(B)$ stands for a layer of material with refractive index $n_a(n_b)$ and thickness $d_A(d_B)$. As a result, we can easily write down formulae of elements and their corresponding structures. Figure 2.19 shows the first Fibonacci elements:

Note that most of the work on these 1D quasiperiodic PhC structures has focused on the properties of the different Fibonacci elements S_j, and the thicknesses of the layers A and B are chosen typically on the order of the wavelength of interest λ (usually λ, $\lambda/2$ or $\lambda/4$). Here, we consider an entirely new sequencing rule for the dielectric layers based on the above Fibonacci elements:

$$F_j = S_1S_2 \cdots S_{j-1}S_j. \tag{2.33}$$

Here, S_j are defined as in (2.31) and figure 2.19. We show in figure 2.20 as an example of the fifth order Fibonacci chain.

In other words, the jth-order Fibonacci structure is the orderly chain sequence of all Fibonacci elements up to the jth-order, S_j. Furthermore, the optical thicknesses of the two layers A and B can be on the order of multiples of a quarter-wavelength

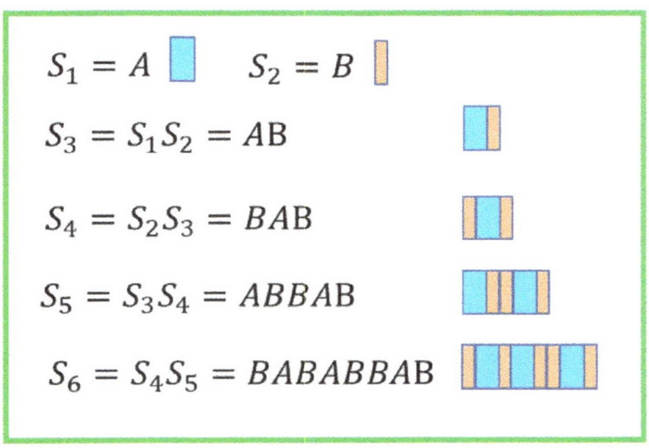

Figure 2.19. First Fibonacci elements of quasi-periodic multilayer dielectric structures.

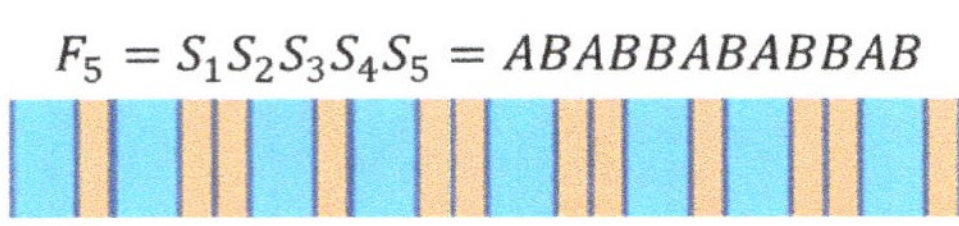

Figure 2.20. Fifth order Fibonacci chain of multilayers of dielectric.

depending on the number of reflection windows needed. It is worth noting that although the structures are simple and straightforward to make, they have very rich optical spectra. For examples, self-similarity of spectra of chains with different orders, and by changing the thickness ratio between the two material layers, mirrors with different numbers of windows at the same Fibonacci order can be created [33, 37, 38]. More interestingly, even in higher order chains with complicated structures, there are spectral regions where the transmission is ~100% which can be considered as localization of light or transmitted modes in the quasi-crystals [33].

Although the quasi-periodic Fibonacci chains are more complicated than the periodic multilayered structures that we have studied earlier, however, in terms of TMM calculation, it is quite similar in programming. As in previous examples, we will define all transfer matrixes of the program, such IMs $B(n_l|n_r)$ and PMs $A(d, n)$, then we can easily write TM for each order structure. For example, the fourth order chain F_4 can be written as:

```
% 4th-order FC
% Air ->-| na |-| nb |-| na |--| nb |--| nb |---| na |---| nb |->-
%        B0a*Aa*Bab*Ab*Bba*Aa*Bab*Ab*Bbb*Ab*Bba*Aa*Bab*Ab*Bbt
%        B0a*Aa*Bab*Ab*Bba*Aa*Bab*     A2b   *Bba*Aa*Bab*Ab*Bbt
F4 = B0a*Aa*Bab*Ab*Bba*Aa*Bab*A2b*Bba*Aa*Bab*Ab;
M4 = F4*Bbt;
R4 = abs(M4(2,1)/M4(1,1))^2;
T4 = n_out/n0*abs(1/M4(1,1))^2;
F5 = F4*Bba*Aa*Bab*A2b*Bba*Aa*Bab*Ab;
M5 = F5*Tbt;
r5 = abs(M5(2,1)/M5(1,1))^2;
t5 = ng/n0*abs(1/M5(1,1))^2;
```

Note that we can replace the two layers 'BB' by one double-thickness layer with the same material. In the program, the PM for that layer is defined as $A(2d_B, n_b)$.

With the simple rule for Fibonacci chain, e.g., equation (2.33), we can design MRWMs with relatively small numbers of layers. More interestingly, at the same Fibonacci order, by changing the thickness ratio between the two material layers, we can create mirrors with different numbers of windows. Figure 2.21 shows the fifth order chain F5 consisting of only 10 layers of A and B, since the two layers BB in figure 2.21 can be deposited as one layer of material B with thickness $2d_B$. We have made several mirrors based on the fifth order chain, F5, having two, and seven windows whose reflection spectra agree very well with theoretical predictions. The mirrors are made using fused silica substrates ($n \sim 1.446$) with coatings, which are formed by ion sputtering alternate layers of silicon dioxide (SiO_2) and tantalum pentoxide (Ta_2O_5). The index contrast between two materials plays a significant role in manipulation of reflection spectra, as shown in figures 2.22–2.24 below.

Our numerical calculations show that if the thicknesses of the layers are kept the same, but the order of the Fibonacci chain is increased, the number of windows does not change, but the magnitude of the reflection increases with the Fibonacci order. Figure 2.23 shows the reflectivity of the fifth, sixth and seventh order chains with the same fundamental thicknesses d_A and d_B. As can be seen from figure 2.5, it is possible to create MPBGs using high order chains with quite low index contrast

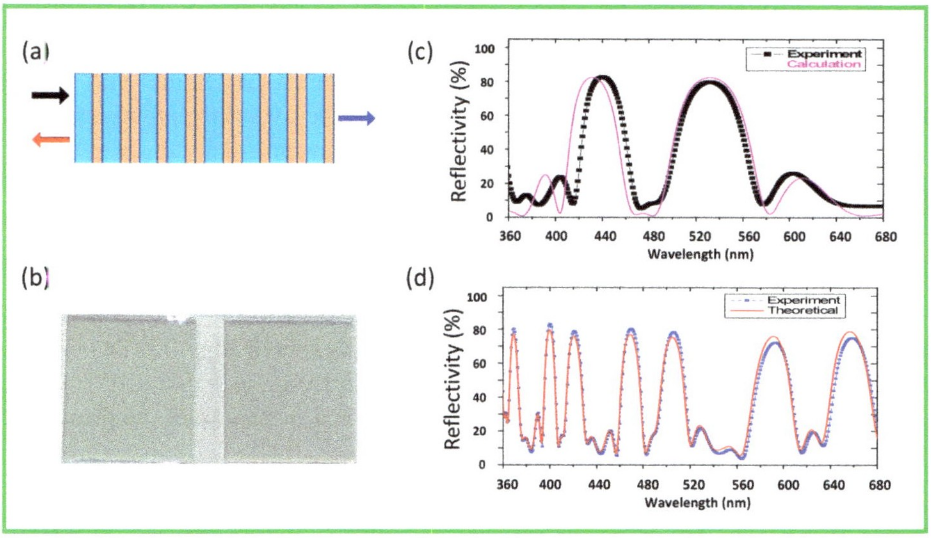

Figure 2.21. (a) Schematic of F5 with A = SiO_2 and B = Ta_2O_5 layers, (b) the mirrors fabricated with F5 structures. (c) Reflectivity of mirrors made by F5 with two windows with thickness $d_A = 81$ nm, $d_B = 165.7$ nm, ($m_A = 1$, $m_B = 3$, $\lambda_0 = 474$ nm), and (d) seven windows with $d_A = 81$ nm, $d_B = 497$ nm ($m_A = 1$, $m_B = 9$). Theory (solid), experiment (dotted). Reprinted from [34], copyright (2010), with permission from Elsevier.

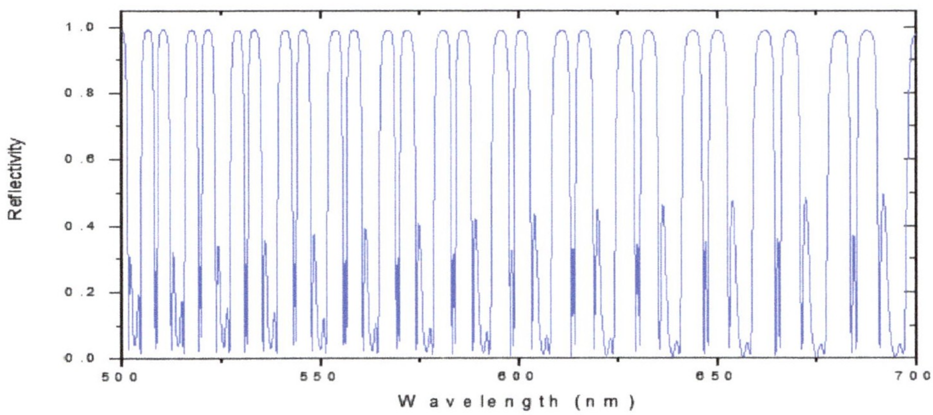

Figure 2.22. Reflectivity of mirror made by F5 with big index contrast $\Delta n = 1.6$, $m_A/m_B = 40$. Reprinted from [34], copyright (2010), with permission from Elsevier.

between the two materials. For example, MPBGs can be generated in seventh order Fibonacci chains (26 layers) with index contrast as low as $\Delta n = 0.68$. Notably, the positions of the main reflection windows are not changed when the Fibonacci chain order is changed. However, the spectra of higher order chains consist of some windows with narrower bandwidths and lower reflectivity.

In this part we present the results of TMM calculations of reflection of quasi-periodic Fibonacci mirrors, a new concept for the design of multiple reflection

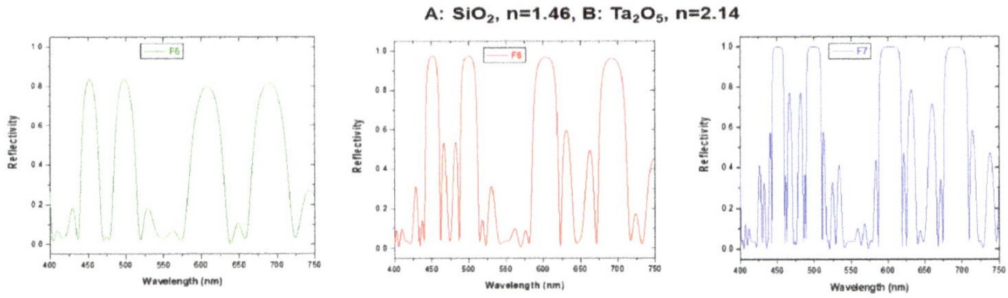

Figure 2.23. Reflectivity of mirrors made by F5 (blue), F6 (red) and F7 (green). A = SiO_2 and B = Ta_2O_5 with thickness d_A = 81.06 nm, d_B = 386 nm (m_A = 1, m_B = 7 and λ_0 = 474 nm). Reprinted from [34], copyright (2010), with permission from Elsevier.

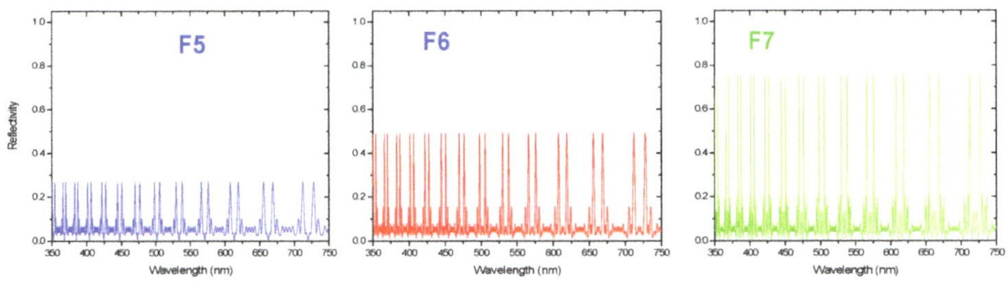

Figure 2.24. Reflectivity of mirrors made by F5 (blue), F6 (red) and F7 (green). A = SiO_2 and B = Al_2O_3 with thickness m_A = 1, m_B = 20. Reprinted from [34], copyright (2010), with permission from Elsevier.

window mirrors in which simple structures are based on a Fibonacci chain of dielectric layers. The new concepts have been demonstrated experimentally with A = SiO_2 and B = Ta_2O_5, having indices 1.46 and 2.14, respectively. The unique Fibonacci chain approach described above has the potential for many applications such as optical limiting in which the MRWM can be used for enhancing nonlinear transmission effects in cavities over a broad region, such as the entire visible [36]. Nonlinear dynamic filters based on MRWM can also be used for optical limiting and sensor protection. Simple MRWMs can also be very effective as frequency combs with many closely spaced reflection/transmission windows. Finally, the discovery of multiple photonic band gaps in such simple structures presents intriguing opportunities for future applications.

2.3 Nonlinear transfer matrix method formalism

Up to this point, all TMM models and calculations are carried out with an assumption that all optical properties of the materials are intensity-independent. Normally, the assumption can be considered as a good approximation when light intensity is not very strong. In the linear regime, all material parameters such as refractive indexes, absorption coefficients etc, can be treated as constants as we did in the previous linear TMM formalism. It is well-known that under irradiation by strong intensity light, such as strong laser beams, especially the laser pulses with high peak intensities, the intensity-dependence of material parameters become significant and cannot be ignored in the modelling and calculations. In such situations, calculations in the linear regime become inaccurate, and even completely wrong if the nonlinear effects—the intensity-dependence of material properties—are strong.

In this part, we will present a new TMM formalism to calculate nonlinear response of multilayered structures, or nonlinear TMM formalism. As can be seen, the nonlinear formalism is much more complicated than the linear one. We follow closely the nonlinear TMM formalism that has been developed by Dankaert *et al* [9]. Readers are encouraged to find a very good theory of this method in the reference.

2.3.1 General equations

Let us consider the multilayered structure with the forward (red arrows) and backward (blue arrows) fields as presented in figure 2.25 below.

Let us assume that each layer is characterized by its thickness $d_j = z_j - z_{j-1}$, its linear refractive index n_{0j} related to the real part of the linear susceptibility $\chi_j^{\prime(1)}$ (j is indexes of the jth layer),

$$n_{0j} = \sqrt{1 + \chi_j^{\prime(1)}}, \tag{2.34}$$

and its attenuation coefficient α_j related to the imaginary part of the linear susceptibility $\chi_j^{\prime\prime(1)}$:

$$\alpha_j = \frac{\omega \chi_j^{\prime\prime(1)}}{n_{0j} \cdot c}. \tag{2.35}$$

Here, ω is frequency, c is speed of light in vacuum. For simplicity, we will restrict ourselves to considering an isotropic medium, and as a result of that the third-order susceptibility $\chi_j^{(3)}$ can be simply considered as a scalar. The electric field amplitude E_j of the normally incident, monochromatic wave with frequency ω inside each layer j can be decomposed in a forward and a backward propagating beam [9]:

$$E_j = E_{Fj} e^{-ik_j z + i\phi_{Fj}(z) - \alpha_j(z - L_j)/2} + E_{Bj} e^{+ik_j z + i\phi_{Bj}(z) + \alpha_j(L_j - z)/2}. \tag{2.36}$$

Here $k_j = \frac{\omega}{c} n_{0j}$ is the wave number, and $\varphi_{F(B)j}(z)$ are the phase shifts due to the nonlinearity (inside the layer j).

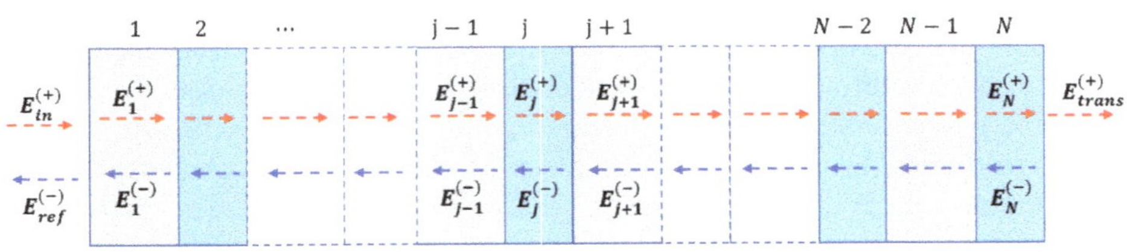

Figure 2.25. Schematic of multilayered structure with forward and backward electric fields in the structure.

In the transfer matrix formalism, the amplitudes of electric fields can be expressed as

$$\begin{pmatrix} E_{Fj} \\ E_{Bj} \end{pmatrix} = M_{j,j+1} \begin{pmatrix} E_{Fj+1} \\ E_{Bj+1} \end{pmatrix}. \tag{2.37}$$

The nonlinear phases can be determined by substituting (2.36) into the nonlinear wave equation

$$\frac{\partial^2 E_j}{dz^2} + \frac{\omega^2}{c^2}(1 + \chi_j^{'(1)} - i\chi_j^{''(1)} + \chi_j^{(3)}|E_j|^2)E_j = 0. \tag{2.38}$$

Using the slowly varying envelope approximation (SVEA) (see appendix B), we have

$$\partial_z\phi_{Fj}(z) = -\frac{\omega\chi_j^{(3)}}{2n_{0j}c}\{|E_{Fj}|^2 e^{-\alpha_j(z-L_{j-1})} + 2|E_{Bj}|^2 e^{-\alpha_j(L_j-z)}\}, \tag{2.39}$$

$$\partial_z\phi_{Bj}(z) = +\frac{\omega\chi_j^{(3)}}{2n_{0j}c}\{2|E_{Fj}|^2 e^{-\alpha_j(z-L_{j-1})} + |E_{Bj}|^2 e^{-\alpha_j(L_j-z)}\}. \tag{2.40}$$

The SVEA condition merely states that the phase variation due to the nonlinear refraction and the amplitude variation due to linear absorption are small compared to the wave vector inside the layer:

$$\alpha_j, \quad \frac{\omega\chi_j^{(3)}|E|^2}{c\, n_{0j}} \ll k_j \text{ or, equivalent } \chi_j^{''(1)}, \quad \chi_j^{(3)}|E|^2 \ll n_{0j}^2 \tag{2.41}$$

The boundary conditions express the continuity of the electric and magnetic fields across each interface. The magnetic induction fields inside layer j :

$$B_{Fj}(z) = -\frac{k_{Fj}(z)}{\omega}E_{Fj}, \quad B_{Bj}(z) = +\frac{k_{Bj}(z)}{\omega}E_{Bj}. \tag{2.42}$$

Where we introduced a compact notation for the generalized wave numbers:

$$k_{Fj}(z) = k_j - \partial_z\phi_{Fj}(z) - i\frac{\alpha_j}{2}, \tag{2.43}$$

$$k_{Bj}(z) = k_j + \partial_z\phi_{Fj}(z) - i\frac{\alpha_j}{2}, \tag{2.44}$$

The magnetic field in layer j then can be written as

$$\begin{aligned} B_j &= B_{Fj}e^{-ik_jz+i\phi_{Fj}(z) - \alpha_j(z-L_j)/2} + B_{Bj}e^{+ik_jz+i\phi_{Bj}(z) + \alpha_j(L_j-z)/2} \\ &= -\frac{k_{Fj}}{\omega}E_{Fj} \cdot e^{-ik_jz+i\phi_{Fj}(z) - \alpha_j(z-L_j)/2} + \frac{k_{Bj}}{\omega}E_{Bj} \cdot e^{+ik_jz+i\phi_{Bj}(z) + \alpha_j(L_j-z)/2}. \end{aligned} \tag{2.45}$$

Using the boundary condition for both EM-fields, we have

$$E_j(L_j) = E_{j+1}(L_j), \tag{2.46}$$

or

$$\begin{aligned} &E_{Fj}e^{-ik_jL_j + i\phi_{Fj} - \alpha_j (L_j - L_{j-1})/2} + E_{Bj}e^{+ik_jL_j + i\phi_{Bj}(L_j)} \\ &= E_{Fj+1}e^{-ik_{j+1}L_j + i\phi_{Fj+1}} + E_{Bj+1}e^{+ik_{j+1}L_j + i\phi_{Bj+1}(L_j) - \alpha_{j+1} (L_{j+1} - L_j)/2} \end{aligned} \tag{2.47}$$

and

$$B_j(L_j) = B_{j+1}(L_j), \tag{2.48}$$

or

$$\begin{aligned} &- k_{Fj}E_{Fj}e^{-ik_jL_j + i\phi_{Fj} - \alpha_j (L_j - L_{j-1})/2} + k_{Bj}E_{Bj}e^{+ik_jL_j + i\phi_{Bj}(L_j)} \\ &= -k_{Fj}E_{Fj+1}e^{-ik_{j+1}L_j + i\phi_{Fj+1}} + k_{Bj}E_{Bj+1}e^{+ik_{j+1}L_j + i\phi_{Bj+1}(L_j) - \alpha_{j+1} (L_{j+1} - L_j)/2} \end{aligned} \tag{2.49}$$

The boundary condition at interface $z = L_j$ can be written in matrix form as below

$$A_j\begin{pmatrix} E_{Fj} \\ E_{Bj} \end{pmatrix} = A_{j+1}\begin{pmatrix} E_{Fj+1} \\ E_{Bj+1} \end{pmatrix} \Rightarrow \begin{pmatrix} E_{Fj} \\ E_{Bj} \end{pmatrix} = A_j^{-1}A_{j+1}\begin{pmatrix} E_{Fj+1} \\ E_{Bj+1} \end{pmatrix} = M_{j,j+1}\begin{pmatrix} E_{Fj+1} \\ E_{Bj+1} \end{pmatrix}. \tag{2.50}$$

where

$$A_j(z) = \begin{bmatrix} e^{-ik_jz - \alpha_j (z - L_{j-1})/2}e^{i\phi_{Fj}(z)} & e^{+ik_jz + \alpha_j (z - L_j)/2}\, e^{i\phi_{Bj}(z)} \\ - k_{Fj}(z)e^{-ik_jz - \alpha_j (z - L_{j-1})/2}e^{i\phi_{Fj}(z)} & k_{Bj}(z)e^{+ik_jz + \alpha_j (z - L_j)/2}\, e^{i\phi_{Bj}(z)} \end{bmatrix}. \tag{2.51}$$

The transfer matrix has the form

$$M_{j,j+1} = e^{-i\delta_j}\begin{pmatrix} m_{11} & m_{12} \\ m_{21} & m_{22} \end{pmatrix}, \tag{2.52}$$

Here, $\delta_j = \phi_{Fj}(L_j) + (k_{j+1} + k_j)L_j$, and

$$m_{11} = \left(\frac{k_{Bj} + k_{Fj+1}}{k_{Bj} + k_{Fj}}\right) \times e^{2ik_jL_j + \alpha_j(L_j - L_{j-1})/2}, \tag{2.53}$$

$$m_{12} = \left(\frac{k_{Bj} - k_{Fj+1}}{k_{Bj} + k_{Fj}}\right) \times e^{2i(k_{j+1} + k_j)L_j + \alpha_j(L_j - L_{j-1})/2 + \alpha_{j+1}(L_j - L_{j+1})/2}, \tag{2.54}$$

$$m_{21} = \left(\frac{k_{Fj} - k_{Fj+1}}{k_{Bj} + k_{Fj}}\right) \times e^{-i[\phi_{Bj}(L_j) - \phi_{Fj}(L_j)]}, \tag{2.55}$$

$$m_{22} = \left(\frac{k_{Fj} + k_{Bj+1}}{k_{Bj} + k_{Fj}}\right) \times e^{-i[\phi_{Bj}(L_j) - \phi_{Fj}(L_j)] + 2ik_{j+1}L_j + \alpha_{j+1}(L_j - L_{j+1})/2}. \tag{2.56}$$

The nonlinear TMM formalism is the system of equations (2.50–2.55) which is much more complicated than the linear one that has been used for calculation of linear systems in the previous sections 2.1 and 2.2. In the following, we will apply the nonlinear TMM formalism to investigate several problems with emphasis on applications. Note that in derivation of the above equations, we followed Danckaert *et al* [9], but readers can find more details of the calculation in the reference.

2.3.2 Modeling example: nonlinear defected PhC structures

In this part, we will apply the nonlinear TMM formalism to study the nonlinear properties of 1D PhC structures under assumption of nonlinear regime, in which optical properties of materials are considered as intensity-dependent. More specifically, we will restrict ourselves to the cases in which the material refractive index can be expressed as $n = n_0 + n_2 I$, where I is light intensity, n_0 and n_2 are linear and nonlinear RIs, respectively. Furthermore, for simplicity we assume these materials have only linear absorption or absorption coefficient is a constant. Nonlinear absorption mechanisms that are intensity-dependent such as two-photon absorption, saturable absorption etc are not considered here. In general, the assumption that materials have only nonlinear refraction indexes $n = n(I)$, but only linear absorption α = constant, is not a very good approximation, especially if laser wavelengths are close to nonlinear absorption resonances of the materials. However, in most situations when wavelengths of lasers are far from the nonlinear absorption resonances, the assumption is considered a reasonably good approximation. We will see that the nonlinear reflectivity and transmissivity of a defected PhC structure become dependent on light intensity. Moreover, nonlinear effects including mode shifting, optical bistability (OB) etc, that have been experimentally realized for applications, can be calculated and analyzed by nonlinear TMM formalism. Note that, there are different approaches to achieve OB [39, 40], but in principle OB can occur only in systems that have nonlinear effect and optical feedback [41]. Here, from the following example, readers could have learned how OB can occur in a simple structure of nonlinear defect PhC. They may come up with new ideas and can design themselves new devices of all-optical switching using the nonlinear TMM formalism and MATLAB® program provided from the following example.

In section 2.2.4, we have calculated the linear spectra of several defected PhC (DPhC) structures. We know from those examples that there are inter-bands within the band gap of the defected PhC that are completely transmitted and are considered as resonant modes of the structures. Before going to investigate the nonlinear DPhCs, let us re-examine the linear spectra of a special PhC structure that can be used to realize all-optical switching or OB. The structure is similar to the DPhCs that we studied in sections 2.2.2 to 2.2.4 in which the thickness thicknesses d_a and d_b of material layers A and B, respectively are defined as previously as

$$d_a = \lambda_0/(4n_a),\ d_b = \lambda_0/(4n_b), \tag{2.57}$$

where n_a and n_b are linear refractive indexes of layers A and B of the grating, respectively. However, the thin defect layer is assumed to be an optically nonlinear material having nonlinear refractive index $n_c = n_{0c} + n_{2c}I$. The thickness d_c and linear refractive index n_{0c} are related by

$$d_c = m\lambda_0/(4n_{0c}), \tag{2.58}$$

where m is an integer number that we will determine later, λ_0 is Bragg resonance of 1D PhC of two material layers A and B. In the linear regime when light intensity is week, the nonlinear term in RI $n_{2c}I \ll n_{0c}$ and can be neglected. In such situations, we can use linear TMM formalism to calculate the spectra of the structure, and the results are shown in figure 2.26 below.

Spectra on the left and the right plots in figure 2.26 show resonant modes of DPhC where m is odd and even, respectively. The results can be obtained by running Program 2.2 provided earlier in this chapter with A = Ti_2O_5 ($n_a = 2.14$), B = SiO_2 ($n_b = 1.46$) and $n_c = 1.60$, and $\lambda_0 = 550$ nm. Now, we can see that the resonant modes of the DPhC can be manipulated in a very simple way. More interestingly, in the case with m even, while the band gap is narrower with increasing m, the central mode has not moved, and is exactly at the Bragg resonant wavelength. That simple observation will be very useful for designing devices with desired resonant modes. Furthermore, when m is even and larger than 4 (m = 6, 8…) or the defect layer is thicker than λ_0/n_{0c} then the DPhC has more than one resonant mode. Note that, the transmission curves with m = 2 (red) and m = 4 (blue) have only a single resonant mode, although the spectra have 100% transmission at the edge of the bad gap. In the Exercise at the end of this chapter, we can prove to ourselves that these two high transmissions are not resonance modes, meaning that there is no enhancement inside the cavity or structures for the light at edge of the band gap. Furthermore, we will

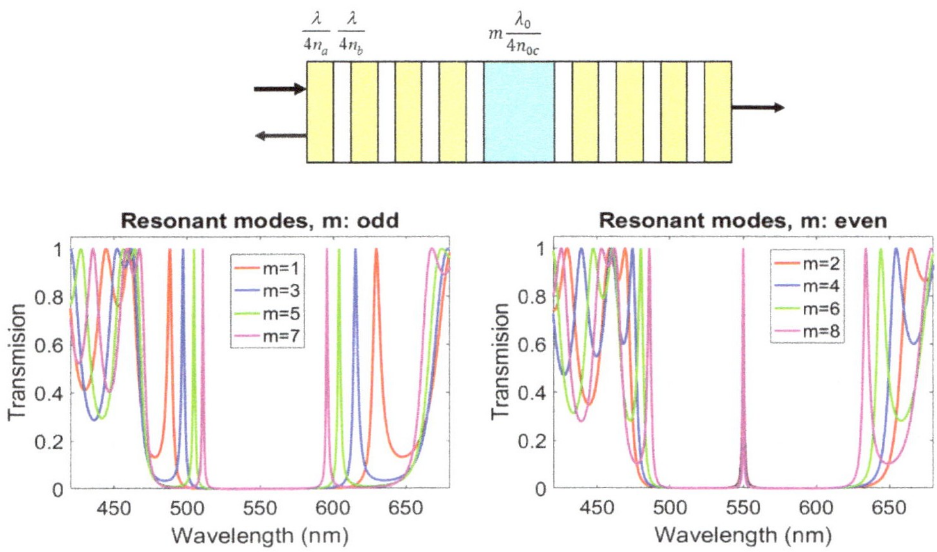

Figure 2.26. Upper: schematic of nonlinear DPhC, in which the defect layer thickness and linear refractive index are related as $d_c = m\lambda_0/(4n_{0c})$. Lower: linear transmission of DPhC: m is odd number (left), and m is even number (right).

learn in that Exercise that the most important and interesting feature of the resonant modes is that the modes are strongly enhanced in these resonant cavities. The main reason for the enhancement of the resonant modes is that the light is strongly confined and reflected multiple times by the two mirrors/reflectors that are embedded in the cavity or the defect layer in DPhC. The resonant light circulates multiple times inside the structure, meanwhile the incident field continuously enters the structure/cavity. As a result, the resonant light is accumulated inside the cavity and its intensity increases significantly compared with the input intensity of the incident field. Understanding the effect of enhancement of resonant light in resonant cavities is very important for optimizing the design of devices.

In general, optical materials typically have very small values of nonlinear refractive index n_2, and therefore in practice, usually very strong laser intensity is required to realize the nonlinear effects. In the case with resonant cavity or DPhC, the light intensity at resonant modes can get enhanced tens or even hundreds of times depending on cavity design, as can be seen in the Exercise. As a result, nonlinear effects are much more strongly enhanced for the resonant modes, and the effects can be realized without the need for very strong intensity lasers, as in the cases of non-resonant modes with and without cavity.

Let us now consider a nonlinear DPhC, in which a thin layer of optical material having nonlinear refractive index n_{2c} is embedded by two gratings that are two Bragg reflectors made by two material layers A = Ti_2O_5 and B = SiO_2. Note that both materials Ti_2O_5 and SiO_2 have negligible weak nonlinear RIs, therefore, we can consider them as linear materials with linear RIs n_a and n_b, respectively. The number of pairs A and B of layers for each of the gratings are N_1 and N_2. The thickness of the defect layer $d_c = m\lambda_0/(4n_{0c})$ is chosen with $m = 4$. In the following, a whole MATLAB® program for calculating nonlinear reflectivity R and transmissivity T of a nonlinear DPhC with $N_1 = N_2 = 8$. The total number of layers in the DPhC including the defect layer is 17. Note that for the nonlinear TMM calculations, an algorithm for backward calculations is used, instead of forward calculations in linear TMM Programs (2.1)–(2.4). In the linear formalism, there is a linear relationship between input and output: each input value of incident field provides a single solution of output reflected and transmitted fields. However, the situation is much more complicated in the nonlinear formalism: there may be multiple output solutions for each input value of the incident field. For such situations, one of the most effective ways to solve the TMM equations is the backward algorithm as presented in Program 2.5. In the program, we start from an assumed value of output transmitted intensity, in this case it is normalized intensity ET at the final surface of the 17th layer. Because layers from the 17th to the 10th, and also from the 8th to the 1st are linear materials, we can simply solve the TMM equations using the linear relationship in TMM between input and output fields. However, for the defect layer—the 9th layer of nonlinear material, first we have to use a trial value of output field to estimate the input value, which is then used to calculate the output. The new output value is used to estimate the input value again and the recursion loop is repeated until convergence of the solutions is achieved.

As usual, the first comment box includes all general information of the model, including the name of the program and the date the program was first written and modified if possible.

MATLAB® Program 2.5

```
%xxxxxxxxxxxxxxxxxxxxxxxxxxxxxxxxxxxxxxxxxxxxxxxxxxxxxxxxxxxx
%  NDPhC_01.m    Nonlinear TMM for DPhC                       x
%                                                             x
%      air -->|A|B|...|A|B|===C===|B|A|...|B|A|--> air        x
%                    N1      nonlinear     N2                 x
%  N1=8, N2 = 8, M = N1+N2+1 = 17 layers                      x
%  A-layer: Linear na, da                                     x
%  B-Layer: Linear nb, db                                     x
%  C-layer: Nonlin nc = n1 + n2*I, dc                         x
%                                                             x
%  No absorption & Normalized Intensity                       x
%  by Dan Nguyen 10/22/2018 modified from version 10/29/2009  x
% xxxxxxxxxxxxxxxxxxxxxxxxxxxxxxxxxxxxxxxxxxxxxxxxxxxxxxxxxxx

format long
n0 = 1.0;                    % RI air (0)
na = 2.14;                   % linear RI layer (A)
nb = 1.46;                   % Linear RI layer (B)
nc = 1.60;                   % Linear RI C-layer
n2 = 1e-6;                   % n2 (cm^2/W)
bt = nc*1.32e-3;             % coef I = |E|^2 in kW/cm^2
bn = n2*bt;                  % normalized coeff
nL = 1.0;                    % Air/substrate (0/S)
Lamb0 = 550.0;               % Wavelength in nm
da = 1*Lamb0/(4*na);         % Thickness, A-layer in nm
db = 1*Lamb0/(4*nb);         % Thickness, B-layer in nm
dc = 4*Lamb0/(4*nc);         % cavity length in nm

L1=da;  L2=L1+db;  L3=L2+da;  L4=L3+db;  L5=L4+da;  L6=L5+db; L7=L6+da;...
L8=L7+db; L9=L8+dc; L10=L9+db; L11=L10+da; L12= L11+db; L13=L12+da;...
L14=L13+db; L15=L14+da; L16=L15+db;L17=L16+da;

NDPhC = fopen('I1e3.dat','w+');     %

wavelength = 450:0.25:700;
R1 = zeros(1,length(wavelength));
T1 = zeros(1,length(wavelength));
R2 = zeros(1,length(wavelength));
T2 = zeros(1,length(wavelength));

ET=1000.0;                          % Normalized Input light intensity
for k = 1:length(wavelength)
    lambda = wavelength(k);
    k0 = n0*2*pi/lambda;
    ka = na*2*pi/lambda;
    kb = nb*2*pi/lambda;
    kc = nc*2*pi/lambda;

    % Backward calculation w assumed ET
    % Linear matrix for A(17) and Air (0 = output)
    m17A011 = (ka + k0)/(2*ka)*exp(2*1j*ka*L17);
    m17A012 = (ka - k0)/(2*ka)*exp(2*1j*(ka+k0)*L17);
    m17A021 = (ka - k0)/(2*ka);
    m17A022 = (ka + k0)/(2*ka)*exp(2*1j*k0*L17);
    EF17 = m17A011*ET;
    EB17 = m17A021*ET;
```

```
M170 = [m17A011, m17A012; m17A021, m17A022];
% Linear matrix for B(16) and A(17)
m16BA11 = (kb + ka)/(2*kb)*exp(2*1j*kb*L16);
m16BA12 = (kb - ka)/(2*kb)*exp(2*1j*(kb+ka)*L16);
m16BA21 = (kb - ka)/(2*kb);
m16BA22 = (kb + ka)/(2*kb)*exp(2*1j*ka*L16);
EF16 = m16BA11*EF17 + m16BA12*EB17;
EB16 = m16BA21*EF17 + m16BA22*EB17;

M16_17 = [m16BA11, m16BA12; m16BA21, m16BA22];

% Linear matrix for A(15) and B(16)
m15AB11 = (ka + kb)/(2*ka)*exp(2*1j*ka*L15);
m15AB12 = (ka - kb)/(2*ka)*exp(2*1j*(ka+kb)*L15);
m15AB21 = (ka - kb)/(2*ka);
m15AB22 = (ka + kb)/(2*ka)*exp(2*1j*kb*L15);
EF15 = m15AB11*EF16 + m15AB12*EB16;
EB15 = m15AB21*EF16 + m15AB22*EB16;

M15_16 = [m15AB11, m15AB12; m15AB21, m15AB22];

% Linear matrix for B(14) and A(15)
m14BA11 = (kb + ka)/(2*kb)*exp(2*1j*kb*L14);
m14BA12 = (kb - ka)/(2*kb)*exp(2*1j*(kb+ka)*L14);
m14BA21 = (kb - ka)/(2*kb);
m14BA22 = (kb + ka)/(2*kb)*exp(2*1j*ka*L14);
EF14 = m14BA11*EF15 + m14BA12*EB15;
EB14 = m14BA21*EF15 + m14BA22*EB15;

M14_15 = [m14BA11, m14BA12; m14BA21, m14BA22];

% Linear matrix for A(13) and B(14)
m13AB11 = (ka + kb)/(2*ka)*exp(2*1j*ka*L13);
m13AB12 = (ka - kb)/(2*ka)*exp(2*1j*(ka+kb)*L13);
m13AB21 = (ka - kb)/(2*ka);
m13AB22 = (ka + kb)/(2*ka)*exp(2*1j*kb*L13);
EF13 = m13AB11*EF14 + m13AB12*EB14;
EB13 = m13AB21*EF14 + m13AB22*EB14;

M13_14 = [m13AB11, m13AB12; m13AB21, m13AB22];

% Linear matrix for B(12) and A(13)
m12BA11 = (kb + ka)/(2*kb)*exp(2*1j*kb*L12);
m12BA12 = (kb - ka)/(2*kb)*exp(2*1j*(kb+ka)*L12);
m12BA21 = (kb - ka)/(2*kb);
m12BA22 = (kb + ka)/(2*kb)*exp(2*1j*ka*L12);
EF12 = m12BA11*EF13 + m12BA12*EB13;
EB12 = m12BA21*EF13 + m12BA22*EB13;

M12_13 = [m12BA11, m12BA12; m12BA21, m12BA22];

% Linear matrix for A(11) and B(12)
m11AB11 = (ka + kb)/(2*ka)*exp(2*1j*ka*L11);
m11AB12 = (ka - kb)/(2*ka)*exp(2*1j*(ka+kb)*L11);
m11AB21 = (ka - kb)/(2*ka);
m11AB22 = (ka + kb)/(2*ka)*exp(2*1j*kb*L11);
EF11 = m11AB11*EF12 + m11AB12*EB12;
EB11 = m11AB21*EF12 + m11AB22*EB12;
```

```
    M11_12 = [m11AB11, m11AB12; m11AB21, m11AB22];

    % Linear matrix for B(10) and A(11)
    m10BA11 = (kb + ka)/(2*kb)*exp(2*1j*kb*L10);
    m10BA12 = (kb - ka)/(2*kb)*exp(2*1j*(kb+ka)*L10);
    m10BA21 = (kb - ka)/(2*kb);
    m10BA22 = (kb + ka)/(2*kb)*exp(2*1j*ka*L10);
    EF10 = m10BA11*EF11 + m10BA12*EB11;
    EB10 = m10BA21*EF11 + m10BA22*EB11;

    M10_11 = [m10BA11, m10BA12; m10BA21, m10BA22];

    MN2 = M10_11*M11_12*M12_13*M13_14*M14_15*M15_16*M16_17*M170;

    % NonLinear matrix for C(9) and B(10)
    % Linear matrix - dummy vary
    mCB11 = (kc + kb)/(2*kc)*exp(2*1j*kc*L9);
    mCB12 = (kc - kb)/(2*kc)*exp(2*1j*(kc+kb)*L9);
    mCB21 = (kc - kb)/(2*kc);
    mCB22 = (kc + kb)/(2*kc)*exp(2*1j*kb*L9);
    JF9 = abs(mCB11*EF10 + mCB12*EB10)^2;
    JB9 = abs(mCB21*EF10 + mCB22*EB10)^2;

eps=1.0;
while eps > 1e-5
    deltaPhi = 3*k0*bn*dc*(JF9 + JB9);
    kFc = kc + k0*bn*(JF9 + 2*JB9);
    kBc = kc + k0*bn*(2*JF9 + JB9);

    m9CB11 = (kBc + kb)/(kBc + kFc)*exp(2*1j*kc*L9);
    m9CB12 = (kBc - kb)/(kBc + kFc)*exp(2*1j*(kc+kb)*L9);
    m9CB21 = (kFc - kb)/(kBc + kFc)*exp(-1j*deltaPhi);
    m9CB22 = (kFc + kb)/(kBc + kFc)*exp(-1j*deltaPhi + 2*1j*kb*L9);
    EF9 = m9CB11*EF10 + m9CB12*EB10;
    EB9 = m9CB21*EF10 + m9CB22*EB10;
    eps = abs((abs(EF9)^2 - JF9))/JF9;
    JF9 = abs(EF9)^2;
    FB9 = abs(EB9)^2;
    End

    M9_10 = [m9CB11, m9CB12; m9CB21, m9CB22];

    % Linear matrix for B(8) and C(9)
    m8BC11 = (kb + kc)/(2*kb)*exp(2*1j*kb*L8);
    m8BC12 = (kb - kc)/(2*kb)*exp(2*1j*(kb+kc)*L8);
    m8BC21 = (kb - kc)/(2*kb);
    m8BC22 = (kb + kc)/(2*kb)*exp(2*1j*kc*L8);
    EF8 = m8BC11*EF9 + m8BC12*EB9;
    EB8 = m8BC21*EF9 + m8BC22*EB9;

    M89 = [m8BC11, m8BC12; m8BC21, m8BC22];

    % Linear matrix for A(7) and B(8)
    m7AB11 = (ka + kb)/(2*ka)*exp(2*1j*ka*L7);
    m7AB12 = (ka - kb)/(2*ka)*exp(2*1j*(ka+kb)*L7);
    m7AB21 = (ka - kb)/(2*ka);
    m7AB22 = (ka + kb)/(2*ka)*exp(2*1j*kb*L7);
    EF7 = m7AB11*EF8 + m7AB12*EB8;
    EB7 = m7AB21*EF8 + m7AB22*EB8;

    M78 = [m7AB11, m7AB12; m7AB21, m7AB22];
```

```
% Linear matrix for B(6) and A(7)
m6BA11 = (kb + ka)/(2*kb)*exp(2*1j*kb*L6);
m6BA12 = (kb - ka)/(2*kb)*exp(2*1j*(kb+ka)*L6);
m6BA21 = (kb - ka)/(2*kb);
m6BA22 = (kb + ka)/(2*kb)*exp(2*1j*ka*L6);
EF6 = m6BA11*EF7 + m6BA12*EB7;
EB6 = m6BA21*EF7 + m6BA22*EB7;
M67 = [m6BA11, m6BA12; m6BA21, m6BA22];

% Linear matrix for A(5) and B(6)
m5AB11 = (ka + kb)/(2*ka)*exp(2*1j*ka*L5);
m5AB12 = (ka - kb)/(2*ka)*exp(2*1j*(ka+kb)*L5);
m5AB21 = (ka - kb)/(2*ka);
m5AB22 = (ka + kb)/(2*ka)*exp(2*1j*kb*L5);
EF5 = m5AB11*EF6 + m5AB12*EB6;
EB5 = m5AB21*EF6 + m5AB22*EB6;
M56 = [m5AB11, m5AB12; m5AB21, m5AB22];
% Linear matrix for B(4) and A(5)
m4BA11 = (kb + ka)/(2*kb)*exp(2*1j*kb*L4);
m4BA12 = (kb - ka)/(2*kb)*exp(2*1j*(kb+ka)*L4);
m4BA21 = (kb - ka)/(2*kb);
m4BA22 = (kb + ka)/(2*kb)*exp(2*1j*ka*L4);
EF4 = m4BA11*EF5 + m4BA12*EB5;
EB4 = m4BA21*EF5 + m4BA22*EB5;
M45 = [m4BA11, m4BA12; m4BA21, m4BA22];

% Linear matrix for A(3) and B(4)
m3AB11 = (ka + kb)/(2*ka)*exp(2*1j*ka*L3);
m3AB12 = (ka - kb)/(2*ka)*exp(2*1j*(ka+kb)*L3);
m3AB21 = (ka - kb)/(2*ka);
m3AB22 = (ka + kb)/(2*ka)*exp(2*1j*kb*L3);
EF3 = m3AB11*EF4 + m3AB12*EB4;
EB3 = m3AB21*EF4 + m3AB22*EB4;
M34 = [m3AB11, m3AB12; m3AB21, m3AB22];

% Linear matrix for B(2) and A(3)
m2BA11 = (kb + ka)/(2*kb)*exp(2*1j*kb*L2);
m2BA12 = (kb - ka)/(2*kb)*exp(2*1j*(kb+ka)*L2);
m2BA21 = (kb - ka)/(2*kb);
m2BA22 = (kb + ka)/(2*kb)*exp(2*1j*ka*L2);
EF2 = m2BA11*EF3 + m2BA12*EB3;
EB2 = m2BA21*EF3 + m2BA22*EB3;
M23 = [m2BA11, m2BA12; m2BA21, m2BA22];

% Linear matrix for A(1) and B(2)
m1AB11 = (ka + kb)/(2*ka)*exp(2*1j*ka*L1);
m1AB12 = (ka - kb)/(2*ka)*exp(2*1j*(ka+kb)*L1);
m1AB21 = (ka - kb)/(2*ka);
m1AB22 = (ka + kb)/(2*ka)*exp(2*1j*kb*L1);
EF1 = m1AB11*EF2 + m1AB12*EB2;
EB1 = m1AB21*EF2 + m1AB22*EB2;
M12 = [m1AB11, m1AB12; m1AB21, m1AB22];

% Linear matrix for Air (0) and A(1)
m0A11 = (k0 + ka)/(2*k0);
m0A12 = (k0 - ka)/(2*k0);
m0A21 = (k0 - ka)/(2*k0);
m0A22 = (k0 + ka)/(2*k0);
EF0 = m0A11*EF1 + m0A12*EB1;
EB0 = m0A21*EF1 + m0A22*EB1;
M01 = [m0A11, m0A12; m0A21, m0A22];
MN1 = M01*M12*M23*M34*M45*M56*M67*M78*M89*M9_10;
```

```
        R1 = abs(EB0/EF0)^2;
        T1 = abs(ET/EF0)^2;

        MT = MN1*MN2;
        R2 = abs(MT(2,1)/MT(1,1))^2;
        T2 = abs(1/MT(1,1))^2;

        fprintf(NLC,'%f %12.8f %12.8f %12.8f %12.8f\n', lambda, R1, T1, R2,
    T2);

    end
      fclose(NLC);

      %%========================== end ================================
```

Let us make some comments on the above program.

First, there are different ways to save the data after calculations. Here we use the way of opening a file with the name T1e3.dat. The file name should be short but still meaningful so that we can conveniently use it later. For example, 'I1e3.dat' is the data of the case with normalized intensity 1e3 = 1000.

```
NDPhC = fopen('I1e3.dat','w+');     % open a file I1e3.dat to save data
```

After calculation, the data is saved in the form of multiple columns for λ, R1, T1 etc.

```
fprintf(DPhC,'%f %12.8f %12.8f %12.8f %12.8f\n', lambda, R1, T1, R2, T2);
fclose(NLC);
```

Second, after running the program with different input intensities and saving data with different names, for example ET = 10 and saving data in file T1e1.dat, 100 and T1e2.dat, and so on, we then write another file to plot the figures with reflections and transmissions for different intensities. The file for plotting is:

```
%xxxxxxxxxxxxxxxxxxxxxxxxxxxxxxxxxxxxxxxxxxxxxxxxxxxxxxxxxxxxx
%  PlotRT.m   plotting R & T figures                          x
%  loading data file T1e1, T1e2, T1e3 etc                     x
%  Plot (lambda, R1, lambda,R2 …)                             x
% xxxxxxxxxxxxxxxxxxxxxxxxxxxxxxxxxxxxxxxxxxxxxxxxxxxxxxxxxxxx
load T1e1.dat
load T1e2.dat
load T1e3.dat
load T2e3.dat
load T3e3.dat
load T4e3.dat
load T5e3.dat
load T6e3.dat
load T7e3.dat

atick1 = 450:50:700;
  figure(1)
  plot(T1e1(:,1),T1e1(:,4),'k', T1e2(:,1),T1e2(:,4),'b',...
       T1e3(:,1),T1e3(:,4),'c', T2e3(:,1),T2e3(:,4),'g',...
       T3e3(:,1),T3e3(:,4),'m', T4e3(:,1),T4e3(:,4),'r','linewidth', 2)
```

```
set(gca,'FontSize',18);
axis([450 700 0 1.2]);
set(gca,'XTick',atick1);
set(gca,'YTick',0:0.2:1);
grid minor
xlabel('Wavelength (nm)');
ylabel('Reflection R');
title( 'Nonlinear Reflection');

figure(2)
plot(T1e1(:,1),T1e1(:,5),'k', T1e2(:,1),T1e2(:,5),'b',...
     T1e3(:,1),T1e3(:,5),'c', T2e3(:,1),T2e3(:,5),'g',...
     T3e3(:,1),T3e3(:,5),'m', T4e3(:,1),T4e3(:,5),'r','linewidth', 2)
set(gca,'FontSize',18);
axis([450 700 0 1.2]);
set(gca,'XTick',atick1);
set(gca,'YTick',0:0.2:1);
grid minor
xlabel('Wavelength (nm)');
ylabel('Transmission T');
title( 'Nonlinear Transmission');

figure(3)
plot(T4e3(:,1),T4e3(:,5),'b', T5e3(:,1),T5e3(:,5),'g',...
     T6e3(:,1),T6e3(:,5),'r','linewidth', 2)
set(gca,'FontSize',18);
axis([450 700 0 1.2]);
set(gca,'XTick',atick1);
set(gca,'YTick',0:0.2:1);
grid minor
xlabel('Wavelength (nm)');
ylabel('Transmission T');
title( 'Nonlinear Transmission');

figure(4)
plot(T4e3(:,1),T4e3(:,4),'b', T5e3(:,1),T5e3(:,4),'g',...
     T6e3(:,1),T6e3(:,4),'r','linewidth', 2)
set(gca,'FontSize',18);
axis([450 700 0 1.2]);
set(gca,'XTick',atick1);
set(gca,'YTick',0:0.2:1);
grid minor
xlabel('Wavelength (nm)');
ylabel('Reflection R');
title( 'Nonlinear Reflection');
%%========================== end ================================
```

Figure 2.27 shows the figures of reflection and transmission of the nonlinear DPhC with different intensities.

Note that in the example the gratings are designed with the Bragg resonance $\lambda_0 = 550$ nm. When intensity is weak, for instance up to ET = 100, the resonant modes are exactly at the Bragg resonant wavelength 550 nm of the two Bragg reflectors. The spectra for $I = 10$ (black curve) and $I = 100$ (blue curve) are almost the same as the linear spectra that are calculated with linear TMM with resonant mode at 550 nm,

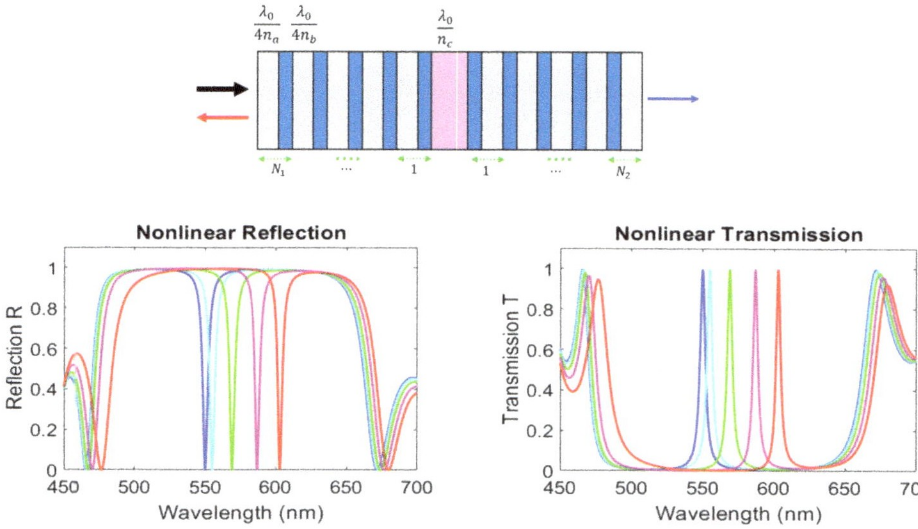

Figure 2.27. Upper: schematic of nonlinear DPhC, in which the defect layer thickness and linear refractive index are related as $d_c = \lambda_0/(n_{0c})$, n0c = 1.60, and nonlinear refractive index $n_2 = 10^{-3}$ cm^2 kW^{-1}. A: Ti_2O_5 ($n_a = 2.14$), B = SiO_2 ($n_b = 1.46$), Lower: nonlinear reflections (left) and nonlinear transmission (right) with different normalized input intensity: ET = 10 (black), 100 (blue), 1000 (cyan), 2000 (green), 3000 (pink) and 4000 (red). The DPhC has $N_1 = N_2 = 8$ pairs of A, B layers, and $\lambda_0 = 550$ nm.

as shown in figure 2.27 above for the case with $m = 4$. We want to stress that the results strongly depend on the nonlinear refractive index n_{2c} of the material in the defect layer. In the above example, we use `n2 = 1e-6; % n2(cm^2/W)`. Readers can easily check that the higher the n_2 values, the stronger the nonlinear shift of the resonant modes. We use this high value n_2 to make the nonlinear effect clearer for ease of study. In practice, we should use the experiment data for all parameters for modeling and simulation of real devices. In our example, when intensity increases up to 4000 the nonlinear effect is simply shifting the resonant mode to longer wavelength with increasing intensity. Increasing intensity further, we will see interesting phenomena that are described in figure 2.28 below.

In figure 2.28 we plot the nonlinear spectra with higher intensity $I = 4000$ (blue), 5000 (green) and 6000 (red). There are two phenomena that occur: (i) mode shifting and (ii) new resonant modes appear in higher intensities. Let us analyze the results in more detail. First, the blue curve in figure 2.28 is for $I = 4000$ and in this case the resonant mode is ~600 nm as compared with 550 nm in the case of weak intensities. The green curve is for intensity 5000 and there are two modes: the first mode is on the right of the mode with $I = 4000$ (blue); this first (green) mode can be considered as the shifting of the blue mode as we know that happened with increasing intensity. The second mode of the green curve is a new one with higher intensity, and in this case ($I = 5000$) this mode appears at 550 nm, which is the resonance in the linear regime. Increasing intensity further, the red curve is for $I = 6000$, and we also see there are two modes, but now both modes are shifting to the longer wavelength as

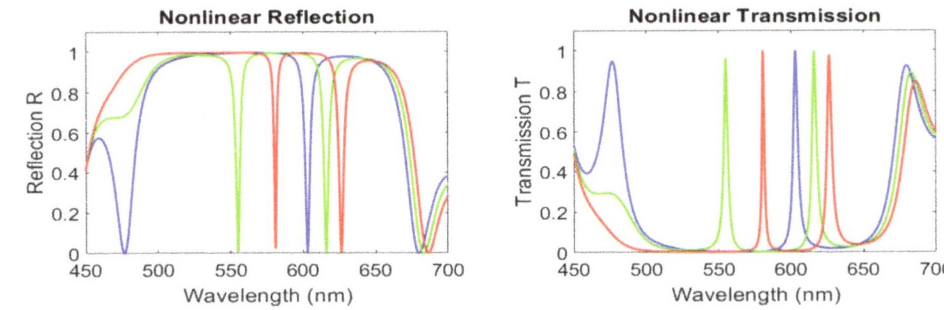

Figure 2.28. Nonlinear reflections (left) and nonlinear transmission (right) with different normalized input intensity: 4000 (blue), 5000 (green) and 6000 (red). The DPhC has $N_1 = N_2 = 8$ pairs of A, B layers, and $\lambda_0 = 550$ nm.

compared to the two green modes (with $I = 5000$). We will see how this analysis will be very useful for understanding the OB effect that we will study later.

Up to this point, we have ignored the absorption of the material in the calculations, although we mentioned absorption in the introduction. Figure 2.29 shows normalized nonlinear transmissions with absorption to compare with no-absorption results in the right panels of figures 2.27 and 2.28.

The normalized spectra with absorption in figure 2.29 clearly show the resonant modes get stronger absorption compared with non-resonant light. To estimate quantitatively the absorption let us show in figure 2.30 the un-normalized spectra with absorption effect in the DPhC.

In the program we assume a linear absorption coefficient $\alpha = 2e{-}4\ \mathrm{nm}^{-1}$, and the thickness of defect layer is about 343 nm. Therefore, the absorption for the single path in the material layer is αL ~0.068 or 0.29 dB or ~6%. Meanwhile, the transmission of the resonant mode at 550 nm transmission is ~57% (blue curve on the left figure in figure 2.31) and the reflection (red curve) is ~5% so that total absorption for the resonant mode is about ~38% or ~2.1 dB. How can that have happened, and is there any inaccuracy in the calculation?

First, let us examine the absorption of non-resonant light outside the band gap, for example the light at 450 nm in figure 2.31. The transmission is about 55% and the reflection is ~39% or absorption is ~6% which is consistent with the absorption $\alpha L = 0.298$ dB for the single-pass absorption in the defect layer. That means, outside the band gap the light can pass the layer without any circulation due to reflection by Bragg reflectors. The absorption calculation is therefore very accurate. Now, let us examine the light at the left edge of the band gap. The transmission is about ~89% with no reflection ($R = 0$) or about 11% absorption. In that case, the light may get partial circulation inside the structure and increase absorption as compare with outside the band gap with only a single pass. In the case of resonant mode, the light can get multiple circulation inside the structure. As a result, multi-pass through the absorption layer increases significantly the absorption for the resonant mode. The effect of nonlinear transmission has been investigated theoretically and experimentally in a nonlinear absorption material filling a cavity [34].

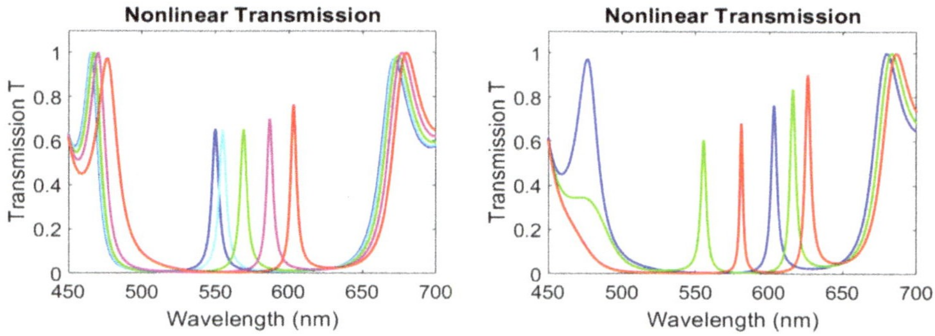

Figure 2.29. Normalized nonlinear transmission with absorption effect α = 1e−4 nm^{-1} and different input intensity. Left: 10 (black), 100 (blue), 1000 (cyan), 2000 (green), 3000 (pink) and 4000 (red). Right: 4000 (blue), 5000 (green) and 6000 (red). The DPhC has $N_1 = N_2 = 8$ pairs of A and B layers, and $\lambda_0 = 550$ nm.

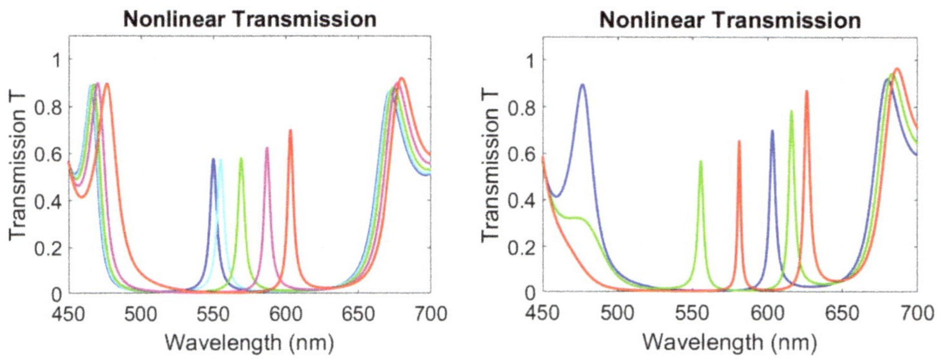

Figure 2.30. Un-normalized nonlinear transmission with absorption effect α = 1e−4 nm^{-1} and different normalized input intensity. Left: 10 (black), 100 (blue), 1000 (cyan), 2000 (green), 3000 (pink) and 4000 (red). Right: 4000 (blue), 5000 (green) and 6000 (red). The DPhC has $N_1 = N_2 = 8$ pairs of A and B layers, and $\lambda_0 = 550$ nm.

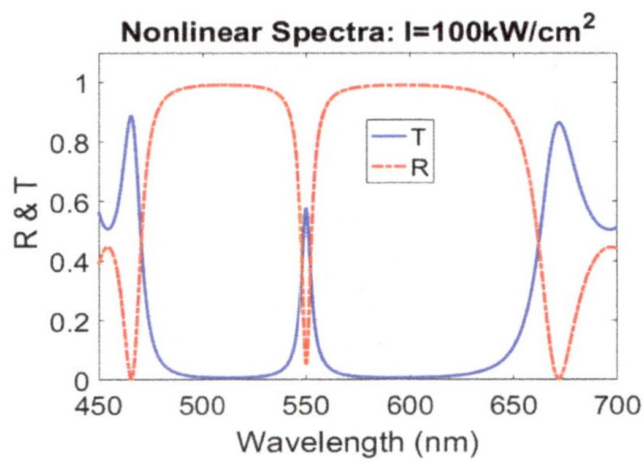

Figure 2.31. Un-normalized nonlinear transmission (blue) and reflection (red) with absorption effect α = 1e−4 nm^{-1} and $I = 100$ (blue).

In order to calculate the absorption effect in nonlinear TMM, the only part that can be used to replace the nonlinear TMM in Program 2.5 above is:

```
%%======================nonlinear TMM with absorption ==============
% NonLinear matrix for C(9) and B(10)
    % Linear matrix - dummy vary
    mCB11 = (kc + kb)/(2*kc)*exp(2*1j*kc*L9);
    mCB12 = (kc - kb)/(2*kc)*exp(2*1j*(kc+kb)*L9);
    mCB21 = (kc - kb)/(2*kc);
    mCB22 = (kc + kb)/(2*kc)*exp(2*1j*kb*L9);
    JF9 = abs(mCB11*EF10 + mCB12*EB10)^2;
    JB9 = abs(mCB21*EF10 + mCB22*EB10)^2;

 eps=1.0;
 while eps > 1e-5
    deltaPhi = 3*k0*bn*dc*(JF9 + JB9);
    kFc = kc + k0*bn*(JF9 + 2*JB9);
    kBc = kc + k0*bn*(2*JF9 + JB9);

    m9CB11 = (kBc + kb)/(kBc + kFc)*exp(2*1j*kc*L9);
    m9CB12 = (kBc - kb)/(kBc + kFc)*exp(2*1j*(kc+kb)*L9);
    m9CB21 = (kFc - kb)/(kBc + kFc)*exp(-1j*deltaPhi);
    m9CB22 = (kFc + kb)/(kBc + kFc)*exp(-1j*deltaPhi + 2*1j*kb*L9);
    EF9 = m9CB11*EF10 + m9CB12*EB10;
    EB9 = m9CB21*EF10 + m9CB22*EB10;
    eps = abs((abs(EF9)^2 - JF9))/JF9;
    JF9 = abs(EF9)^2;
    FB9 = abs(EB9)^2;
    end
    M9_10 = [m9CB11, m9CB12; m9CB21, m9CB22];
%%===============================end===================================
```

Up to this point we have studied the reflection and transmission in a broad wavelength region with different incident light intensity or input intensity. Let us us now calculate the nonlinear transmission of a laser beam with fixed wavelength but intensity changes. We will see the interesting effect of OB in the nonlinear DPhC structure.

In figure 2.32 we show the intensity output–input characterization of a nonlinear DPhC with parameters as described earlier. In the figure, the red and green arrows show the hysterical behavior of the output intensity I_{out} as input intensity I_{in} changes. Starting from $I_{in} = 0$ when intensity is weak, the output I_{out} increases in monotony with I_{in}. When I_{in} increases and reaches a critical point, I_{out} will jump to the upper branch of the OB curve and from there it will go up with increasing I_{in}. However, if the state of the system is in the upper branch and the I_{in} is decreased then the output I_{out} will follow the green arrows. The optically hysterical characterization, OB, is the mechanism for all-optical switching and operation of optical transistors.

It is very important to stress that the OB curve in figure 2.32 is obtained numerically from Program 2-5 above with some changes. The numerical results do not dictate how the input–output would behave hysterically as the reds and green arrows indicate. In Exercise 2 we will use the results that we have already obtained earlier for nonlinear transmission with different intensity to understand the OB characterization.

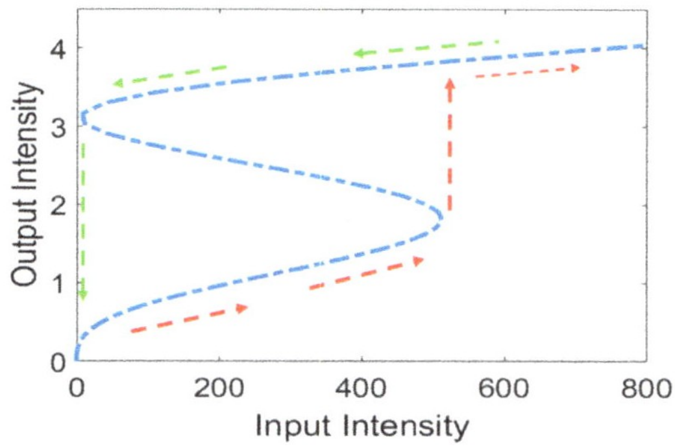

Figure 2.32. Dimensionless input–output characterization of a nonlinear DPhC. The red and green arrows show the hysterical behavior output intensity as a function of input intensity or the OB phenomenon.

To make the change in Program 2.5 for obtaining the OB curve, we just fix the wavelength of the input light instead of the broad band as before and running with normalized input intensity I_{in0}. We can convert input and output intensity by multiple with coefficient b_0 in the program. Another important point is that the program calculates nonlinear TMM in a backward way. First, with each initial value of I_{in0}, a value of output I_{out} is used to calculate backwardly the correct value of I_{in} that corresponds to the output I_{out}.

```
lambda=550;
for Iin0 = 100:100:10000,
    Iout = b0*ET^2;            %Output intensity kW/cm^2
```

And we save data for plotting as:

```
fprintf(NPhC,'%f  %12.8f\n', Iin,  Iout);
```

***Exercise* 1.**

Prove by TMM formalism that light at resonant modes can be strongly enhanced inside a resonant cavity or a DPhC. For example, the light intensities inside the structures are calculated for the light with different wavelengths. The results will show strong intensity inside cavity/DPhC for the resonant mode, or the resonant enhancement, but no enhancement for the non-resonant. Figure 2.33 shows the calculated intensities inside DPhC for the light with two different wavelengths; graph (1) corresponds to light at the edge of the band gap (pink arrow), and graph (2) corresponds to the inter-band inside the band gap (green arrow 2). Both graphs are plotted in the same scale of the forward and backward light intensity. The results show the inter-band mode has intensity more than 20 times the input light intensity. Meanwhile, the light at the band edge also gets small enhancement. This example is about the important characterization of the resonant modes.

It is worth noting that the curves in the right panels are not light intensity distribution inside structure. The graphs in figure 2.33 are plots of $|E^{(+)}|^2$, $|E^{(-)}|^2$ and $|E^{(+)}|^2 + |E^{(-)}|^2$, but the intensity in medium with RI n is $n|E|^2$. When the refractive indexes of the material are taken into account, the intensity will be higher. Moreover, the total field is the superposition of backward and forward fields inside

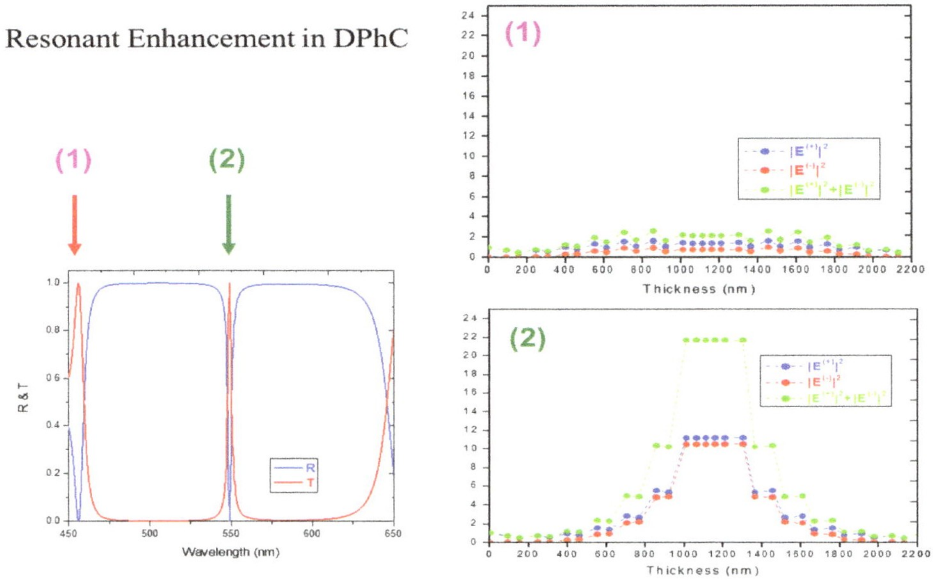

Figure 2.33. Left: transmission of light at the edge of band gap (1) and resonant wavelength (2) are almost 100%, but only the resonant mode has strong enhancement, as shown in the panels on the right.

the structure, $E(z) = E^{(+)}(z)e^{-ikz} + E^{(-)}(z)e^{+ikz}$ and therefore intensity in the jth layer should be $I_j = n_j|E(z_j)|^2$. Due to the superposition of forward and backward field inside structure, standing waves are formed, as shown in figure 2.34 below.

The field distribution inside the structure could give us some insight information of the system that is useful for designing and optimizing operation of the device. For instance, in some situations, we may want to have very strong enhancement, but in other cases we may try to reduce the standing waves inside the defect. Before writing the program to calculate field distribution, we should have a model that describes the physics and mathematics of the problem. In the following an instruction to develop that model is presented.

First, let us assume there is no incident wave from the right of the structure, as described in figure 2.35.

And using TMM formalism as presented above, we can easily calculate both reflection field $E^{(-)}(0)$ and transmission $E^{(+)}(L)$. If we want to calculate the fields at one layer with coordinate z_p inside the structure, then let us consider figure 2.36.

From TMM formalism, we can write the equation at z_p

$$\begin{pmatrix} E^{(+)}(z_p) \\ E^{(-)}(z_p) \end{pmatrix} = \begin{pmatrix} M_{p11} & M_{p12} \\ M_{p21} & M_{p22} \end{pmatrix} \begin{pmatrix} E^{(+)}(L) \\ 0 \end{pmatrix} = \begin{pmatrix} M_{p11} & M_{p12} \\ M_{p21} & M_{p22} \end{pmatrix} \begin{pmatrix} t \\ 0 \end{pmatrix}, \tag{2.59}$$

Or,

$$E^{(+)}(z_p) = M_{p11} \cdot E^{(+)}(L) = \frac{M_{p11}}{M_{11}}, \quad E^{(-)}(z_p) = M_{p21} \cdot E^{(+)}(L) = \frac{M_{p21}}{M_{11}}. \tag{2.60}$$

$$E_j = E_j^{(+)} e^{-ik_j z} + E_j^{(-)} e^{+ik_j z}$$

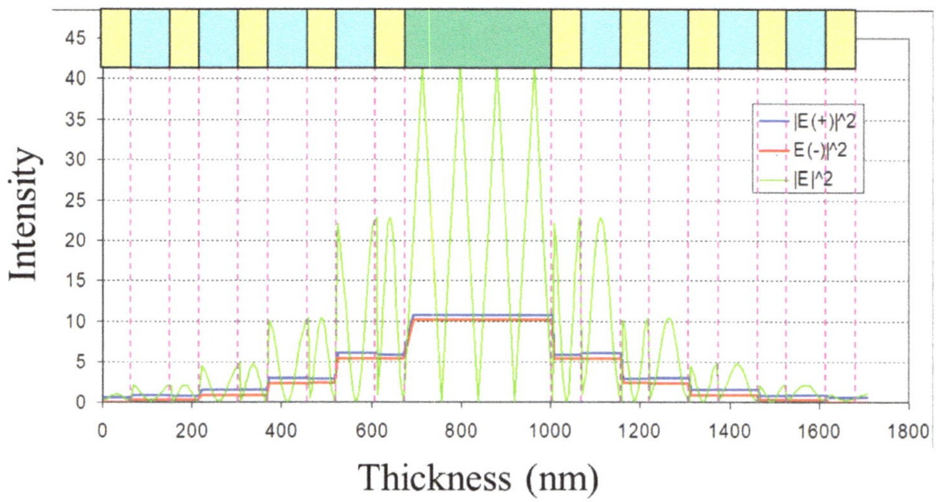

Figure 2.34. Light intensity distribution inside the DPhC. The intensity inside the defect can be much higher (~40 times in this case) as compared with the input intensity.

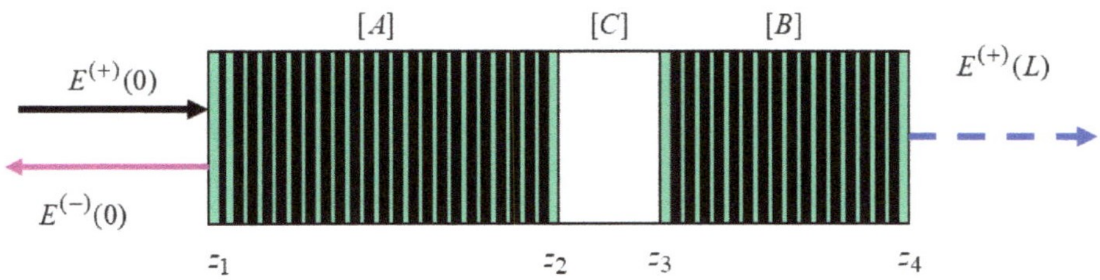

Figure 2.35. Diagram of TMM model for calculating the field distribution inside the DPhC.

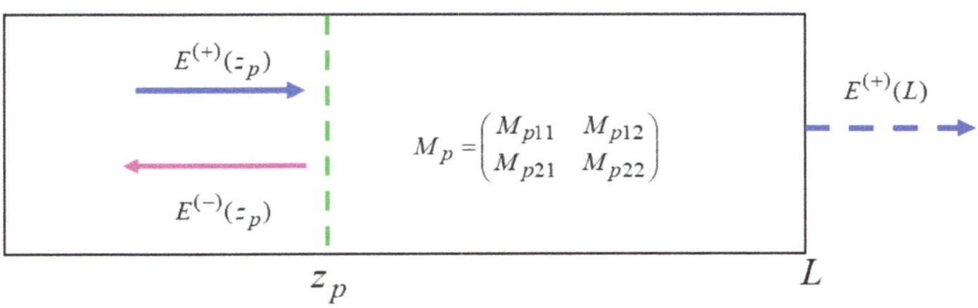

Figure 2.36. Diagram of TMM model for calculating the field distribution at coordinate z_p inside the DPhC.

The elements of TM of the whole structure M_{ij} can be easily calculated as shown in the examples, especially in MATLAB® Programs 2.1–2.4. Once we know z_p, we can also calculate M_p and its elements M_{pij}. Therefore, we can calculate backward and forward fields at any position z_p inside the structure. The total field superposition of those fields can be easily calculated.

***Exercise* 2**

Using the results of nonlinear transmission from the above example for the nonlinear DPhC above, prove the hysterical behavior of the output–input characterization in figure 2.32.

Hint: Calculate spectra of nonlinear transmission with different input intensities. Investigate the mode changes, and we will see the output intensity will jump at the critical point, and that is due to the second mode appearing when intensity reaches the critical point c in figure 2.37.

Resonant modes shift to the long wavelength when intensity increases and are represented by transmission curves a, b c, in the figure on the left. While resonant modes shift to the left up to the c-curve, the transmission values at the fixed wavelength 550 nm are reduced when input intensity increases. However, the absolute values of output intensity still increase up to point c in the figure on the right. From point c, increasing input intensity further to point d the second mode appears in the nonlinear transmission (left figure) due to strong nonlinear effect induced by high intensity, and transmission at 550 nm is very high. As a result, the output intensity jumps to very high value (point d) in the figure on the right. Readers should do the calculation and see themselves when the critical point is reached in both figures.

The hysterical characterization of the OB curve can be analyzed more rigorously by using linearized stability theory [39–41]. The theory can prove mathematically that with a small perturbation, states of the system corresponding to the middle branch of the OB curve are unstable or virtual states that may be mathematical solutions, but do not exist. Only states in lower and upper branches would be stable and exist as real states of the system. However, under high nonlinearity conditions states in those two branches can be instabilities and several effects like self-oscillations and chaos can occur. The analysis above is a simple approach to see how hysterical characterization can happen, but does not predict and explain the nonlinear dynamics of the systems.

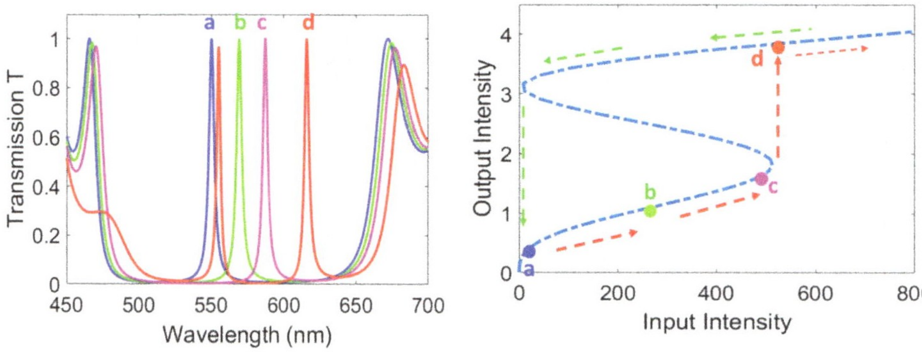

Figure 2.37. Resonant modes shift to the long wavelength when intensity increases (a, b, c in both figures). The critical point d on the left graph of transmission corresponds to the jump from c to d in input–output hysterical characterization in the figure on the right.

Problem 2.1

From the boundary conditions for the electric fields at the boundary coordinate z_m between two materials, using equation (2.9) for situation $\theta_l = \theta_r = 0$, derivation of relationship of electric fields at the boundary as in equation (2.10)

$$E^{(+)}(z_m^-) = \frac{n_l + n_r}{2n_l} E^{(+)}(z_m^+) + \frac{n_l - n_r}{2n_l} E^{(-)}(z_m^+)$$
$$E^{(-)}(z_m^-) = \frac{n_l - n_r}{2n_l} E^{(+)}(z_m^+) + \frac{n_l + n_r}{2n_l} E^{(-)}(z_m^+).$$

Hint:

Using equation (2.9) for situation $\theta_l = \theta_r = 0$, we can write the amplitude of refection and transmission when the light propagates from the left to right of the interface as

$$r_{lr} = \frac{n_l - n_r}{n_l + n_r}, \quad \text{and} \quad t_{lr} = \frac{2n_l}{n_l + n_r}, \tag{2.61}$$

and for the light propagating from right to left as

$$r_{rl} = -r_{lr} = \frac{n_r - n_l}{n_l + n_r} =, \quad \text{and} \quad t_{rl} = t_{lr} = \frac{2n_l}{n_l + n_r}, \tag{2.62}$$

From the continuity condition for the electric field at the boundary at the interface, we have relationships between the fields as

$$E^{(-)}(z_m^-) = r_{lr}E^{(+)}(z_m^-) + t_{rl}E^{(-)}(z_m^+), \tag{2.63}$$

$$E^{(+)}(z_m^+) = t_{lr}E^{(+)}(z_m^-) + r_{rl}E^{(-)}(z_m^+), \tag{2.64}$$

And from (2.63) and (2.64) we can easily get equation (2.10).

References

[1] Joannopoulos J D, Johnson S G, Winn J N and Meade R D (ed) 2008 *Photonic Crystals: Molding the Flow of Light—Second Edition* 2nd edn (Princeton, NJ: Princeton University Press)

[2] Claus D 2013 *Photonic Crystals: Technology, Theory and Challenges (Materials Science and Technologies)* (New York: Nova Science Pub Inc.)

[3] Usanov D A, Nikitov S A, Skripal A V and Ponomarev D V 2019 *One-Dimensional Microwave Photonic Crystals: New Applications* 1st edn (Boca Raton, FL: CRC Press)

[4] Knittl Z 1976 *Optics of Thin Films: An Optical Multilayer Theory* (London: Wiley)

[5] Yeh P 1988 *Optical Waves in Layered Media* (Wiley Series in Pure and Applied Optics) (New York: Wiley)

[6] Yeh P, Yariv A and Hong C-S 1977 Electromagnetic propagation in periodic stratified media. I. General theory *J. Opt. Soc. Am.* **67** 423–38

[7] Byrnes S J 2018 Multilayer optical calculation. arXiv:1603.02720v3 [physics.comp-ph]

[8] Lekner J 1994 Light in periodically stratified media *J. Opt. Soc. Am.* A **11** 2892–9

[9] Danckaert J *et al* 1991 Dispersive optical bistability in stratified structures *Phys. Rev.* B **44** 8214

[10] Yang Z S, Kwong N H, Binder R and Smirl A L 2005 Stopping, storing, and releasing light in quantum-well Bragg structures *J. Opt. Soc. Am.* **B22** 2144–56

[11] Nguyen D T, Kwong N H, Binder R and Smirl A L 2007 Mechanism of all-optical spin-dependent polarization switching in Bragg-spaced quantum wells *Appl. Phys. Lett.* **90** 18116

[12] Bragg W H and Bragg W L 1913 The reflexion of x-rays by crystals *Proc. R. Soc. Lond. A.* **88** 428–38

[13] Cowley J M 1975 *Diffraction Physics* (Amsterdam: North-Holland)

[14] Siegman A E 1986 *Lasers* (Mill Valey, CA: University Science Books)

[15] Svelto O 1998 *Principles of Lasers* 4th edn (New York: Plenum Publishing Corporation)

[16] Saleh B E A and Teich M C 1991 *Fundamental of Photonics* (Wiley Series in Pure and Applied Optics) (New York: Wiley)

[17] Snyder A W and Love J D 1983 *Optical Waveguide Theory* (New York: Chapman and Hall)

[18] Marcuse D 1974 *Theory of Dielectric Optical Waveguides* (New York: Academic)

[19] Thyagarajan K and Ghatak A 1998 *An Introduction to Fiber Optics* (Cambridge: Cambridge University Press)

[20] Spiegelberg C, Geng J, Hu Y, Kaneda Y, Jiang S and Peyghambarian N 2004 *J. Lightwave Technol.* **22** 57

[21] Digonnet M J F (ed) 1993 *Rare Earth Doped Fiber Lasers and Amplifiers* (New York: Marcel Dekker, Inc.)

[22] Desurvire E 1994 *Erbium-Doped Fiber Amplifiers* 1st edn (New York: Wiley-Interscience)

[23] NP Photonics Inc. www.npphotonics.com

[24] Schülzgen A, Li L, Nguyen D, Spiegelberg C, Matei Rogojan R, Laronche A, Albert J and Peyghambarian N 2008 Distributed feedback fiber laser pumped by multimode laser diodes *Opt. Lett.* **33** 614–6

[25] Li L, Schülzgen A, Temyanko V L, Spiegelberg C, Nguyen D T, Zhu X, Moloney J V, Albert J and Peyghambarian N 2008 Cladding-pumped distributed feedback phosphate glass fiber lasers *CLEO/QELS* paper CMA1

[26] Kashyap R 2010 *Fiber Bragg Gratings* 2nd edn (New York: Academic)

[27] Othonos A 1997 Fiber Bragg gratings *Rev. Sci. Instrum.* **68** 4309

[28] Jin L, Tan Y N, Quan Z, Li M P and Guan B O 2012 Strain-insensitive temperature sensing with a dual polarization fiber grating laser *Opt. Express* **20** 6021–8

[29] Ahmad H, Latif A A, Zulkifli M Z, Awang N A and Harun S W 2012 Temperature sensing using frequency beating technique from single-longitudinal mode fiber laser *IEEE Sensors J.* **12** 2496–500

[30] Pan S and Yao J 2009 A wavelength-switchable single-longitudinal-mode dual-wavelength erbium-doped fiber laser for switchable microwave generation *Opt. Express* **17** 5414

[31] Levine D and Steinhardt P J 1984 Quasicrystals: a new class of ordered structures *Phys. Rev. Lett.* **53** 2477–80

[32] Gellermann W, Kohmoto M, Sutherland B and Taylor P C 1994 Localization of light waves in Fibonacci dielectric multilayers *Phys. Rev. Lett.* **72** 633–36

[33] Kohmoto M, Sutherland B and Iguchi K 1987 Localization in optics: quasiperiodic media *Phys. Rev. Lett.* **58** 2436–8

[34] Nguyen D T, Norwood R and Peyghambarian N 2010 Multiple spectral window mirrors based on Fibonacci chains of dielectric layers *Opt. Commun.* **283** 4199–202

[35] Hendrickson J, Richards B C, Sweet J, Khitrova G, Poddubny A N, Ivchenko E L, Wegener M and Gibbs H M 2008 Excitonic polaritons in Fibonacci quasicrystals *Opt. Express* **16** 15382–7
[36] Nguyen D T, Sheng C, Thomas J, Norwood R, Kimball B, Steeves D M and Peyghambarian N 2008 Observation of nonlinear transmission enhancement in cavities filled with nonlinear organic materials *Appl. Opt.* **47** 5777
[37] Macia E, Maciá and Domínguez-Adame F 2000 *Electrons, Phonons and Excitons in Low Dimensional Aperiodic Systems* (Madrid: Editorial Complutense)
[38] Maciá E 2012 Exploiting aperiodic designs in nanophotonic devices *Rep. Prog. Phys.* **75** 036502
[39] Nguyen B A and Nguyen T D 1989 A new mechanism for optical bistability *Phys. Lett.* A **136** 71–2
[40] Nguyen T D 1994 Nonlinear self-oscillations in exciton-biexcitons system *Phys. Lett.* A **193** 462–6
[41] Haug H (ed) 1988 *Optical Nonlinearities and Instabilities in Semiconductors.* (New York: Academic)

Dan T Nguyen

Chapter 3

Beam propagation method for modeling multimode cladding-pumped fiber amplifiers

3.1 Modeling problems for multimode cladding-pumped fiber amplifiers

Rare-earth-doped fiber lasers and fiber amplifiers have emerged from research laboratories during the 1970s and earlier 1980s to become indispensable components of many of today's technologies in communication, medical equipment, materials processing, and military applications etc. Until today, fiber lasers and fiber amplifiers have continued to attract great efforts of research and development due to their potential for various applications. Although there are many different configurations, fiber lasers and fiber amplifiers can be divided into two categories based on the optical pumping approach: single-mode (SM) or multimode (MM) pump schemes. In the SM pump scheme (SMP), SM laser sources are usually used to couple the pump power to the fiber core, which is typically in the range of a few μm to a few tens of μm for single-mode fibers (SMFs), as shown in figure 3.1(a). As a result, there are limitations in pump power, and a lot of effort for coupling in an SMP scheme. To overcome these problems, a different pumping method has been developed in which pump power from a high-power MM laser sources is coupled to a large cladding of double-clad fiber (DCF) surrounding a rare-earth-doped core. The pump power is gradually coupled to the core and is absorbed by rare-earth doped materials. The method therefore is called MM cladding-pumped or MMP scheme since MM laser sources are used, as shown in figure 3.1(b).

As can be seen, pump power in the MMP scheme can be as high as tens of watts and much higher when pump-combined couplers are used. Moreover, couplings would be much more flexible than in the SMP method as the pump beams are now coupled to a large inner cladding of DCF, as shown in figure 3.1(b). In the MMP, laser sources with very high power, but not very high quality beams can be used for pump sources. As an example, figure 3.1(b) shows a top-hat MM pump profile of a

doi:10.1088/978-0-7503-2272-0ch3

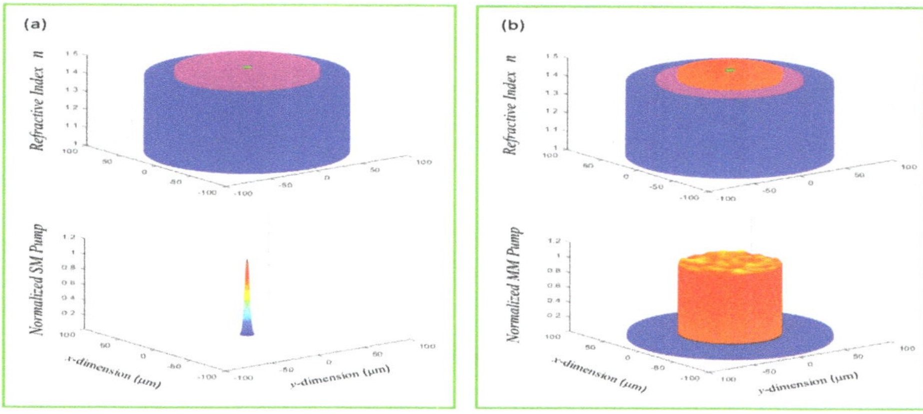

Figure 3.1. Upper: MATLAB® plot of index profile: (a) Single-clad fiber for SM pumped fiber amplifiers with clad is the pink area, (b) double-clad fiber for MMP scheme: first clad (red), second clad (pink). Lower: (a) SMP scheme with SM pump profile, and (b) MM pump profile for MMP scheme. Both fibers have SM core (green) and jacket (blue).

high-power diode beam having both amplitude and phase noises that is coupled into the inner cladding of the fiber. In contrast, the SMP scheme requires SM beams of high quality, as shown in figure 3.1(a). Nowadays, an MMP scheme is the most common method for high-power fiber lasers and fiber amplifiers.

It is worth noting that although the MMP scheme has been widely used in practice for fiber lasers and fiber amplifiers, modeling MMP rare-earth doped fibers has been very challenging. Modeling problems of MMP can be generally stated as how to describe effectively the coupling of the MM beam of the pump from a large clad to a small rare-earth doped core, and the effects of cladding shapes and fiber structures to the pump absorption and in turn to the laser and amplifier performance. Clearly, these questions are important not only for understanding the process of amplification in the fiber, but also for optimizing designs of both active fibers and configurations which play decisive roles for system performance. At the same time, modeling methods that are based on the propagation and rate equations model (PREM) have been proven to be very accurate and reliable for engineers and scientists to predict and optimize SMP fiber amplifier performances [1–3] but are not capable for an MMP scheme. Commercial software for modeling and simulation of SMP fiber lasers and fiber amplifiers is widely available. It would take a whole book to present a comprehensive theory of rare-earth-doped fiber amplifiers, and readers are encouraged to study those fields in excellent books, such as references [1–3] mentioned earlier. Here, again we focus mostly on the problem of modeling an MMP scheme whose details have been lacking in literature. The example of modeling will be examined by MATLAB® programs.

3.1.1 PREM modeling SM pumped fiber amplifiers

In this section we will describe the modeling problems of MMP fiber amplifiers (MMPFAs). We will see why the method that has been so successfully used for modeling and simulation of SMP fiber amplifiers (SMPFAs) cannot provide a good solution for MMPFAs. First, let us describe briefly the modeling and simulation method for SMPFAs. As mentioned earlier, the most effective and widely used method for SMPFAs is based on the PREM. After understanding the SMP scheme, we will see that the PREM could only provide qualitative solutions for the performance of MMPFAs. However, that is not good enough in most of the cases. In brief, PREM cannot be used to describe accurately the multimode nature in MMPFAs. Readers with fundamental knowledge of fiber optics would be able to estimate approximately that the number of modes guiding the inner clad could be well in the range of thousands of modes. That is because the diameter of inner clads for guiding MM pump beams are quite large, typically in the range of 100 μm and even much larger for very high power systems. Those guiding modes have very different spatial distributions in different fiber structures, especially cladding shapes. Worse, we cannot treat them as a bunch of independent modes because they are coupled and the mode couplings are strongly dependent on fiber structures. Due to the mode couplings, the power carried by modes can be transferred among the modes when propagating along the fiber. Therefore, spatial distribution of the MM beam is not uniform along the fiber. The variation of spatial distribution is strongly dependent on the fiber structure, and it is difficult to describe the whole MM beam by an effective overlap factor along the fiber. The dependence is very complicated so there is no method that can describe the propagation of an MM beam with such a high number of modes.

For simplicity, let us start with a two-level system in the description of PREM for an SMPFA. The two-level system is perfectly applied for Yb-doped fiber amplifiers (YDFAs) [4], but it can also be used for other rare-earth systems in the simplest approximations. For example, the system of 1480 nm-pumped Er-doped fiber amplifiers (EDFAs) and 1570-nm-pumped Tm-doped fiber amplifiers have very low dopant concentrations [1–3]. In such conditions, all processes including up-conversion and quenching that are dependent on dopant concentrations are negligible. However, the description of PREM for SMPFA in this part can be applied to other rare-earth doped fibers with some modifications. In PREM, both pump and signal powers are SM fields and each of them can be described by propagation equations coupled with a system of equation—the rate equations that describe the amplification processes in active materials. In general, several pump beams with different wavelengths can be launched into the fiber at the same or different input sides of the fiber. In other words, different pump beams can propagate in the same or opposite direction from the signal beam, the so called co-pump or counter-pump schemes, respectively. In practice, YDFAs can be pumped effectively in broad bands from 915 to 980 nm. The amplification process of the signal can be described as follows: first, pump photons are absorbed and excite ions from ground state $F_{7/2}$ to up-level state $F_{5/2}$, and is represented by a red arrow in figure 3.2. From the excited state, ions can emit stimulated (blue arrow) and spontaneous (black arrows) photons that amplify signal and generate amplified spontaneous emission (ASE), which is the noise source of the amplifier. Figure 3.2 shows examples of cross sections of Yb-ions doped glass.

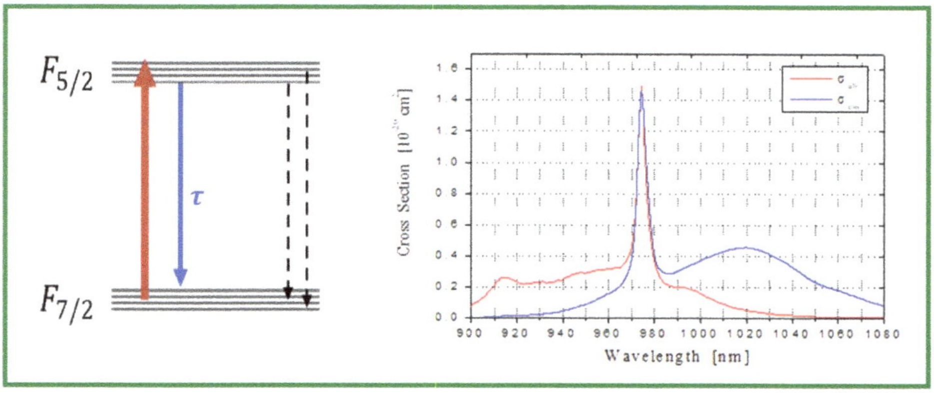

Figure 3.2. Left: energy diagram of Yb^{+3} ions doped in glasses. Pump photons are absorbed by ions in ground state $F_{7/2}$ and excite these ions to excited state $F_{5/2}$, both states have several sub-level states (red arrow). Stimulated and spontaneous emissions of photons by excited ions amplify the signal and generate amplified spontaneous emission (ASE) in the medium, and are presented by blue and black arrows, respectively. Right: absorption (red) and emission (blue) cross-sections of Yb^{+3} ions doped in glass.

In general, modeling and simulation of sing mode pumped YDFA using PREM can be described as follows. First, the propagations of SM pump and SM signal fields are described by propagation equations coupled with the rate equations that describe the absorption and emission, as presented in figure 3.2. Note that single or multiple SM signals with different wavelengths can each also be described by a propagation equation. The method can also describe very well situations in which a single pump and multiple pumps with different wavelengths are pumped at one or both ends of the fiber. Pump and signal fields can propagate in the same direction (co-pump scheme) or in opposite direction (counter-pump). Second, in addition to signal and pump fields, excited Yb^{+3} ions can also emit photons spontaneously which generate random noises that propagate in both directions along the fiber. The noise also gets amplified in propagation along the fiber. These noises are called amplified spontaneous emission (ASE) which has very negative impacts on the noise properties of fiber amplifiers. ASE can be also modeled and simulated very well in the PREM by treating a broad band of ASE as multiple channels of forward $ASE^{(+)}$ and backward $ASE^{(-)}$ that propagate in opposite directions along the fiber.

For simplicity, let us consider a simple but popular configuration of YDFAs in which both SM pump and SM signal are co-propagation (co-pump scheme). The propagations of the pump and signal field along the fiber can be described by propagation equations as [1]

$$\frac{dP_P(\lambda_P, z)}{dz} = -\left\{\int_{core} [\sigma_{abs}(\lambda_P) N_1(r, z) - \sigma_{emis}(\lambda_P) N_2(r, z)]\, \Psi_P(r) r dr d\phi - \alpha_P\right\} \times P_p(\lambda_P, z), \tag{3.1}$$

$$\frac{dP_S(\lambda_S, z)}{dz} = -\left\{\int_{core} [\sigma_{abs}(\lambda_S)N_1(r, z) - \sigma_{emis}(\lambda_S)N_2(r, z)]\, \Psi_S(r) r dr \cdot d\phi - \alpha_S\right\} \times P_S(\lambda_S, z). \tag{3.2}$$

Here, in equations (3.1) and (3.2) $P_{P(S)}$ stand for power of pump (P) and signal (S) which are varied with propagation distance z, $\sigma_{abs(emis)}(\lambda_P)$ and $\sigma_{abs(emis)}(\lambda_S)$ are absorption (*abs*) and emission (*emis*) cross sections at pump (λ_P) and signal (λ_S) wavelengths, which are experimental parameters from figure 3.2. $N_1(r, z)$ and $N_2(r, z)$ are populations of ions in ground and excited states, respectively, and their values will be determined by rate equations that will be presented later. α_S and α_P are propagation losses for signal and pump, respectively. $\Psi_S(r)$ and $\Psi_P(r)$ are profiles of signal and pump fields which are both SM guided in the fiber core. These SM profiles can be determined by solving the mode equations [1, 5] in the fiber and are normalized as

$$\iint \Psi_{S(P)}(r, \varphi) r\, dr \cdot d\varphi = 1. \tag{3.3}$$

The populations of ions in ground state N_1 and exited state N_2 are satisfied by the condition

$$N_{Yb}(r, z) = N_1(r, z) + N_2(r, z). \tag{3.4}$$

Here N_{Yb} is total population of Yb^{+3} ions in the medium (population unit is cm^{-3}).

In general, equations (3.1) and (3.2) are three-dimensional equations, which are quite complicated to solve. However, if we assume the populations are not varied across fiber core $N_{1,2}(r, z) = N_{1,2}(z)$ which is a good approximation for SM fibers with step-index cores. In such situations, equations (3.1) and (3.2) can be integrated and they become 1D propagation equations as

$$\frac{dP_P(z)}{dz} = -\{\Gamma_P[\sigma_{abs}(\lambda_P)N_1(z) - \sigma_{emis}(\lambda_P)N_2(z)] - \alpha_P\}\, P_p(z), \tag{3.5}$$

$$\frac{dP_S(z)}{dz} = \{\Gamma_S[\sigma_{ems}(\lambda_S)N_2(z) - \sigma_{abs}(\lambda_S)N_1(z)] - \alpha_S\}\, P_S(z). \tag{3.6}$$

In equations (3.5) and (3.6), $\Gamma_{P(S)}$ stands for overlap factor of the pump(signal) mode profile $\Psi_{S(P)}(r, \varphi)$ with the fiber core, defined as

$$\Gamma_{P(S)} = \iint_{core} \Psi_{S(P)}(r, \varphi) r\, dr \cdot d\varphi. \tag{3.7}$$

For simplicity, we will omit the index $\lambda_{P(S)}$ in pump and signal powers $P_{P(S)}$ in equations (3.5) and (3.6). We also change the order of the terms in equations for signal, which is useful for later discussions. For example, from equation (3.5) we will see the pump power is reduced when propagating due to absorption. However, equation (3.6) shows the signal can get gain with power increases if the whole term in brackets on the right hand side is positive. Therefore, the signal gain of fiber at any coordination z in the fiber can be defined as

$$g(\lambda_S, z) = \Gamma_S[\sigma_{ems}(\lambda_S)N_2(z) - \sigma_{abs}(\lambda_S)N_1(z)] - \alpha_S. \tag{3.8}$$

The total gain of the fiber with the length L can be calculated by integral

$$G(\lambda_S) = \int_0^L \{[\sigma_{ems}(\lambda_S)N_2(z) - \sigma_{abs}(\lambda_S)N_1(z)] - \alpha_S\}\, dz. \tag{3.9}$$

From equation (3.8) we can see that signal gain is a function of cross sections and populations. The higher the inversion population the higher the gain. From equation (3.8), we have

$$g(\lambda_S, z) = N_{Yb}\sigma_{ems}(\lambda_S)[n_2(z) - \eta(\lambda_S)n_1(z)] - \alpha_S, \tag{3.10}$$

where $n_{1,2} = N_{1,2}/N_{Yb}$ and $\eta(\lambda_S) = \sigma_{abs}(\lambda_S)/\sigma_{ems}(\lambda_S)$.

And the pump absorption is defined from equation (3.5) as

$$\alpha = \Gamma_P[\sigma_{abs}(\lambda_P)N_1(z) - \sigma_{emis}(\lambda_P)N_2(z)] + \alpha_P. \tag{3.11}$$

As described earlier, ASE can be also modeled and simulated in the PREM by treating a broad band of ASE as M channels of forward $\sum_{i=1}^{M} ASE_i^{(+)}$ and backward $\sum_{i=1}^{M} ASE_i^{(-)}$, where each channel carries a power $P_i^{(\pm)}$ of bandwidth of $\Delta\lambda_i$ that propagates in opposite directions along the fiber and are described by propagation equations as below [1–3, 5]:

$$\begin{aligned}\frac{dP_i^{(+)}(z)}{dz} &= \{\Gamma_i[\sigma_{ems}(\lambda_i)N_2(z) - \sigma_{abs}(\lambda_i)N_1(z)] - \alpha_i\} \\ &\quad P_i^{(+)}(z) + h\nu_i\Delta\nu_i\Gamma_i\sigma_{ems}(\lambda_i)N_2,\end{aligned} \tag{3.12}$$

$$\begin{aligned}\frac{dP_i^{(-)}(z)}{dz} &= -\{\Gamma_i[\sigma_{ems}(\lambda_i)N_2(z) - \sigma_{abs}(\lambda_i)N_1(z)] + \alpha_i\}P_i^{(-)}(z) \\ &\quad - h\nu_i\Delta\nu_i\Gamma_i\sigma_{ems}(\lambda_i)N_2.\end{aligned} \tag{3.13}$$

Here, h is plank constant, $\nu_i = 2\pi c/\lambda_i$ is frequency of ASE of bandwidth $\Delta\lambda_i$ centered at λ_i. Γ_i is the overlap of mode profile of ASE light of wavelength λ_i. α_i is loss of ASE and is assumed equal to signal loss α_S. Readers are encouraged to find a very comprehensive description of ASE and how to establish the propagation equations for forward and backward ASEs in textbook [1].

From these above descriptions, we can present generally the problem of modeling and simulation of YDFA with propagation equations by figure 3.3

Usually, a small bandwidth of $\Delta\lambda_i$ is chosen for one channel of ASE, and the smaller $\Delta\lambda_i$ the more accurate the noise calculation—the noise figure of fiber amplifiers. However, if $\Delta\lambda_i$ is too small then the number of ASE equations becomes too big and it could be a problem for solving the system of equations of propagation. A bandwidth of $\Delta\lambda_i = 0.5$ nm is usually chosen to produce good accuracy results of noise figure and reasonably large number of equations. For example, in a popular scheme of YDFA pumped at 980 nm the ASE band can be as broad as from 1000 to 1100 nm, therefore the model will include 400 propagation equations for forward

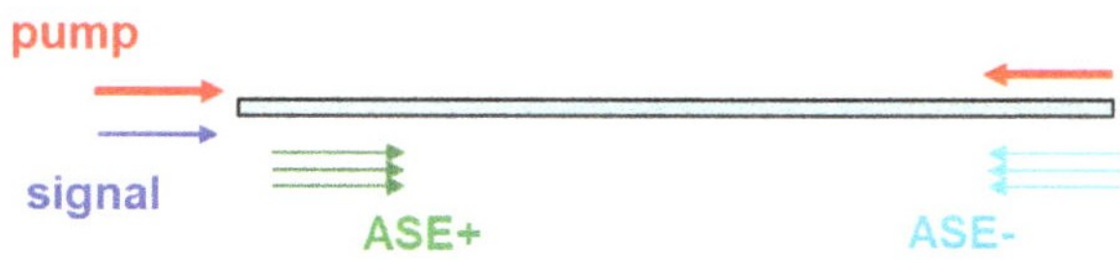

Figure 3.3. General diagram of modeling problem of YDFA by propagation equations. In general, pumps (red arrows) can be launched from both ends of the fiber, one pump is co-pump and the other is counter-pump with the signal (blue arrow). There are broad bands of amplified spontaneous emission (ASE) that propagate in the same (forward—ASE$^+$) and opposite (backward—ASE$^-$) directions of signal. These ASEs can be treated as multiple ASE channels, each carried a power of ASE within a small bandwidth $\Delta\lambda_i$ with the center wavelength λ_i. Each of these fields: pumps, signal and ASE is described by a propagation equation of an SM field that has overlap Γ_P, Γ_S and Γ_i, respectively.

and backward ASEs. At this point, it is clear that all propagation equations (3.5)–(3.6) and (3.11)–(3.12) are ready to be solved if populations $N_{1,2}$ are known. Unfortunately, the populations have complicated relationships with pump, signal and also ASE powers. Furthermore, the values of populations are not constant along the fibers as pump, signal and ASE powers are changed with propagation distance. Therefore, simulation of gain (signal), noise (ASE) and pump absorption require knowing populations at each step of propagation. In the following, there is a brief description of how to solve populations and then equations of propagation for all fields down to the fiber length of fiber amplifiers.

In general, rate equations are the mathematical model for the rate of change of populations in the amplification processes described in figure 3.2. Let us describe the changes for up-level population N_2 in the diagram in figure 3.2 in which ions in ground and exited state are N_1 and N_2, respectively. First, pump photons get absorbed by ions in ground state and excite these ions to excited state. Therefore, the rate of change of up-level population N_2 is proportional to $R_{abs}N_1$ with R_{abs} the pump absorption rate that will be defined later. The excited ions can transit to ground state by emission of stimulated and spontaneous photons which can be characterized by pump emission rate R_{ems} and lifetime τ of the excited ions. Therefore, the rate of change of N_2 is opposite with both $R_{ems}N_2$ and N_2/τ. Similarly, the signal and ASE photons can be absorbed by ground-state ions and excite these ions to the up-level state with an absorption rate W_{abs}. As a result rate of change of N_2 is proportional to $W_{abs}N_1$. By the same reason as described for the pump, the rate of change of N_2 is opposite with both $W_{ems}N_2$ with W_{ems} the emission rate for signal and ASE. Now, the rate equations of the up-level population N_2 in YDFA can be written as [1, 5]:

$$\frac{dN_2}{dt} = (R_{abs} + W_{abs})N_1 - \left(R_{ems} + W_{ems} + \frac{1}{\tau}\right)N_2, \tag{3.14}$$

and the rate equation for N_1 is determined by the relationship $N_{Yb} = N_1 + N_2 = const$.

The rates absorption and emission are defined as [1, 5]

$$R_{abs} = \Gamma_P\frac{\sigma_{abs}(\lambda_P)P_P(z)}{h\nu_P A_{core}}; \quad R_{ems} = \Gamma_P\frac{\sigma_{ems}(\lambda_P)P_P(z)}{h\nu_P A_{core}}, \tag{3.15}$$

$$W_{abs} = \frac{\Gamma_S \sigma_{abs}(\lambda_S)}{h\nu_S A_{core}} P_S(z) + \sum_i \frac{\Gamma_i \sigma_{abs}(\lambda_i)}{h\nu_i A_{core}} \left\{ P_i^{(+)}(\lambda_i, z) + P_i^{(-)}(\lambda_i, z) \right\}, \tag{3.16}$$

$$W_{ems} = \frac{\Gamma_S \sigma_{ems}(\lambda_S)}{h\nu_S A_{core}} P_S(z) + \sum_i \frac{\Gamma_i \sigma_{ems}(\lambda_i)}{h\nu_i A_{core}} \left\{ P_i^{(+)}(\lambda_i, z) + P_i^{(-)}(\lambda_i, z) \right\}. \tag{3.17}$$

Here, A_{core} is area of the core, and other parameters are already defined earlier.

As we are interested in the steady operation of the fiber amplifier, we restrict ourselves to the steady-state solutions of the rate equation for populations:

$$(R_{abs} + W_{abs})N_1 - \left(R_{ems} + W_{ems} + \frac{1}{\tau}\right)N_2 = 0. \tag{3.18}$$

Using $n_i = N_i/N_{Yb}$ and relation $n_1 + n_2 = 1$ for any two-level system, we have

$$n_2 = \frac{\tau(R_{abs} + W_{abs})}{1 + \tau(R_{abs} + R_{ems} + W_{abs} + W_{ems})}. \tag{3.19}$$

Now, we can lay out the algorithm for simulating the PREM of a YDFA as follows. For simplicity, let us ignore the ASE for the moment, and restrict ourselves to the case of co-pump scheme YDFA. At the input end of the fiber $z = 0$ where both SM pump and SM signal enter the fiber core, we know the initial conditions input pump and signal powers $P_{P0} = P_P(z = 0)$ and $P_{S0} = P_S(z = 0)$ can be easily determined. Then, the population N_2, and therefore N_1, are determined at position $z = 0$ by equation (3.18), which required values of signal and pump powers at that position from equations (3.14)–(3.16). Once populations N_1 and N_2 are known at z, we can easily integrate propagation equations (3.11) and (3.12) to $z + \Delta z$ in the propagation direction, and obtain values of $P_P(z + \Delta z)$ and $P_S(z + \Delta z)$, respectively. The calculation process is continued by calculation population at $z + \Delta z$ and then population $N_i(z + \Delta z)$ and so on until we reach the output-end, where we obtain output signal power $P_S(L)$ and the more useful values of gain defined as ratio between output and input signal power. For more convenience, the gain is defined in dB unit as

$$G = 10 \cdot \log_{10}\left(\frac{P_S(L)}{P_S(0)}\right) \text{ (dB)} \tag{3.20}$$

The problem we just considered is a simplified situation when ASEs are ignored. In fact, as can be seen from equations (3.16)–(3.17), ASEs' power appears in both absorption W_{abs} and emission W_{ems} rates, and because ASE bands are very broad, the total power of ASEs is actually quite high and therefore it has a strong impact on the population inversion, and therefore ASEs' role in the gain and noise of fiber amplifier should not and cannot be ignored. When ASEs are taken into account, the situation becomes more complicated. Now, the system of propagation equations becomes a two-boundary condition problem. As described in figure 3.3, the boundary conditions for forward and backward ASEs are known as

$P_i^{(+)}(z=0) \equiv 0$ and $P_i^{(-)}(z=L) \equiv 0$, respectively. However, if we want to integrate the propagation equations from z to L we need to know boundary conditions for backward ASE at $z=0$, and those are the unknown values initially. The two-boundary conditions are as below,

$$P_S(z=0) = P_{S0},\ P_P(z=0) = P_{P0}, \tag{3.21a}$$

$$P_i^{(+)}(z=0) = 0,\ P_i^{(-)}(z=L) = 0, \quad i \in [i, M] \tag{3.21b}$$

In order to solve the propagation equations (3.6)–(3.7) and (3.12)–(3.13) with two-boundary conditions, usually trial values are chosen for unknown boundary conditions. Then the whole system of equations is integrated forwardly from $z=0$ to $z=L$ with the known initial values, P_{S0}, P_{P0}, $P_i^{(+)}(z=0)=0$ and trial values $P_i^{(-)}(z=0)=\varepsilon$ for unknown $P_i^{(-)}(z=0)$. It is assumed that solutions of the first forward integration obtained at the end of propagation are $P_{SL}^1 = P_S(z=L)$, $P_{PL}^1 = P_P(z=L)$, $P_{iL}^{(+)1} = P_{iL}^{(+)}(z=L)$ and $P_{iL}^{(-)1} = P_{iL}^{(-)}(z=L)$. Then, the system of equations will be integrated backwardly from $z=L$ to $z=0$, but now the initial values of backward ASEs are known $P_i^{(-)}(z=L)=0$, and the exact values for other variations are unknown. In such situations, the first forward solutions P_{SL}^1, P_{PL}^1 and $P_{iL}^{(+)1}$ that we have just obtained together with known values $P_i^{(-)}(z=L)=0$ will be chosen as initial conditions for backward integration from $z=L$ to $z=0$. The solution of first backward integration $P_i^{(-)1} = P_i^{(-)1}(z=0)$ will then be used as initial values for unknown backward ASE and together with known values of the other variations for second forward integration. The processes are repeated until the solutions are in convergence with a chosen accuracy. This process of solving propagation equations with two-boundary conditions is presented in [1–3, 5].

3.1.2 Problems of modeling MM pumped fiber amplifiers by PREM

For clarity, we want to repeat again here that the PREM-based method of modeling and simulation of YDFA and other rare-earth doped fiber amplifiers is very accurate and reliable for calculating, analyzing and even designing fibers and experimental configurations for almost all SM pumped fiber amplifiers, and fiber lasers that will be discussed later in this chapter. However, when applied for modeling MMP fiber amplifiers the PREM has very limited successful, and we will point out the main reason for that in the following.

The most important reason why PREM-based methods fail to model and simulate MMP fiber amplifiers is that the methods are not able to capture accurately the multimode nature of the MM pump scheme, in which a multimode beam propagates in a highly multimode waveguide—a large inner cladding *and* coupled to a smaller core. Why is that? First, as stated earlier, if we try to use the PREM method that describes each mode by an equation of propagation, then we end up with thousands of equations for the pump beam. Solving that problem is difficult but still solvable by supercomputers with huge memory. However, even with these efforts, the method is unable to describe the behavior of the whole beam in the structure because it cannot model the couplings among all the thousand modes that are strongly dependent on not only the size but also the shape of structure. And if the method cannot describe the most important feature of the MMP fiber amplifiers—the MM and cladding pumped scheme—then it is not able to produce a good solution for the problem. To avoid that stated problem, the vast majority of engineers and scientists have to rely on a different method—it is still the PREM-based one, but instead of describing each mode of the MM beam, they try to use a *much simpler concept: the effective overlap factor* [6–8]. Let us discuss why this simple approximation can be used for modeling MMP fiber amplifiers qualitatively in some cases, but fails in most of the cases, even qualitatively.

In the approximation of effective overlap factor (EOF), the whole MM beam, in this case the MM pump is described by the effective overlap of the MM beam with the core by a factor—the EOF. The pump is assumed to fill the whole inner cladding but only a small part of the beam is overlapping with the core. As a result, EOF is defined as the ratio between core and inner clad areas $\Gamma_{eff} = A_{core}/A_{inner}$ [6–8]. The argument for the use of that approximation is that only the core-overlap part of pump beam gets absorbed by the core, and takes part in the amplification processes. That part of the pump beam can be represented by the EOF. Therefore, the argument goes on, we can still use the equation of propagation (3.5) for modeling MM pump replacing the overlap factor Γ_P of a SM pump by the effective overlap factor EOF Γ_{eff}. It is possible to see at best the EOF approximation can provide qualitative solutions for the MMP fiber amplifiers in some situations. In fact, if the operation of the fiber amplifiers is in the following conditions, the EOF approximation can be applied. First, if the fibers are perfectly rounded without any imperfection, the mode couplings, which are sensitive to perturbations, could be negligibly small and in that case the spatial distribution of the beam is almost the same for the whole length of the fiber. However, fibers are never perfect without any imperfections and it is almost impossible to avoid mode couplings in such a waveguide having structure like inner cladding around a core. Examples are in passive MM fibers where the core itself is as large as 50–62.5 μm. Without an amplification process in the passive MM fibers, mode couplings always occur and are the main reasons for limiting performance of MM fibers as compared with SM fibers. Secondly, the fiber amplifiers can be in the saturated regime where the pump is strong enough to saturate absorption for the whole fiber length. In such a situation, populations of active materials are almost the same for the whole fiber length. If populations vary along the fiber, their variations play a role as perturbation source, and increase mode couplings which change spatial power distribution and in return change absorption and populations. Fiber amplifiers with low dopant concentration can operate in a saturated regime in the case that both fiber ends are pumped strongly by co- and counter-pumps. Even if these two conditions are satisfied, the EOF approximation is still unable to describe the so-called skew rays in a perfectly rounded fiber. The skew rays are the rays that start from inner cladding and would never hit the center core as they propagate skewedly around the core, see for example [9–11]. Therefore, there is a part of MM pump beam, the skew part, which will never hit and get absorbed by the core. As a result, pump residues in experiments are always higher then for EOF modeling which predicts the pump will get absorbed eventually. Note that bending fibers can help reducing skew ray effects and all pump power can eventually be absorbed. However, different bending provides different absorption and we cannot describe that effect by EOF approximations. That is why in several cases, PREM-based methods with EOF approximation can sometimes only qualitatively prove modeling results for high quality fibers with low dopant concentrations.

The most serious problem for the PREM-based method with EOF approximation is that even in the best situation when the two above conditions are satisfied, the method still cannot be used to predict fiber amplifiers in different fiber structures. Therefore, the modeling results are not useable for optimizing the fiber amplifiers for better performance. As can be seen, the pump absorption is always characterized by the EOF, and as a result, the larger the overlap the better pump absorption, which usually leads to higher population inversion and better performance in both gain and noise figure (NF). In other words, because higher ratio between core and inner cladding areas or EOF $\Gamma_{eff} = A_{core}/A_{inner}$ leads to better performance, the only solution for optimization is to increase core size and reduce inner clad. It is true that the larger the EOF the higher absorption if fibers have the same structure. However, that ratio has certain limits: SM fiber has core size limited and cannot be larger than that determined by SM condition $V = \pi a\sqrt{n_{core}^2 - n_{clad}^2}/\lambda \leqslant 2.405$ (a is core radius). At the same time, inner cladding should not be reduced too much as the benefit of using a high power beam for pumping that couples to the inner cladding is diminished with reduced inner cladding size. More important, the PREM-based method cannot explain why the fiber with *different cladding shapes or different fiber structures provides different results even if these fibers have similar or even the same EOF*. Therefore, that is the key for designing rare-earth doped fibers for high performance systems including very high power fiber amplifiers and fiber lasers [9–12].

In fact, from general principles of optical physics, double-clad or even multiple-clad fibers with different cladding shapes have been manufactured to provide much better performance than the circularly rounded-clad fibers that are originally designed for MM cladding-pumped fiber amplifiers, as shown in figure 3.1(b). The main reason for better performance: the fiber structures break the circularly rounded symmetry of the original double clad fibers and that can reduce significantly the skew-ray effects and increase core absorption in MM cladding-pumped fiber amplifiers. Figure 3.4 shows some examples of double-clad fibers for MMP fiber amplifiers with different cladding shapes of elliptical, D-shape and rectangular shapes, and figure 3.5 illustrates the effects of skew rays in circular rounded and D-shape fibers.

As can be seen from figure 3.4, the common feature of those fibers is that their structure and cladding are modified to break the symmetry of circularly rounded shape, which is the origin of skew ray effect. As a result, absorption in those fibers with modified cladding is usually much stronger than the originally double-clad fiber with rounded shape shown in figure 3.1(b).

It is well known that many special fiber designs provide extraordinary performance, much greater than that of an originally circular double-clad fiber [9–12].

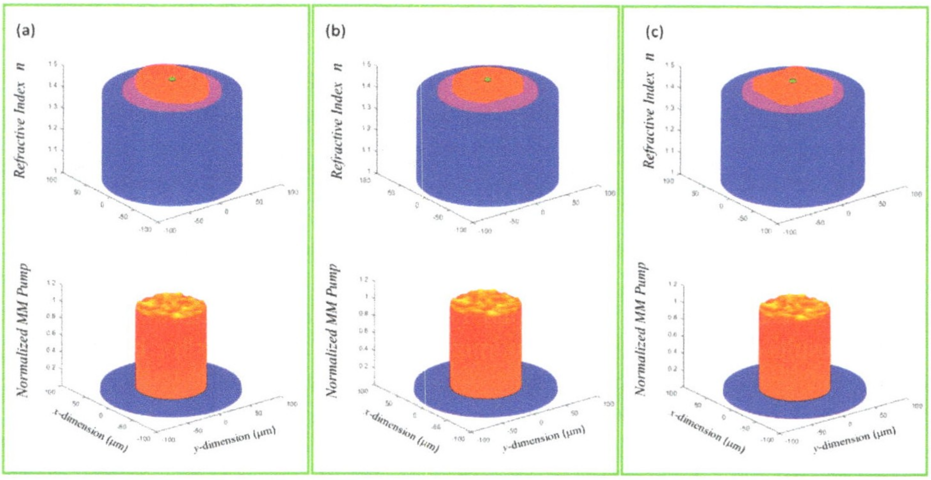

Figure 3.4. Upper: MATLAB® plot of index profile of double-clad fibers with different shapes of inner cladding (red): (a) elliptical shape, (b) D-shape, and (c) rectangular-shape. Lower: MM pump profile of top-hat in MMP scheme for all cases. All fibers have SM core (green) and and jacket (blue).

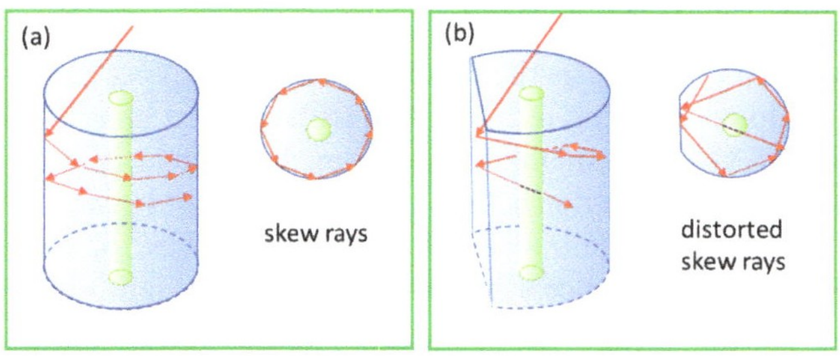

Figure 3.5. Schematic of skew rays in circularly symmetric clad fiber (a), and distorted skew rays in D-shape fiber. Fiber core is in green color, and size of core and clad are not scaled.

Therefore, fiber structures play very crucial roles for amplification processes and that important feature cannot be modeled by methods that are based on an EOF approximation that depends only on the ratio between core and clad areas.

In the next section, we will present a modeling method that has been proved to be very good and effective for MMP fiber amplifiers.

3.2 Beam propagation method for modeling multimode cladding-pumped fiber amplifiers

As presented above, the modeling methods that are based on PREM for MM cladding-pumped fiber amplifiers suffered not just a technical problem, but a real problem of principle. It is very natural to find a new computational method that can overcome the description of multimode nature in the cladding pumped scheme of fiber amplifiers, since modeling became a key for the design of fiber amplifiers, especially high power fiber amplifiers. It is well-known that one of the most powerful methods of simulating light propagation in complicated structures—the beam propagation method (BPM) is very effective simply to implement for simulation multimode beam propagation in multimode fibers. The BPM has been proposed dating back to the 1970s [13–15]. Since then, it has been proven to be a very good modeling method for multimode light propagation in optical waveguides. The BPM has been widely used in the optics industry as well as in scientific applications—with commercial software available. However, most software and publications have been about using BPM for modeling passive waveguides in which loss is constant along the propagation length.

We will call that a standard BPM to distinguish with our cases where we will use the BPM to deal with active fibers, specifically rare-earth doped fiber amplifiers. In short, our model of the effective BPM is the incorporation of the standard BPM and the rate equations to model and simulate cladding-pumped fiber amplifiers. By doing that, the method has good features of both the standard BPM and PREM, at the same time avoiding the weaknesses of both methods. For instance, we will use the BPM to simulate the multimode propagation of pump beam—the most difficult problem of MM cladding-pumped fiber amplifiers. Note that if we want to restrict ourselves to the case of SM signal simplification, which has the most important applications, we will consider the fiber satisfied SM condition or the fiber core will support only one guiding mode, which is the fundamental mode of the fiber. Because of that, we can still use the propagation equation to describe the signal instead BMP. In doing that, we can save nearly half of computing memory and computing time but the results are as good as using the BPM for modeling signal propagation. With this in mind, we now present how to do both the BPM calculation for MM pump propagation and integration of the propagation equation for the signal at the same time.

Let us start the calculation process from the input entrance $z = 0$ of the fiber, where the power of the pump and signal P_{P0} and P_{S0}, respectively, are known as input values. For simplicity, ASEs are ignored at this time.

Because the signal is an SM beam overlapping with the core by factor Γ_S we can treat the signal similarly in the PREM method to what is presented in section 3.1.1. The overlap factor for the pump at the entrance of the fiber can be represented by EOF $\Gamma_{eff} = A_{core}/A_{Inner}$ since the pump power that couples to the core at the entrance is proportional to the ratio. At $z = 0$, we can calculate populations $N_i(0) = N_i(z = 0)$ by using the rate equations, e.g., equations (3.14)–(3.17) for a YDFA. The, pump absorption $\alpha[N_i(0)]$ and signal gain $g[N_i(0)]$ as functions of populations are determined by equations (3.8) and (3.11), respectively. The pump absorption is then used in BPM to propagate the pump beam to the next step with increment Δz. At the same time, signal gain can be used to integrate the propagation equation for a SM signal. The schematic of the effective BMP for MM a cladding-pump fiber amplifier is presented in figure 3.6.

The processes are repeated until the end of the fiber length. Note that, in general, we can use the BPM to propagate both pump and signal beams, however, it is much better to use the propagation equation to describe the signal. By doing that, we can

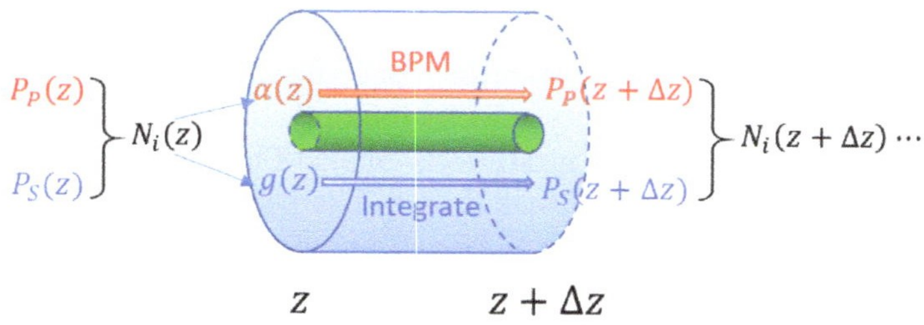

Figure 3.6. Schematic of effective BPM for MM cladding-pumped fiber amplifiers. At the fiber entrance, z pump and signal powers $P_P(z)$, $P_S(z)$ are assumed to be known as input powers. The rate equations are solved to determine populations $N_i(z)$ and pump absorption $\alpha[N_i(z)]$ and signal gain $g[N_i(z)]$ at that position. The pump absorption is then used in the BPM to propagate the pump beam with a step Δz. Meanwhile, an SM signal can be described by propagation equation and can be integrated with a step Δz. The processes are repeated until the end of the fiber length.

save nearly half the computing memory and time as compared to using the BPM for both pump and signal.

Solving the propagation equation is well known and will not be repeated here. Let us describe details of the BPM in general and then apply it for the pump in MM cladding-pumped fiber amplifiers. First, the slowly varying electric field envelop $E(x, y, z)$ of light propagating along the z-axis in a waveguide can be written by the paraxial wave equation as follows [16]:

$$\begin{aligned}\frac{\partial E(x, y, z)}{\partial z} &= \frac{i}{2k}\nabla_{\perp}^{2}E(x, y, z) + ik_0\Delta n(x, y, z)E(x, y, z)\\ &\quad - \frac{\alpha(x, y, z)}{2}E(x, y, z)\\ &= \left\{\frac{i}{2k}\left(\frac{\partial^2}{\partial x^2} + \frac{\partial^2}{\partial y^2}\right) + ik_0\Delta n(x, y, z) - \frac{\alpha(x, y, z)}{2}\right\}\\ &\quad E(x, y, z).\end{aligned} \tag{3.22}$$

where $k = n_0\omega/c = n_0k_0 = n_0 2\pi/\lambda$ with $k_0 = \omega/c$ and refractivity index (RI) profile $n(x, y, z) = n_0 + \Delta n(x, y, z)$ relative to reference refractive index n_0, and α is the power absorption/loss of the waveguide. where n_0 is reference index and can be chosen differently. Figure 3.7 shows as an example RI $n(x, y, z)$ with n_0 being the RI of the second clad in double-clad fiber.

The paraxial wave equation (3.22) can be re-written as

$$\frac{d}{dz}E(x, y, z) = \left(\hat{D} + \hat{V}\right)E(x, y, z), \tag{3.23}$$

where operators $\hat{D}$ and $\hat{V}$ are given by

$$\hat{D} = \frac{i}{2k}\left(\frac{\partial^2}{\partial x^2} + \frac{\partial^2}{\partial y^2}\right), \quad \text{and} \quad \hat{V} = \{ik\Delta n(x, y) - \alpha(x, y)\}. \tag{3.24}$$

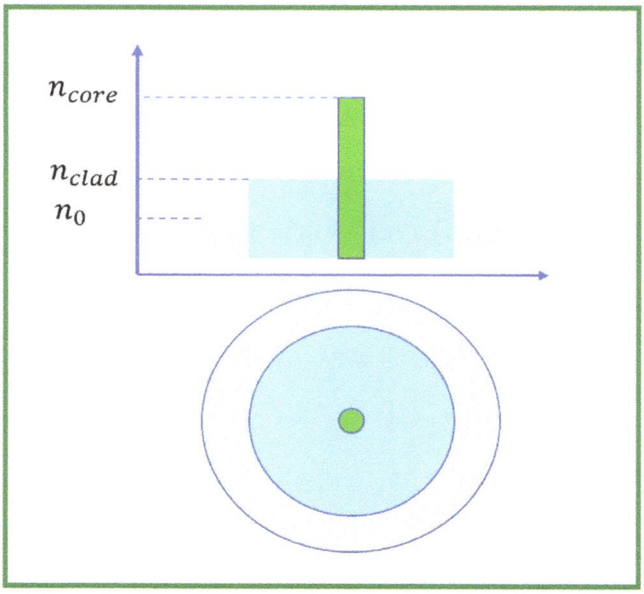

Figure 3.7. Schematic of refractive index $n(x, y, z) = n(r)$ with n_0, n_{clad} and $n_0 = n_{clad2}$ are refractive indexes of fiber core, first and second clad.

Note that because these operators $\hat{D}$ and $\hat{V}$ do not commute, $\hat{D}\hat{V} - \hat{V}\hat{D} \neq 0$, we cannot simply write the solution of equation (3.23) as

$$E(x, y, z + \Delta z) = e^{(\hat{D}+\hat{V})\Delta z}E(x, y, z) = e^{\hat{D}\Delta z}e^{\hat{V}\Delta z}E(x, y, z), \tag{3.25}$$

However, because $[\hat{D}\hat{V} - \hat{V}\hat{D}]\Delta z \to 0$ if $\Delta z \to 0$, therefore for a small propagation step Δz the solution of (3.23) can be approximated as [13–16]

$$E(x, y, z + \Delta z) = e^{(\hat{D}+\hat{V})\Delta z}E(x, y, z) \approx e^{\hat{D}\Delta z}e^{\hat{V}\Delta z}E(x, y, z). \tag{3.26}$$

Note that the approximation in equation (3.26) is second-order accurate in the increment step Δz of propagation, meaning the accuracy of the calculation is proportional to $(\Delta z)^2$.

A further approximated solution with third-order accurate $(\Delta z)^3$ can also be obtained in a small propagation step using the following re-arrangement [16]:

$$E(x, y, z + \Delta z) = e^{(\hat{D}+\hat{V})\Delta z}E(x, y, z) \approx e^{\hat{D}\frac{\Delta z}{2}}e^{\hat{V}\Delta z}e^{\hat{D}\frac{\Delta z}{2}}E(x, y, z), \tag{3.27}$$

where $\exp(\hat{D} \cdot \Delta z/2)$ means take a half step of diffraction alone, and $\exp(\hat{V} \cdot \Delta z/2)$ means take the whole step of linear propagation alone. This calculation is third-order accurate in the step length and requires that the change produced by each step is small compared to unity. Equations (3.26) or (3.27) can be solved very effectively by a fast Fourier transformation (FFT) [13–16].

Equation (3.26) can be solved by FFT as follows:

$$\begin{aligned} E(x, y, z + \Delta z) &= \exp(\hat{D}\Delta z)\exp(\hat{V}\Delta z)E(x, y, z) \\ &= \exp(\hat{D}\Delta z)\left\{\exp(\hat{V}\Delta z)E(x, y, z)\right\} \\ &= \exp(\hat{D}\Delta z)\left\{\tilde{E}(x, y, z)\right\}. \end{aligned} \tag{3.28}$$

Applying $f(x, y) = fft2\{ifft2[f(x, y)]\}$ where *fft2* and *ifft2* operators are the fast Fourier and inversed fast Fourier transforms, respectively, in two-dimensional x,y-space, we have

$$\begin{aligned} E(x, y, z + \Delta z) &= fft2\left\{ifft2\left[\exp(\hat{D}\Delta z)\left\{\tilde{E}(x, y, z)\right\}\right]\right\} \\ &= fft2\left\{\exp\left[-\left(k_x^2 + k_y^2\right)\Delta z\right]ifft2\left\{\tilde{E}(x, y, z)\right\}\right\}. \end{aligned} \tag{3.29}$$

The benefit of using formulae (3.29) is that the difficulty of calculating diffraction term is replaced by simple function $\exp[-(k_x^2 + k_y^2)\Delta z]$, and other operations are replace by very fast operations, especially if we use built-in functions fft2 and ifft2 in MATLAB®. Then, our algorithm for having solution $E(x, y, z + \Delta z)$ can be described by the following steps:

1. $$\tilde{E}(x, y, z) = \exp[ik_0\Delta n(x, y, z)\cdot\Delta z]\cdot\exp\left[-\frac{1}{2}\alpha(x, y, z)\cdot\Delta z\right] \times E(x, y, z), \tag{3.30a}$$

2. $$\zeta(x, y, z) = iff2[\tilde{E}(x, y, z)], \tag{3.30b}$$

3. $$E(x, y, z + \Delta z) = fft2\left\{\exp\left[-\left(k_x^2 + k_y^2\right)\Delta z\right]\cdot\zeta(x, y, z)\right\}, \tag{3.30c}$$

4. $$z = z + \Delta z \tag{3.30d}$$

Those are the most important expressions of our program. Once we have electric field we then have intensity and power. For example, power of the laser field in the fiber can be written as

$$P_P(z) = \iint dxdy\,|E(x, y, z)|^2. \tag{3.31}$$

The method has been successfully applied to simulate Yb-doped multicore fiber lasers [17, 18]. We have developed our own MATLAB® codes, and the simulation results of QWs are presented in the following.

3.3 Modeling example: effective BPM modeling MM cladding-pumped Yb-doped fiber amplifiers

In this section, we will apply the general description above of the effective BPM to perform modeling of MM cladding-pump YDFA. Let us briefly describe the problem with some important expressions of the formalism. For simplicity, at the very first stage of studying let us first ignore the ASEs. We will describe later in this chapter how to deal with ASEs in MM cladding-pumped rare-earth doped fiber amplifiers. With that in mind, the main equations for MM pump and SM signal beams in the effective BMP can be re-written as

$$P_P(x, y, z + \Delta z) \doteq BMP\{\alpha\{n_i(P_P(z), P_S(z))\}\Delta n(x, y, z), P_P(x, y, z)\}, \quad (3.32)$$

$$\frac{dP_S(z)}{dz} = \{\Gamma_S N_{Yb}[\sigma_{ems}(\lambda_S)n_2(z) - \sigma_{abs}(\lambda_S)n_1(z)] - \alpha_S\}\, P_S(z), \quad (3.33)$$

$$n_2 = \frac{\tau(R_{abs} + W_{abs})}{1 + \tau(R_{abs} + R_{ems} + W_{abs} + W_{ems})}. \quad (3.34)$$

where, $BMP\{\alpha\{n_i(P_P(z), P_S(z))\}\Delta n(x, y, z), P_P(x, y, z)\}$ in equation (3.32) is just an abbreviated expression of the BPM processes that depend on evaluation of populations at each step of increment step Δz, and is described in equation (3.30).

At the entrance of the fiber, the transverse power distribution of the pump $P_P(x, y, z_0)$ (and therefore the pump power in the core $P_{P0} = \Gamma_{eff} P_P(x, y, z_0)$) and signal power $P_{S0} = P_S(x, y, z_0)$ are known and are taken as input parameters. The steady state solution (3.34) of rate equations is solved numerically providing values of the local ground and excited state populations at z_0, and therefore the values of pump absorption $\alpha(N_i(z_0))$

$$\alpha = N_{Yb}\ [\sigma_{abs}(\lambda_P)n_1(z_0) - \sigma_{emis}(\lambda_P)n_2(z_0)] + \alpha_P, \quad (3.35)$$

and the local signal gain

$$g(\lambda_S, z) = \Gamma_S N_{Yb}\sigma_{ems}(\lambda_S)[n_2(z) - \eta(\lambda_S)n_1(z)] - \alpha_S, \quad (3.36)$$

are computed numerically at place z_0.

Having given the value of the pump absorption $\alpha(N_i(z_0))$, the BPM then computes the transverse distribution of the pump power $P_P(x, y, z = z_0 + \Delta z)$, and then pump power coupled to the core $P_{Pcore}(x, y, z = z_0 + \Delta z)$ is determined where Δz is a step increment in the calculation. At the same time, the single-mode propagation equation for the signal can be integrated numerically with the known local gain $g(N_i(z_0))$ to produce the signal power $P_S(x, y, z = z_0 + \Delta z)$. With the knowledge of the transverse distributions of the pump power $P_P(x, y, z)$, $P_{Pcore}(x, y, z)$, and the signal power $P_S(x, y, z)$ at $z = z_0 + \Delta z$, the rate equations are solved numerically determining local populations, pump absorption, and signal gain at $z = z_0 + \Delta z$. The pump and signal propagation are calculated by BPM and integration, respectively, to the next step increment Δz. The processes are repeated until the desired fiber length is reached. The effective BPM outputs the pump and signal powers that are dependent not only on the size of core and clad, but more importantly on fiber structures. In BPM calculation, the spatial distribution of the pump power along the propagation distance and the pump power in the doped core $P_{Pcore}(x, y, z)$ can also be determined. Therefore, the effective BPM can describe the interaction of the MM beam with the fiber structure. As a result, the method can model the effects of cladding shapes as well as fiber structure on amplifier operation.

It is worth noting that, in the cladding-pumped scheme, the initial pumping conditions such as the intensity and phase distribution can also play a role in amplifier performance. In our model, a noisy input beam is used to simulate spatial incoherence

$$E_{P0} = E_P(x, y, z = 0) = \Xi(x, y)(1 + \rho_A \xi(x, y))e^{2\pi i \rho_\varphi \varphi(x, y)}, \tag{3.37}$$

where $\Xi(x, y)$ represents the ideal coherent input beam, $\rho_{A,\varphi}$ controls the amplitude (A) and phase (φ) noise sources, $\xi(x, y)$ gives the spatial distribution of the amplitude noise, and $\varphi(x, y)$ gives the spatial distribution of the phase noise. Both noise sources are assumed to have a finite spatial coherence length ω_{coh}. In our calculation, the noise sources $\xi(x, y)$ and $\varphi(x, y)$ are normalized so that their maximum magnitude is unity, so that $\rho_{A,\varphi}$ controls the amplitude of the noise sources.

The phase fluctuations are probably the most important factor in the initial pumping conditions, and we discuss them here. Consider phase fluctuations of amplitude ρ_A and coherence ω_{coh}. Then, the characteristic wave vector introduced into the field is

$$\delta k \simeq \frac{2\pi\rho_\varphi}{\omega_{coh}} = k_0 \rho_\varphi \frac{\lambda}{\omega_{coh}} \tag{3.38}$$

or $\delta k/k_0 \simeq \rho_\varphi(\lambda/\omega_{coh})$. Consider $\rho_\varphi = 0.1$, $\omega_{coh} = 4\lambda$ then $\delta k/k_0 \simeq 0.025$, and we do not expect the spatial frequency content of the beam to be greatly modified from the ideal coherent beam. However, for $\rho_\varphi = 0.1$, $\omega_{coh} = 4\lambda$, then $\delta k/k_0 \simeq 0.25$ (or 25%), and the spatial frequency content of the beam is greatly affected with the result that many incident rays could leave the fiber as opposed to being trapped. In many experiments, the coherent length of the pump lasers is measured to be about 10–12 μm (~10λ in a YDFA).

In the following example, we will study a simple problem of modeling an MM cladding-pumped YDFA without ASE using a MATLAB® program. We want to stress again that the program is written in a way that follows closely the physical and mathematical model, not in an optimized computing way. Our focus is to understand the modeling method, and not the writing programs. In order to have a model that can simulate problems with ASE we have to develop the model further.

3.3.1 Example with MATLAB®: cladding-pumped YDFA

Let us study an example of modeling an MM cladding-pumped YDFA with parameters of fiber structure that are: core diameter d = 5 μm, inner cladding R_1 = 50 μm, outer cladding R_2 = 62 μm and the fiber jacket with R_{jack} = 100 μm. Other parameters of fiber structures such as indexes, core position, background loss are provided in the following program. Other important parameters for fiber amplifiers are dopant concentrations, in this case Yb- and Er-concentrations for Yb–Er co-doped system. As usual, some normalized parameters are used in the program, and their derivation is provided in the appendix and explanations at the end of the chapter.

MATLAB® Program 3.1

```
%xxxxxxxxxxxxxxxxxxxxxxxxxxxxxxxxxxxxxxxxxxxxxxxxxxxxxxxxxxxxxxxxx
%  MMP_Yb.m - MM Propagation in Yb-doped fiber Amplifier          x
%  Elliptical Cladding                                            x
%  Multi-Mode Propagation of Pump, SM signal                      x
%                                                                 x
%  by Dan Nguyen  12/04/2008                                      x
%xxxxxxxxxxxxxxxxxxxxxxxxxxxxxxxxxxxxxxxxxxxxxxxxxxxxxxxxxxxxxxxxx

    function [zar,rval,Pz,Pabs,Pc,Psig,Irz] = MMP_Yb

    hbar = 6.626e-34;            % planck constant, JxS
      c0 = 3e+14;                % light velocity in um/S
%  ******* FIBER STRUCTURE ***************************************
%
       d  = 5.0;                 % core diameter in um
       A  = pi*(d/2)^2;
   xcore  = 0.00;                % Core position
   ycore  = 0.00;                %
                                 % Indices
    ncore  = 1.465;              % core index
    nclad1 = 1.460;              % inner clad index
    nclad2 = 1.440;              % inner clad index
     njack = 1.400;              % jacket index
     deln = 000e-4;              % refractive-index fluctuations

%
   Rclad1   = 50.00;             % 1st cladding radius [in um]
   Rclad2   = 62.0;
   Rjack    = 200.0;             % jacket radius [in um]

   alfa     = 0.01;              % propagation loss in dB/cm
   linabs   = alfa*1e-4/4.343;   % cladding intensity loss per micron
%
   xymax    = 512.0;             % grid size in micron
    nxy     = 512;               % number of transverse points
    L       = 50.0;              % Fiber length in cm
    zmax    = L*1e+4;            % progation distance in micron
    nz      = L*1e+4;            % number of propagation points
    ncycle  = L*1e+2;            % points between outputs

 %**** end of fiber structute ***********************************

%********** PUMPING CONDITIONS *********************************
   lambdaP = 0.975;            % Pump wavelength in um
   lambdaS = 1.030;            % Signal wavelength in um

   w0x      = 45.0;            % Pump spot size in um
   w0y      = 45.0;
   nxsup    = 12;              % supper-gaussian order
   nysup    = 12;
   Rpump    = 50.0;            % Pumping Area Radius
```

```
   ycen     = 00.0;           % beam displacement from center
   inrand   = -1;             % type of input
%                          = 0   Gaussian or modal input

%                          = +1  Random amplitude fluctuations on a top hat
%                                that fills the cladding
%                          = -1  Random amplitude fluctuations on a top hat
%                                of radius rtop

   noiseA  = 0.1;             % amplitude noise level
   noiseP  = 0.1;             % phase noise level
   wcorX   = 4.0;             % spatial correlation length in X, in micron
   wcorY   = 4.0;             % spatial correlation length in Y, in micron

%**** end of pumping condition ********

%**** Input Material and Normalized Coefficients************************


     s12  = 0.05e-12;         % signal absorption cross section, in um^2
     s21  = 0.40e-12;         %    **  emission
     sap  = 1.45e-12;         % pump absorption cross section, in um^2
     sep  = 1.40e-12;         %    **  emission
     nyb  = 10.0e+08;         % Yb concentration, in ions/um^3
     t21  = 1.50e-03;         % lifetime t21, in second (S)

                              % Normalized Coefficients
      GS  = 0.80;             % Overlap factor of SM Signal
      etS = s12/s21;          % signal eta = s12/s21
      etP = sap/sep;          % pump   eta = sap/sep
      beS = GS*nyb*s21;       % signal
      beP = nyb*sep;          % pump

      % Normalized power
      qs  = hbar*c0*A/(t21*s21*lambdaS*GS); % Normalized signal in W
      qp  = hbar*c0*A/(t21*sep*lambdaP);    % MM Pump   in W
%****** end of material input ********

%****** SET UP INPUT FOR BPM ****************************************
%
   nout    = nz/ncycle;
   dz      = zmax/nz;
   dx      = xymax/nxy;
   n1      = round(d/dx);
   dx      = d/n1;
   xymax   = nxy*dx;
    kmax    = 2*pi/dx;
    dk      = kmax/nxy;
    z       = 0;
   nmid    = floor(nxy/2);

%***Set Up Input Beam
%
   efield      = zeros(nxy,nxy);
   pcore       = zeros(nxy,nxy);
   pclad1      = zeros(nxy,nxy);
   pclad2      = zeros(nxy,nxy);
   pjack       = zeros(nxy,nxy);
   ppump       = zeros(nxy,nxy);
```

```
         % Grid in x-y Space
v        = [0:nxy-1];
rval     = v*dx - xymax/2;
[x,y]    = meshgrid(v,v);
x        = x*dx - xymax/2;
y        = y*dx - xymax/2;

          % Grid in K-Space
 p        = find(v > nmid);
 v(p)     = nxy-v(p);
 v        = v*dk;
 [k1,k2]  = meshgrid(v,v);


          % Core Profile
rcore     = sqrt((x-xcore).^2+(y-ycore).^2);
p         = find(rcore <= d/2);
pcore(p)  = 1;

           % Inner clading profile
p          = find(sqrt(x.^2+y.^2) <= Rclad1);
pclad1(p)  = 1;
pclad1     = pclad1  - pcore;


          % Fiber Jacket Profile
p          = find(sqrt(x.^2+y.^2) <= Rclad2);
pclad2(p)  = 1;
pclad2     = pclad2 - pclad1  - pcore;

          % Fiber Jacket Profile
p         = find(sqrt(x.^2 + y.^2) <= Rjack);
pjack(p)  = 1;

          % Pumping Input Area
r         = sqrt(x.^2+y.^2);
p         = find(r <= Rpump);
ppump(p)  = 1;

          % Normalized Noisy Input Beam
GausC     = exp(-x.^2/wcorX^2-y.^2/wcorY^2);
normC     = dx^2*sum(sum(GausC));
GausC     = GausC/normC;
          % Amplitude Noise
stocA     = ifft2( fft2(rand(nxy)-0.5).*fft2(GausC) );
stocA     = noiseA*stocA/max(max(abs(stocA)));
          % Phase Noise
stocP     = ifft2( fft2(rand(nxy)-0.5).*fft2(GausC) );
stocP     = noiseP*stocP/max(max(abs(stocP)));

efield    = exp( -(x/w0x).^(2*nxsup)-((y-ycen)/w0y).^(2*nysup) ).*...
             (1+stocA).*exp(2*pi*sqrt(-1)*stocP);

   if inrand == +1
     efield = (1-pjack).*(1+stocA).*exp(2*pi*sqrt(-1)*stocP);
   end

  if inrand == -1
    efield = ppump.*(1+stocA).*exp(2*pi*sqrt(-1)*stocP);
  end
```

```
              % Normalized initial field
norm          = dx*dx*sum(sum(abs(efield).^2));
efield        = sqrt(1/norm)*efield;

              % Transfer Function
arg           = -dz*(k1.^2 + k2.^2)/(4*pi*nclad2);
 freq         = exp(sqrt(-1)*arg);
absdz         = dz*linabs.*pjack;

              % absorbing boundary
 fabc          = exp(-(2.0*r/xymax).^16);

%**** END OF SET UP FOR BPM ******

%**********************************************************************
%                      START BEAM PROPAGATION                         *
%**********************************************************************

 Irz         = zeros(nxy,nz/ncycle+1);
 Irz(:,1)    = abs(efield(nmid,:)).^2';
 zout        = zeros(1,nz/ncycle+1)';
 zout(1)     = 0;
 Psum        = 0;
 count       = 1;

 dphi0       = 2*pi*dz*((ncore^2  - nclad2^2)/(2*nclad2))*pcore...
             + 2*pi*dz*((nclad1^2 - nclad2^2)/(2*nclad2))*pclad1...
             + 2*pi*dz*((njack^2  - nclad2^2)/(2*nclad2))*pjack;

           % Input Pump and Signal Power
   Pin       = 100;                   % Input power in W
    Ps0      = 1.0;                   % Signal Power in W
     pp      = (d/Rpump)^2*Pin/qp;    % Normalized to qp at z=0
     ps      = Ps0/qs;                % normalized to qs


 for k = 1:nz
      dphi   = dphi0 + 2*pi*dz*deln*rand(nxy);

          sj = ps;
          n2 = (etP*pp + etS*ps)/(1 + (1+etP)*pp + (1+etS)*ps );

              % Core absorption per micron & dz
     coreabs = beP*(etP - (1+etP)*n2);
         adz = dz*coreabs*pcore;

             % Signal 1st-intergation per micron
       step1 = (beS*(n2*(1+etS)-etS)-linabs)*ps;
          ps = sj + dz*step1;

             % Norm Pfield w diffraction & absorption in dz
    efield   = efield.*exp(sqrt(-1)*dphi-adz/2-absdz/2);

             % FFT after K-Transfomation
    efield   = fft2(ifft2(efield).*freq); %.*fil;

 %***** Calculation of Pump and Signal Propagation**********

     zar(k)  = z*1e-4;
             % Total Pump Power [%]
     Pz(k)   = 100*dx^2*sum(sum(abs(efield).^2));
             % Pump Power in Core Pcpt [%] Pcore [W]
     Pcpt    = 100*dx^2*sum(sum(abs(efield).^2.*pcore));
      Pc(k)  = Pcpt*Pin/100;

             % Absorption Pump Power [%]
      Psum   = Psum + dz*coreabs*Pcpt;
     Pabs(k)= Psum;

             % Signal Power in W
     Psig(k)= ps*qs;
```

```
%***** OUTPUT DATA ****************************************

    if mod(k+1,ncycle)==0,
       count = count + 1
       Irz(:,count)   = abs(efield(nmid,:)).^2';
       zout(count)    = z*1e-4;
       pump = Pz(k)
       absorp = Pabs(k)
       signal = Psig(k)
     end


    z  = z + dz;
    pp = Pc(k)/qp;
    ps = Psig(k)/qs;
 end

%
% ****  END OF PROPAGASTION ******************

      figure(1)
      plot(zar,Pz,'b', zar,Psig,'g', zar,Pabs,'r');
      set(gca,'FontSize',15);
      legend('P_p', 'P_S', 'P_{abs}')
      axis([ 0 max(zar) 0 100 ]);
      xlabel('Propagation distance (cm)');
      grid on
      ylabel('Power [W]')
      title('P_P=100W, P_S(0)=1W, L=50cm');

%
      figure(2)
      colormap(jet)
      imagesc(zout,rval,Irz);
      set(gca,'FontSize',15);
      axis([ 0 max(zar) -65 65 ]);
      xlabel('Propagation distance (cm)');
      ylabel('Lateral distance (microns)');
      title('Intensity profile');
%===============================================================
```

In the MATLAB® window, just click on a green triangle  'Run' button on the top panel; after a while we get the simulation results for the YDFA length $L = 50$ cm with pump power $P_P = 100$ W and input signal power $P_S = 1$ W, which are shown in figure 3.8. In our PC (Dell 64 GHz processor) it take a little bit more than 1 minute for 1 cm of fiber length, and almost 1 hour for 50 cm in this example.

First, let us have some simple observations from the results in figure 3.8. In the program the propagation loss was assumed as 0.01 dB cm^{-1} or 1 dB m^{-1}. In this example $L = 50$ cm, therefore the propagation loss should be 0.5 dB. The results show at $L = 50$ cm the pump power is about ~52% and absorption power ~(41 ÷ 42)% therefore the propagation loss is ~0.5 dB as expected. The signal power is about ~39 W versus 41 W absorption, which is very close to quantum defect in the system $(1 - \lambda_P/\lambda_S) \times 100 \sim 5\%$. Note that, at the input position $z = 0$, the signal power $P_S(0) = 1$ W and absorption power $P_{abs}(0) = 0$. Note that in the example, we have used some conditions that are not typical, such as a very high

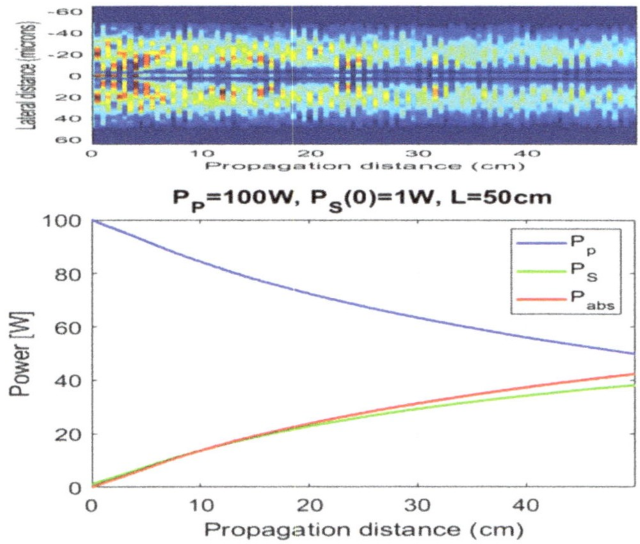

Figure 3.8. Simulation results with $L = 50$ cm, pump power $P_P = 100$ W and input signal power $P_S = 1$ W. Upper: image of MM pump propagation in the fiber. Lower: powers of pump beam (blue), pump absorption (P_{abs}) and amplified signal along the YDFA.

pump power 100 W that is coupled to cladding of $R = 50$ μm. Usually, such a high power beam of pump is typical coupled to much larger cladding. However, using that value is very convenient for us to take quick estimates as above. The Yb concentration in this example is also somewhat higher than the concentrations in Yb-doped silicate glass fiber.

Second, let us take a look more closely at the image of pump propagation in the fiber. The results show clearly that the spatial distribution of pump intensity is not uniform across the fiber and it changes along the fiber. As mentioned earlier, mode coupling is mostly unavoidable in MM beam propagation in MM waveguides, and it is the main reason for the spatially ununiform distribution of the beam intensity. It is well known from waveguide theory that different modes in an MM waveguide have spatial profiles [19, 20]. Power of a beam propagating in MM waveguide is carried by the modes of the MM waveguide. Depending on the launching conditions each mode can carried different amounts of power or different modal powers. Without any mode coupling, the spatial distribution of the whole beam is unchanged during the propagation. Furthermore, if the input beam has spatially uniform distribution, the beam overlapping with the core can be described by unchanged coefficient of effective overlap factor (EOF) $\Gamma_{eff} = A_{core}/A_{inner}$, and that is the basic assumption of the EOF method as described earlier. However, from the mode coupling theory [21, 22] the powers carried by the coupled modes can transfer periodically from one to the other during propagation. As a result, spatial distribution of power is changed periodically along the uniform waveguides. Note that, in our case considered here, an MM beam propagating in inner cladding has hundreds or even thousands of modes of an absorbed core inside. Due the mode coupling the modes can exchange and transfer power among them during the propagation. As a result, spatial distribution of the beam changes along the fiber. The couplings among the modes actually depend strongly on the waveguide structure such as size and shape. It is very hard to describe the beam of hundreds or thousands of modes as in this example when the MM pump beam propagates in an MM waveguide of diameter 100 μm. Meanwhile, the BPM does not treat the MM beam as a combination of modes as in mode couple theory, instead it treats the beam as a whole and uses the wave equation to progress the propagation of the whole beam. Therefore, the BPM is especially effective for this type of problem. All we need is to have spatial profile of waveguides (index, size, shape) and input beam profile.

3.3.2 Explanation of program 3.1

a. Because the program is presented very close to the physics and mathematics model in the text, there are only a few things that need to be explained. It is clear that the preparation of this programming is mostly for making fiber/waveguide structure profiles. After having correct profiles, the effective BPM is incorporating the standard BPM and the rate equation solving populations $N_i(z)$ and, therefore, determined pump absorption $\alpha[N_i(z)]$ and signal gain $g[N_i(z)]$ coefficients for each steps of propagation. Note that in this example, the signal is assumed as an SM beam so that we do not need to use the BPM to propagate the signal beam, which requires more memory and also double computing time. Instead, knowing gain coefficient is good enough to determine the power of the SM signal.
b. The normalized power coefficients: q_s and q_p in Program 3.1. As usual, it is better to normalize the variables in programs, in this case the pump and signal powers. Notice that the rate absorption and emission have unit $[1/T]$ (T: time) therefore, we can normalize these rates using the lifetime of excited Yb-ions by rewriting equations (3.33)–(3.34) as

$$\tau_{21}W_{abs(ems)} = \Gamma_S \frac{\tau_{21}\sigma_{abs(ems)}}{h\nu_S A_{core}} P_S(z) + \sum_i \Gamma_i \frac{\tau_{21}\sigma_{abs(ems)}}{h\nu_i A_{core}} \left\{ P_i^{(+)}(z) + P_i^{(-)}(z) \right\}$$

$$\begin{aligned} \tau_{21}W_{ems} &= \frac{P_S(z)}{q_S} + \sum_i \frac{1}{q_i} \left\{ P_i^{(+)}(z) + P_i^{(-)}(z) \right\} \\ &= p_S + \sum_i \left\{ p_i^{(+)}(\lambda_i, z) + p_i^{(-)}(\lambda_i, z) \right\} \\ \tau_{21}W_{abs} &= \eta_S \frac{P_S(z)}{q_S} + \sum_i \eta_i \frac{1}{q_i} \left\{ P_i^{(+)}(z) + P_i^{(-)}(z) \right\} \\ &= \eta_S p_S + \sum_i \eta_i \left\{ p_i^{(+)}(\lambda_i, z) + p_i^{(-)}(\lambda_i, z) \right\}. \end{aligned} \tag{3.39}$$

Here, $q_S \sim q_i$ are normalized power for signal and ASE power ($\Gamma_S \sim \Gamma_i$).

$$q_S = \frac{h\nu_S A_{core}}{\Gamma_S \sigma_{ems} \tau_{21}} \sim q_i = \frac{h\nu_S A_{core}}{\Gamma_i \sigma_{ems} \tau_{21}} \quad \text{and} \quad \eta_S = \frac{\sigma_{abs}(\lambda_S)}{\sigma_{ems}(\lambda_S)} \sim \eta_i = \frac{\sigma_{abs}(\lambda_i)}{\sigma_{ems}(\lambda_i)}. \tag{3.40}$$

Using the normalized power (3.40), we can rewrite equation (3.33) in the dimensionless formulation that is used in the program

$$\frac{dP_S(z)}{q_S dz} = \left\{ \Gamma_S N_{Yb} \sigma_{ems} \left[n_2(z) - \frac{\sigma_{abs}}{\sigma_{ems}} n_1(z) \right] - \alpha_S \right\} \frac{P_S(z)}{q_S},$$

or

$$\frac{dp_S(z)}{dz} = \{\beta_S[n_2(z) - \eta_S n_1(z)] - \alpha_S\} p_S(z). \tag{3.41}$$

where $\beta_S = \Gamma_S N_{Yb} \sigma_{ems}$ and is '`beS = GS*nyb*s21`' in the program, where Gs is overlap factor for signal.

c. For the cladding pump scheme, the normalized coefficient $q_P = h\nu_P A_{core}/\sigma_{ems}\tau_{21}$ of power for the pump does not include the overlap factor Γ_P as in the case of SM beam. The reason for that is in an MM cladding-pump we will calculate the pump power coupled to the core, and we do not use any overlap factor for the MM beam in the calculation.
d. In this example, we do not include any effect of ASE. The effect will be discussed later.

3.4 Modeling example: effective BPM modeling of MM cladding-pumped Yb–Er do-doped fiber amplifiers

Optical fiber was first developed in 1970 by scientists of Corning Inc. with attenuation $\alpha \sim 20$ dB km^{-1} (currently $\alpha \sim 0.15$ dB km^{-1}). A couple of decades later, optical fibers have revolutionized the communications industry. It is not exaggerated to state that Er-doped fiber amplifiers (EDFAs) have played a crucial role for enabling the communications revolution. For the last decades, optical communication has changed the ways people communicate, from entertainment to business, security and almost everything in our modern society. Among many rare-earth doped fiber amplifiers, the EDFA is unique with a broad band of gain for signal in wavelengths around 1550 nm where the silica fiber exhibits the minimum attenuation, which is the third window of silica glass fiber in optical telecommunications. Note that, there are two types of EDFA. In the first type, the fiber core glass is doped with Er_2O or simply Er-doped *only*, while in the second type the fiber core glass is doped with Er_2O and Yb_2O or Yb–Er co-doped. Correspondingly, we have Er-doped only fiber amplifier and Yb–Er co-doped fiber amplifiers. It is worth noting that the technology of rare-earth doped fiber in general and EDFA in particular is a very big area. We do not intend to provide even a brief review of the technology. Readers are encouraged to find that information in some excellent text books mentioned several times earlier [1–3].

In the following, we will first briefly describe the model of Er-doped only fiber amplifiers with an SM pump scheme. The aim of that part is to provide some basics of EDFAs so that we can study in more detail a more complex system, i.e., the Yb–Er co-doped fiber amplifiers. As will be seen, the latter system plays an important role in high-power fiber amplifiers (HPFAs) and fiber lasers. In practice, most HPFAs in the wavelength region from 1.530–1565 nm (C-band or simply 1.5-μm band) have used Yb–Er co-doped fibers as active fibers in their systems. As stated earlier, in the high-power regime, pump powers are usually high, typically from tens to hundreds or even kW-power levels. In such situations, beam combiners are used to combine different beams from a number of pump diodes. The beam combiner delivers the pump power for the fiber amplifiers. As a result, passive fibers used for beam combiners usually have large core sizes, many with core diameters are larger than 100 μm. In a good design of HPFA, the pump power from the beam combiner will couple to the inner cladding of the fiber amplifier in any general Yb–Er co-doped fiber. The inner cladding should be matched with or slightly larger than the core of the beam combiner. HPFAs are therefore mostly classified as an MM cladding-pump scheme. Therefore, Yb–Er co-doped fiber amplifiers are perfectly fit for the effective BPM that we described above. However, an Er-doped system is simpler, and understanding amplification processes in the system is very helpful for modeling the more complicated Yb–Er co-doped one. Note that, although many if not most Er-doped only fiber amplifiers belong to SM pump schemes, some of them have used the cladding-pumped method. We will discuss these cases later in this chapter.

3.4.1 Modeling 1480 nm-pumped Er-doped fiber amplifiers

It is worth noting that Er-doped fiber amplifiers can be pumped by two different pump wavelengths, 1480 and 980 nm. First, let us consider the simpler scheme with a 1480 nm pump, as described in figure 3.9. We follow the conventional description of amplification processes in Er-ions [16], as shown in figure 3.9. The upper diagram shows a case of EDFA, in which both pump (red arrow) and signal (blue arrow) are co-propagation and couple into the same input of the fiber. The diagram of energy levels shows the main processes of amplification in an EDFA pumped by a 1480 nm pump. In the diagram, the pump photons in the wavelength region ~1480 nm (red-solid arrow) excite Er-ions from ground state $^4I_{15/2}$ to the excited state $^4I_{13/2}$. The excited Er-ions can be relaxed to the ground state $^4I_{15/2}$ by emitting photons in a broad band in two different ways: (i) spontaneous and (ii) stimulated emission. The spontaneously emitted photons generate the noise in the medium, meanwhile the stimulated emission photons provide gain to the signal photons with the same wavelengths as presented by the blue arrow in the diagram. In an amplification medium the spontaneously emitted photons can be amplified and they are called amplified spontaneous emission (ASE) presented by dashed-arrows in the diagram on the left panel of figure 3.9. In general, if two Er-ions under excitation are close enough, the energy of one excited Er-ion can transfer to the other also in excited state. As a result, one Er-ion drops to the ground state and the other jumps to higher

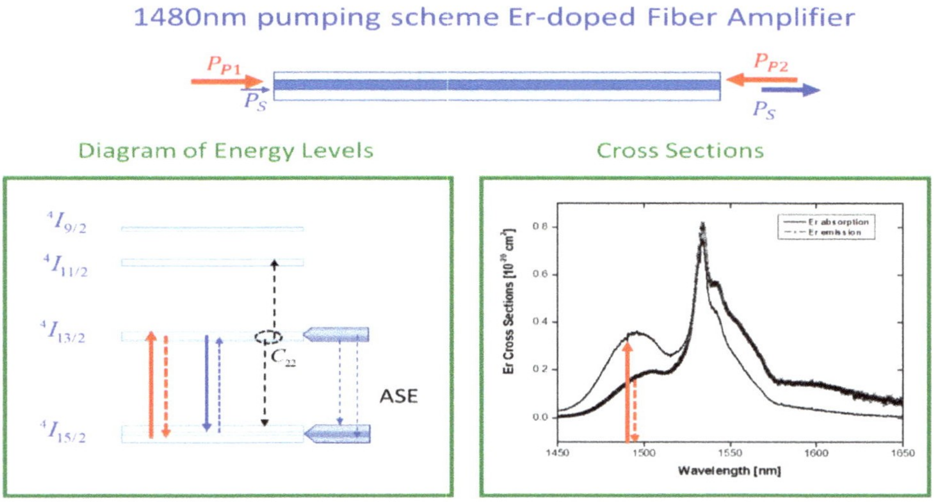

Figure 3.9. Upper: schematic of co-propagation an EDFA with pump (red arrow) and signal (blue arrows). Lower: diagram of energy in Er-ions doped in glass. Red arrow: pump photons, blue arrow: stimulated emission photons, dashed-thin arrows: ASEs. Dashed-bold arrow: the up-conversion characterized by C_{22} coefficient. Cross sections panel: absorption and emission CR of Er-ions in phosphate glass (permission by IEEE). Red arrow indicates a 1480 nm-pump scheme.

level $^4I_{9/2}$. The process is call up-conversion and is characterized by an up-conversion coefficient C_{22}, which strongly depends on Er-concentration doped in glass. In general, if the Er concentration is low, the up-conversion coefficient can be negligibly small. The right panel shows the absorption and emission cross-sections of an Er-ion doped in phosphate glass just as examples. Again, in that figure, the red arrow presents the pump photons in the region 1480 nm, but in fact it can be pumped in a broad band from ~145 to ~1532 nm. As can be seen in the figure of cross section (CR), there is a broad band in which the emission CR is higher than the absorption one, which is the wavelength band in which a signal can get amplified if the population of excited Er-ions is larger than 50% [16].

Note that, in the energy diagram and plot of CRs above, the excited ions can transit to ground state by emission of photons with wavelength of the pump photons indicated by the dashed arrow in the diagram and CR in figure 3.9. The strength of absorption and emission rates of pump photons are determined by the values of absorption and emission CRs, respectively. Similarly, signal photons can be re-absorbed by Er-ions in the ground state presented by the dashed blue arrow in the diagram. The rates of absorption and emission for pump and signal photons (and also ASE) are described by expressions in equations (3.14)–(3.17) in the first part of this chapter. It is clear that from those expressions, the strength of absorption and emission rates are determined solely by the values of absorption and emission CRs, respectively. That is the reason why we should pump the system in the band of wavelength, where pump absorption CR is higher than pump emission CR. For the same reason, it is better to amplify the signal in the region where the emission CR is higher than the absorption one.

Although it looks quite simple, the energy diagram together with CRs in figure 3.9 provides very important information of the amplification processes in 1480 nm-pumped EDFAs. Firstly, the ratio between pump absorption and emission CRs $\eta_P = \sigma_{abs}(\lambda_P)/\sigma_{ems}(\lambda_P)$ peaks around 1480–1490 nm indicating that these are the most efficient wavelengths for pumping. Secondly, the values of absorption CR of Er-ions $\sigma_{abs}(\lambda_P)$, $\lambda_P \in$ [1480 nm − band] is quite low as compared to the Yb-ions in 980 nm band (shown in figure 3.2). In particular, in Er-doped glass the maximum value of absorption CR $\sigma_{abs,Er}(1480\ \text{nm}) \sim 0.4 \times 10^{-22}\ \text{cm}^{-2}$ is much lower than that of the Yb system at 980 nm $\sigma_{abs,Yb}(980\ \text{nm}) \sim 1.5 \times 10^{-22}\ \text{cm}^{-2}$. That fact indicates that absorption at 1480 nm in an Er-system is much lower than absorption in the Yb system at 980 nm. It also opens up the possibility of doping both Yb and Er in the same glass host to significantly increase absorption, as we will see in the next section. Thirdly, as indicated in the energy diagram, the up-conversion process negatively impacts the amplification as it depletes the population of the excited state without contributing to the stimulated emission. The up-conversion process is therefore contributing noise in the fiber amplifier and also generating heat, which is very serious for amplifier operation. Physically, if the Er concentration increases the space between Er-ions is reduced, and the higher the concentration the closer the ions are and the stronger the up-conversion effect (higher up-conversion coefficient). This is a big obstacle for the case of the Er-doped only system. Because absorption CR $\sigma_{abs,Er}(1480\ \text{nm})$ is relatively low, the Er concentration should be high to improve the total absorption. However, the up-conversion in high concentration systems would be very negative to the amplification process as it generates both noise and heat. That is also the main reason why Er-doped only fiber amplifiers are usually in relatively low Er-concentration, which could be good in the low power regime using an SM pump scheme. Therefore, modeling SM pumped Er-doped only fiber amplifiers is quite similar to the SM pumped Yb-doped fiber in section 3.1 earlier. In general, we can pump the EDFA from both ends, one pump is in co-propagation and the other one is counter-propagation with the signal. The pumps, signal and ASEs can be described by propagation equations as below

$$\frac{dP_P(z)}{dz} = \pm\{\Gamma_P[\sigma_{emis}(\lambda_P)N_2(z) - \sigma_{abs}(\lambda_P)N_1(z)] \pm \alpha_P\}\, P_p(z), \tag{3.42}$$

$$\frac{dP_S(z)}{dz} = \{\Gamma_S[\sigma_{ems}(\lambda_S)N_2(z) - \sigma_{abs}(\lambda_S)N_1(z)] - \alpha_S\}\, P_S(z). \tag{3.43}$$

$$\begin{aligned}\frac{dP_i^{(\pm)}(z)}{dz} &= \pm\{\Gamma_i[\sigma_{ems}(\lambda_i)N_2(z) - \sigma_{abs}(\lambda_i)N_1(z)] \mp \alpha_i\}\, P_i^{(\pm)}(z)\\ &\pm h\nu_i\Delta\nu_i\Gamma_i\sigma_{ems}(\lambda_i)N_2,\end{aligned} \tag{3.44}$$

In equations (3.42) and (3.44) the signs ± are for co-pump and counter pump cases.

The rate equation that describes the rate of change for population in the case with up-conversion can be described as [1–4, 16, 24–28]

$$\frac{dN_1}{dt} = -(W_{abs} + R_{abs} + R_{13})N_1 + \left(W_{ems} + R_{ems} + \frac{1}{\tau_{21}}\right)N_2 + C_{22}N_2^2, \tag{3.45}$$

$$\frac{dN_2}{dt} = (W_{abs} + R_{abs})N_1 - \left(W_{ems} + R_{ems} + \frac{1}{\tau_{21}}\right)N_2 + \frac{N_3}{\tau_{32}} - 2C_{22}N_2^2, \tag{3.46}$$

$$N_{Er} = N_1 + N_2 + N_3. \tag{3.47}$$

Here, the populations of states ${}^4I_{15/2}$, ${}^4I_{13/2}$ and ${}^4I_{11/2}$ are denoted N_1, N_2 and N_3, respectively. τ_{21} is the fluorescence lifetime of the second level ${}^4I_{13/2}$, and τ_{32} is nonradiative relaxations from third level ${}^4I_{11/2}$ to lower levels ${}^4I_{13/2}$, respectively. The rate equations (3.45)–(3.46) now include nonlinear terms $C_{22}N_2^2$. In the limit of low concentration we can ignore the nonlinear terms, and also the high-level population $N_3 \sim 0$ since in that case there is no mechanism to excite Er-ions to these states. In that case analytical solutions of populations in steady-state can be easily obtained as below

$$n_1^0 = \frac{N_1^0}{N_{Er}}\frac{1 + \tau_{21}(R_{ems} + W_{ems})}{1 + \tau_{21}(R_{ems} + W_{ems} + R_{abs} + W_{abs})}, \tag{3.48}$$

$$n_2^0 = \frac{N_2}{N_{Er}} = \frac{1 + \tau_{21}(R_{abs} + W_{abs})}{1 + \tau_{21}(R_{ems} + W_{ems} + R_{abs} + W_{abs})}. \tag{3.49}$$

Here, the upper index '0' stands for the system without up-conversion process. In that case, the 1480 nm-pumped EDFAs behave exactly as the two-level system YDFAs presented earlier, as all high-level states in the energy diagram of Er-ions do not take part in the amplification process. However, when the nonlinear terms are included, numerical solutions for the rate equations are necessary.

The rates of absorption and emission for pump and signal have similar expressions as in the case of the SM-pumped YDFA presented in section 3.1.

$$\begin{aligned} R_{abs} &= \Gamma_{P1}\frac{\sigma_{abs}(\lambda_{P1})P_{P1}(z)}{h\nu_{P1}A_{core}} + \Gamma_{P2}\frac{\sigma_{abs}(\lambda_{P2})P_{P2}(z)}{h\nu_{P2}A_{core}}, \\ R_{ems} &= \Gamma_{P1}\frac{\sigma_{ems}(\lambda_{P1})P_{P1}(z)}{h\nu_{P1}A_{core}} + \Gamma_{P2}\frac{\sigma_{ems}(\lambda_{P2})P_{P2}(z)}{h\nu_{P2}A_{core}}. \end{aligned} \tag{3.50}$$

$$W_{abs} = \frac{\Gamma_S\sigma_{abs}(\lambda_S)}{h\nu_S A_{core}}P_S(z) + \sum_i \frac{\Gamma_i\sigma_{abs}(\lambda_i)}{h\nu_i A_{core}}\left\{P_i^{(+)}(\lambda_i, z) + P_i^{(-)}(\lambda_i, z)\right\}, \tag{3.51}$$

$$W_{ems} = \frac{\Gamma_S\sigma_{ems}(\lambda_S)}{h\nu_S A_{core}}P_S(z) + \sum_i \frac{\Gamma_i\sigma_{ems}(\lambda_i)}{h\nu_i A_{core}}\left\{P_i^{(+)}(\lambda_i, z) + P_i^{(-)}(\lambda_i, z)\right\}. \tag{3.52}$$

Simulation of an SM 1480 nm-pumped EDFA using the model that is based on equations (3.42)–(3.45) is very standard, and some commercial software is available. Therefore, we will not present the code for SM pumped EDFA. We will, however study one example of MM 1480 nm cladding-pumped Er-doped only fiber amplifiers. As stated earlier, the Er-concentration could be quite different depending on the host glass. At the earlier time, when EDFAs were mostly based on silica glass with quite low Er-concentration, typically in the range of 0.1–0.2wt%. Due to the structure of silica glass, when Er concentration increases higher than that range, the possibility of clustering of Er-ions is very high, resulting in both up-conversion and quenching effects in which population of the excited level is quickly depleted. However, there are some host glasses where Er concentration is much higher and clustering effects are much weaker than in silica glass. For example, Er_2O can be doped in phosphate glass up to few percent weight with relatively small values of up-conversion coefficient. In such high Er concentration, an MM cladding-pump Er-doped fiber can be used in high power operation. The benefit of that scheme is that pumping 1480 nm has quite a good quantum defect $\eta = (1 - \lambda_P/\lambda_S) \sim 5\%$, which is one of the main sources of heat in the EDFA.

Modeling example 3.2: Multimode cladding-pumped 1480 nm EDFA

We will use MATLAB® program to study an example of MM 1480 nm cladding-pump EDFA. In the case of study, we will again ignore the ASE—that complicated problem will be studied at the end of this chapter. In this example we will consider the case of Er-doped phosphate glass fiber with 3wt% of Er concentration. As mentioned earlier, Er concentration is much higher than with silica glass fiber. Before going into details of the program, we want to discuss several important points in this example which is new compared with the MM cladding pump YDFA in the earlier example with program 3.1.

First, gain and noise which are characterized by the noise figure (NF) are most important in a fiber amplifier. However, in many applications where a laser pulse is amplified, another parameter, the B-integral which is phase change of the pulse due to nonlinear effect $\phi \sim n_2 \int |E(x, y, z)|^2 dz$ accumulated during the propagation is very important requiring optimization of design [30, 31]. In general, the higher the ϕ, the higher the pulse distortion, and usually in the pulse amplification we need to keep ϕ as small as possible, typically less than π to achieve reasonably good beam quality. Note that in many applications a higher power laser beam but completely distorted pulses is not good for the application. In the example, we will calculate B-integral for a pulse with pulse width $\tau = 1$ ns, repetition rate 500 kHz.

Second, in this example we will solve the nonlinear rate equation numerically instead using the analytical solution as in the case of the YDFA. Remember, in the rate equations in the Er-doped fiber amplifier we have nonlinear terms $\sim C_{22}N_2^2$. In this example, we have a function `NewtonSys(pp, ps)` to solve the nonlinear rate equations and return the values of populations at each step of the propagation. As can be seen, the function `NewtonSys(pp, ps)` has two inputs `pp` and `ps` which are normalized pump and signal power. At each step in propagation, the function is called to calculate the pump absorption and signal gain. Finally, in this example, we use the Newton method for solving the nonlinear equations, but readers can modify the code using a different method.

MATLAB® Program 3.2

```
%xxxxxxxxxxxxxxxxxxxxxxxxxxxxxxxxxxxxxxxxxxxxxxxxxxxxxxxxxxxxxxx
%  MMPEr1480.m - 1480 Er-doped Fiber Amplifier                  x
%  Multi-Mode Propagation of Pump, SM signal                    x
%  Big Core d = 10 micron                                       x
%                                                               x
%  by Dan Nguyen  01/24/2020                                    x
%xxxxxxxxxxxxxxxxxxxxxxxxxxxxxxxxxxxxxxxxxxxxxxxxxxxxxxxxxxxxxxx

    function [zar,rval,Pz,Pabs,Pc,Psig,Pheat,Irz,B0] = MMPEr1480
     global t1 t2 alfa;
     global etaS etaP cup;

    hbar = 6.626e-34;          % planck constant, JxS
      c0 = 3e+14;              % light velocity in um/S
% ******* FIBER STRUCTURE ***************************************
%
       d  = 10.0;              % core diameter in um
       A  = pi*d^2/4.0;        % Core Area, in um^2
     MFD  = 16.0;
     Aeff = pi*MFD^2/4.0;
                               % Indices
    ncore = 1.5820;            % core index
    nclad = 1.5810;            % inner clad index
    nclout= 1.5030;            % outer clad index
    njack = 1.0000;            % jacket index
     deln = 001e-4;            % refractive-index fluctuations

     xcore = 0.00;             % Core position
    ycore  = 0.00;             %

%
    Rclad  = 50.0;             % Clad radius in micron
    Rclout = 80.0;             % Outer clad radius
    Rjack  = 100.0;            % Jacket radius

   alfa    = 0.02;             % propagation loss in dB/cm
   linabs  = alfa*1e-4/4.343;  % cladding intensity loss per micron
%
   xymax   = 256.0;            % grid size in micron
    nxy    = 512;              % number of transverse points
    L      = 10;               % Fiber length (cm)
    zmax   = L*1e+4;           % propagation distance in micron
    nz     = L*1e+4;           % number of propagation points
    ncycle = L*1e+2;           % points between outputs

 %**** end of fiber structute ************************************


 %********** PUMPING CONDITIONS **********************************

   lambdaP = 1.480;              % Pump wavelength in um
   lambdaS = 1.550;              % Signal wavelength in um
   tau     = 1e-9;               % pulse width 1ns
   fre     = 500e3;              % Rep Rate 500 kHz
   Trep    = 1/fre;              % T=1/f
```

```
   coef     = 0.88*Trep/tau;
   NonI     = 2.0e-7;              % nonlinear index of phosphate glass
                                   % n2 = 2 10^(-16) cm2/W

   w0x      = 62.0;            % Pump spot size in um
   w0y      = 62.0;
   nxsup    = 12;               % supper-gaussian order
   nysup    = 12;
   Rpump    = 65.0;            % Pumping Area Radius

   ycen     = 00.0;             % beam displacement from center
   theta    = 6;
   inrand   = -1;                % type of input
%                        = 0   Gaussian or modal input
%                        = +1  Random amplitude fluctuations on a top hat
%                              that fills the cladding
%                        = -1  Random amplitude fluctuations on a top hat
%                              of radius rtop

   noiseA   = 0.01;              % amplitude noise level
   noiseP   = 0.01;              % phase noise level
   wcorX    = 4.0;              % spatial correlation length in X, in micron
   wcorY    = 4.0;              % spatial correlation length in Y, in micron

%**** end of pumping condition *********

%**** Input Material and Normalized Coefficients***********************
    %% phostpahe glass
      s12   = 0.263e-12;         % 1550 signal absorption CR, in um^2
      s21   = 0.368e-12;         %    **  emission
      sap   = 0.245e-12;         % 1480-pump absorption CR, in um^2
      sep   = 0.083e-12;         %    **  emission

      ner   = 3.45e+08;          % Er concentration, in ions/um^3 (1.15 =1w%)

      c22   = 4.50e-6;           % C22, um^3/s
      t21   = 7.00e-03;          % lifetime t21, in second (S)
      t31   = 1.00e-07;
      t32   = 2.00e-05;          % Lifetime t32, in s
        t1  = t21/t31;
        t2  = t21/t32;

                                % Normalized Coefficients
      GS   = 0.80;              % Overlap factor of SM Signal
      etaS = s12/s21;           % signal eta = s12/s21
      etaP = sap/sep;           % pump   eta = sap/sep
      beS  = GS*ner*s21;        % signal
      beP  = ner*sep;           % pump
      k0   = 2*pi/lambdaP;
      cup  = c22*ner*t21;

      % Normalized power
      qs   = hbar*c0*A/(t21*s21*lambdaS*GS); % SM Signal in W
      qp   = hbar*c0*A/(t21*sep*lambdaP);    % MM Pump   in W

%% ****** end of material input *********
```

```
%****** SET UP INPUT FOR BPM ******************************************
%
   nout    = nz/ncycle;
   dz      = zmax/nz;
   dx      = xymax/nxy;
   n1      = round(d/dx);
   dx      = d/n1;
   xymax   = nxy*dx;
    kmax    = 2*pi/dx;
    dk      = kmax/nxy;
    z       = 0;
   nmid    = floor(nxy/2);

%***Set Up Input Beam
%
   efield      = zeros(nxy,nxy);
   pcore       = zeros(nxy,nxy);
   pclad       = zeros(nxy,nxy);
   pclout      = zeros(nxy,nxy);
   pjack       = zeros(nxy,nxy);
   ppump       = zeros(nxy,nxy);

               % Grid in x-y Space
    v          = [0:nxy-1];
    rval       = v*dx - xymax/2;
    [x,y]      = meshgrid(v,v);
    x          = x*dx - xymax/2;
    y          = y*dx - xymax/2;

               % Grid in K-Space
    p          = find(v > nmid);
    v(p)       = nxy-v(p);
    v          = v*dk;
    [k1,k2]    = meshgrid(v,v);

               % Fiber Core Profile
   rcore       = sqrt((x-xcore).^2+(y-ycore).^2);
   p           = find(rcore <= d/2);
   pcore(p)    = 1;

               % inner clading profile
   p           = find(sqrt(x.^2+y.^2) <= Rclad);
   pclad(p)    = 1;
   pclad       = pclad - pcore;

               % Fiber Jacket Profile ver 2 (10/2013)
    p           = find(sqrt(x.^2+y.^2) <= Rclout);
    pclout(p)   = 1;
    pclout      = pclout - pcore - pclad;

              % Fiber Jacket Profile ver 2 (10/2013)
    p           = find(sqrt(x.^2+y.^2) <= Rjack);
    pjack(p)    = 1;
    pjack       = pjack - pcore - pclad - pclout;

                % Pumping Input Area
    r           = sqrt(x.^2+y.^2);
    p           = find(r <= Rpump);
    ppump(p)    = 1;
```

```
                % Normalized Noisy Input Beam
   GausC        = exp(-x.^2/wcorX^2-y.^2/wcorY^2);
   normC        = dx^2*sum(sum(GausC));
   GausC        = GausC/normC;
                % Amplitude Noise
   stocA        = ifft2( fft2(rand(nxy)-0.5).*fft2(GausC) );
   stocA        = noiseA*stocA/max(max(abs(stocA)));
                % Phase Noise
   stocP        = ifft2( fft2(rand(nxy)-0.5).*fft2(GausC) );
   stocP        = noiseP*stocP/max(max(abs(stocP)));

   efield       = exp( -(x/w0x).^(2*nxsup)-((y-ycen)/w0y).^(2*nysup) ).*...
                  (1+stocA).*exp(2*pi*sqrt(-1)*stocP);

     if inrand == +1
       efield = pjack.*(1+stocA).*exp(2*pi*sqrt(-1)*stocP); %NEW
     end

     if inrand == -1
       efield = ppump.*(1+stocA).*exp(2*pi*sqrt(-1)*stocP);
     end
                % Normalized initial field
   norm         = dx*dx*sum(sum(abs(efield).^2));
   efield       = sqrt(1/norm)*efield;

                % Transfer Function
   arg          = -dz*(k1.^2 + k2.^2)/(2*k0*nclout);
    freq        = exp(sqrt(-1)*arg);
   absdz        = dz*linabs*(pcore + pclad);

                % absorbing boundary
    fil         = exp(-(2.0*r/xymax).^16);

%**** END OF SET UP FOR BPM ******

%*************************************************************************
%                   START BEAM PROPAGATION                               *
%*************************************************************************
   Irz        = zeros(nxy,nz/ncycle+1);
   Irz(:,1)   = abs(efield(nmid,:)).^2';
   zout       = zeros(1,nz/ncycle+1)';
   zout(1)    = 0;
   Psum       = 0;
   Pscat      = 0;
   BInt       = 0;
   count      = 1;
   dphi0      = 2*pi*dz*((ncore^2 - nclout^2)/(2*nclout))*pcore...
              + 2*pi*dz*((nclad^2 - nclout^2)/(2*nclout))*pclad...
              + 2*pi*dz*((njack^2 - nclout^2)/(2*nclout))*pjack;
            % Input Pump and Signal Power
     Pin      = 100.00;             % Input power in W
      Ps0     =   0.50;             % Signal Power in W
     Geff     = (d/Rclad)^2;        % overlap at the input
      pp      = Geff*Pin/qp;        % Normalized to qp
      ps      = Ps0/qs;             % normalized to qs

 for k = 1:nz
     dphi  = dphi0 + 2*pi*dz*deln*rand(nxy);
        sj = ps;
         x = NewtonSys(pp, ps);
       n2 = x(2);
```

```
          % Core absorption per micron & dz
    coreabs = beP*(etaP-(1+etaP)*n2);
        adz = dz*coreabs*pcore;

          % Signal 1st-intergation per micron
     step1 = (beS*(n2*(1+etaS)-etaS)-linabs)*ps;
        ps = sj + dz*step1;

        % B-Intergral
        BInt = BInt + 2*pi*NonI*dz*ps*qs/(lambdaS*Aeff);
        B0(k)= BInt;

          % Norm Pfield w diffraction & absorption in dz
   efield  = efield.*exp(sqrt(-1)*dphi-adz/2-absdz/2);

          % FFT after K-Transfomation
   efield  = fft2(ifft2(efield).*freq); %.*fil;

%*********end of transform procedure ********************

%***** Calculation of Pump and Signal Propagation*********

    zar(k)  = z*1e-4;
          % Total Pump Power [%]
    Ppump   = 100*dx^2*sum(sum(abs(efield).^2));
    Pz(k)   = 0.01*Pin*Ppump;
          % Pump Power in Core Pcpt [%] Pcore [W]
    Pcpt    = 100*dx^2*sum(sum(abs(efield).^2.*pcore));
     Pc(k)  = Pcpt*Pin/100;

          % Absorption Pump Power [%]
     Psum   = Psum + dz*coreabs*Pcpt;
    Pabs(k)= 0.01*Pin*Psum;

          % Background loss of pump power [%]
    Pscat   = Pscat + dz*linabs*100*dx^2*sum(sum(abs(efield).^2));
    Plos(k)  = 0.01*Pin*Pscat;

           % Signal Power in W
     Psig(k)= ps*qs;
     Pheat(k)= (Psum + Pscat)*Pin*0.01 - (ps*qs - Ps0);


%***** OUTPUT DATA ****************************************
     if mod(k+1,ncycle)==0,
        count = count + 1
        Irz(:,count)  = abs(efield(nmid,:)).^2';
        zout(count)   = z*1e-4;
        pump = Pz(k)
        absorp = Pabs(k)
        signal = Psig(k)
        heat   = Pheat(k)
      bnumber = coef*B0(k)
      end
      z  = z + dz;
      pp = Pc(k)/qp;
      ps = Psig(k)/qs;
  end
```

```
%
% ****  END OF PROPAGASTION *******************
%
    figure
      plot(zar, Pz,'b', zar,Pabs,'g', zar,Pheat,'r');
      set(gca,'FontSize',15);
      axis([ 0 L 0 Pin ]);
      xlabel('Propagation distance (cm)');
      ylabel('Pump Power [W]')
      set(gca,'YTick',0:20:Pin)
      grid on;
      legend('P_{pump}','P_{abs}', 'P_{heat}')
      title('3%Er, \Phi_{in}=130\mu, P=100W@1480nm, S=0.5W@1550nm');
%
      figure
      imagesc(zout,rval,Irz);
      set(gca,'FontSize',15);
      axis([ 0 L -65 65 ]);
      set(gca,'YTick',[-60 -10 0 10 60])
      xlabel('Propagation distance (cm)');
      ylabel('Lateral distance (microns)');
      title('3%Er, \Phi_{in}=130 \mu, P=100W@1480nm, S=0.5W@1550nm');
%
      figure
      plot(zar,Psig,'b');
      set(gca,'FontSize',15);
      axis([ 0 L 0 55]);
      set(gca,'YTick',0:5:55)
      xlabel('Propagation distance (cm)');
      ylabel('Signal Power [W]')
      title('3%Er, \Phi_{in}=130 \mu, P=100W@1480nm, S=0.5W@1550nm');
      grid on;

      figure
      plot(zar,B0,'r');
      set(gca,'FontSize',15);
      axis([ 0 L 0 max(B0)]);
      xlabel('Propagation distance (cm)');
      ylabel('B Integral [rad]')
      title('3%Er, \Phi_{in}=130 \mu, P=100W@1480nm, S=0.5W@1550nm');
     grid on;
%
%==============================================================
% function:
% NewtonSys Newton method for system of nonlinear equations
%==============================================================
function x = NewtonSys(pp, ps)
  global t1 t2;
  global etaS etaP cup;

 tol = 1e-4;
 kmax = 5;
 %initial-value
    wa  = etaS*ps;
    n10 =(1+ ps + pp)/(1 + wa + ps + (1 + etaP)*pp );
    n20 =(wa + etaP*pp)/(1 + wa + ps + (1 + etaP)*pp );
     x0 = [n10, n20];
  xold = x0; iter = 1;

  while (iter <= kmax)
```

```
% system of nonlinear equations

f1=-(wa+etaP*pp)*xold(1)+(1+ps+pp)*xold(2)+t1*(1-xold(1)-xold(2))+ cup*xold(2)^2;
f2=(wa+etaP*pp)*xold(1)-(1+ps+pp)*xold(2)+t2*(1-xold(1)-xold(2))-2*cup*xold(2)^2;

% Jacobian
df11 = -(wa + etaP*pp + t1);
df12 =  (1 + ps + pp - t1) + 2*cup*n20;

df21 = (wa + etaP*pp - t2);
df22 = -(1 + ps + pp + t2) - 4*cup*n20;


F = [f1; f2];

J = [df11, df12;
     df21, df22];

y = - J\F;
xnew = xold + y';

dif = norm(xnew/xold - 1);
  if dif <= tol
     x = xnew; %disp([iter   xnew   dif]);
     return;
  else
      xold = xnew;
  end
    iter = iter + 1;
end
 disp('Newton method did not converge')
    x = xnew;

%end
```

Results:

In figure 3.10 we show the simulation results with the image of MM pump propagating in the fiber (upper) and the pump and signal power along the fiber. The fiber has 3% Er doped phosphate glass fiber which is very high compared with 0.1%–0.2wt% concentration in silica glass fiber.

As can be seen, a 1480 nm pumping scheme has quite low absorption even with very high concentration and quite large core size $d = 16$ μm in this case due to a low absorption cross section. As a result, population inversion and signal gain is low in this case. One may think of compensation of low absorption at 1480 nm by designing the system with a much longer Er-doped fiber so that the fiber can absorb most of the pump power and hopefully to convert it to the signal power. However, it is not as simple as that. First, the power curves in figure 3.10 show that the pump power coupled to the fiber core is significantly reduced after ~10 cm even though the total pump power in the fiber is still very high. That can be understood in a conventional way as attributed to the skew-ray effect in the perfect symmetrical fiber considered in this example. Therefore, even though the total pump power in the fiber is still very high, the pump power in the core is reduced quite significantly leading the

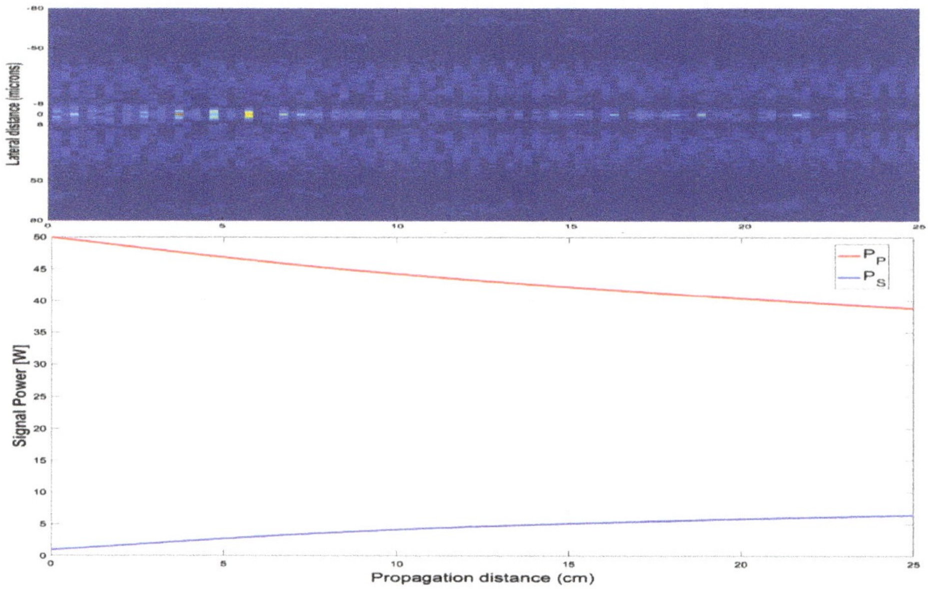

Figure 3.10. Simulation results of 1480 nm-pumped EDFA with 3% Er, $d = 16$ μm, inner clad radius 50 mm and fiber length with $L = 25$ cm, pump power $P_P = 50$ W and input signal power $P_S = 1$ W. Upper: image of MM pump propagation in the fiber. Lower: powers of pump beam (blue), pump absorption (P_{abs}) and amplified signal along the YDFA.

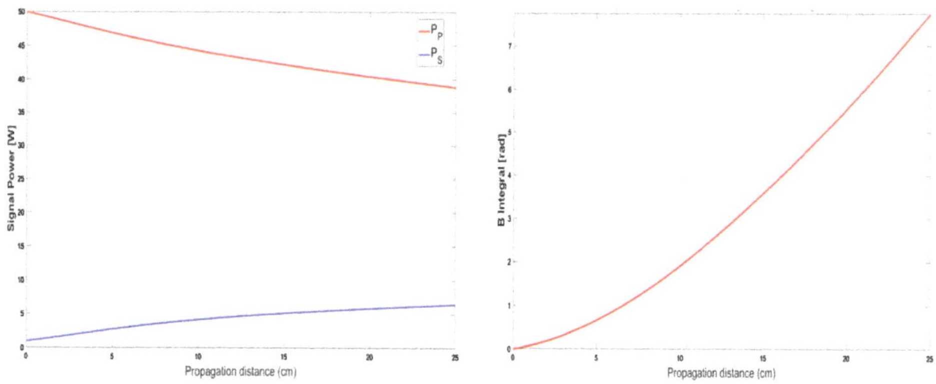

Figure 3.11. Simulation results of 1480 nm-pumped EDFA with 3% Er, $d = 16$ μm, inner clad radius 50 mm and fiber length with $L = 25$ cm, pump power $P_P = 50$ W and input signal power $P_S = 1$ W. Upper: image of MM pump propagation in the fiber. Lower: Powers of pump beam (blue), pump absorption (P_{abs}) and amplified signal along the YDFA.

signal curve to start faltering. Longer fiber may not really help signal gain much more. More importantly, the results of the *B*-integral in figure 3.11 show that the value of *B* increases very fast and reaches high values that are not preferable in a real application, e.g., $B > 3$.

Explanation of Program 3.2

There are some important points of the modelling equations that we want to present clearly so that readers can easily follow and understand program 3.2.

1. Normalization: in solving physics problems, we have to use the units of variables such as time, weight, distance and power etc. Quite often the units make our equations very messy causing confusion. Therefore, it is better transfer the variables to dimensionless ones. For example, we define a new dimensionless pump and signal powers as $pp = P_p/qp$ and $ps = P_s/qs$, respectively, by normalized power coefficients qp and qs. The propagation equation of SM signal becomes

$$\frac{dps(\tilde{z})}{d\tilde{z}} = \{\gamma_s[n_2(\tilde{z}) - \eta_s n_1(\tilde{z})] - \tilde{\alpha}_s\}ps(\tilde{z}), \tag{3.53}$$

in which $\gamma_s = L_0\Gamma_s\sigma_{21}(\lambda_s)N_{Er}$, dimensionless distance $\tilde{z} = z/L_0$ and $\tilde{\alpha} = L_0\alpha_s$ with $L_0 = 1$ cm is chosen as unit length. The gain for SM signal that is estimated in the program at each step of propagation is

$$g(\tilde{z}) = \{\gamma_s[n_2(\tilde{z}) - \eta_s n_1(\tilde{z})] - \tilde{\alpha}_s\}. \tag{3.54}$$

And the BPM propagation for the MM pump beam is

$$\alpha_{BPM} = \left\{\beta[\eta_p n_1(x) - n_2(x)] + \tilde{\alpha}_p\right\}, \tag{3.55}$$

$\beta = L_0\Gamma_p\sigma_{ep}N_{Er}$ and $\tilde{\alpha}_p = L_0\alpha_p$, $\alpha_p \approx \alpha_s$ is background loss of the fiber.

2. Nonlinear rate equations: below is a description of the main equations using the function
   `function x = NewtonSys(pp, ps)`
   a. Dimensionless nonlinear rate equations (3.45)–(3.47) in steady-state can be re-written as

$$-(W_{ab} + \eta_p P)n_1 + (1 + P + W_{em})n_2 + t_1 n_3 + C_{up}n_2^2 = 0 \tag{3.56}$$

$$(W_{ab} + \eta_p P)n_1 - (1 + P + \eta_p P)n_2 + t_2 n_3 + 2C_{up}n_2^2 = 0 \tag{3.57}$$

$$n_1 + n_2 + n_3 = 1. \tag{3.58}$$

   Here, we define:

$$t_1 = \tau_{21}/\tau_{31}; \quad t_2 = \tau_{21}/\tau_{31}; \quad C_{up} = \tau_{21}C_{22}N_{Er}. \tag{3.59}$$

   These nonlinear rate equations (NREs) above can be solved numerically by the Newton method as discussed later.

3. Algorithm for solving NRE:
   the following is the description of the algorithm of `function x= NewtonSys(pp, ps)` in program 3.2. The function `NewtonSys(pp, ps)` solves NREs (3.56)–(3.58) using the Newton method and returns the normalized populations $n_i = N_i/N_{Er}$, $i = 1, 2, 3$. The populations obtained

from the function are used to calculate pump absorption and signal gain at each step in the propagation.

We re-write the nonlinear equations (3.56)–(3.58) in the following forms

$$F_1 = -(W_{ab} + \eta_p P)n_1 + (1 + P + W_{em})n_2 + t_1(1 - n_1 - n_2) + C_{up}n_2^2, \tag{3.60}$$

$$F_2 = (W_{ab} + \eta_p P)n_1 - (1 + P + W_{em})n_2 + t_2(1 - n_1 - n_2) - 2C_{up}n_2^2. \tag{3.61}$$

Then we need to solve the Jacobian equation:

$$\begin{pmatrix} J_{11} & J_{12} \\ J_{21} & J_{22} \end{pmatrix}\begin{pmatrix} n_1 \\ n_2 \end{pmatrix} = -\begin{pmatrix} \dfrac{\partial F_1}{\partial n_1} & \dfrac{\partial F_1}{\partial n_2} \\ \dfrac{\partial F_2}{\partial n_1} & \dfrac{\partial F_2}{\partial n_2} \end{pmatrix} \quad \text{or} \quad \hat{J}(n_1, n_2) \times \hat{n} = -\hat{F}(n_1, n_2). \tag{3.62}$$

Where the element of matrix $F(\mathbf{n})$ can be expressed as follows

$$\begin{aligned} F_{11} &= \frac{\partial F_1}{\partial n_1} = -(W_{ab} + \eta_p P + t_1); \\ F_{12} &= \frac{\partial F_1}{\partial n_2} = (1 - t_1 + P + W_{em}) + 2C_{up}n_2; \end{aligned} \tag{3.63}$$

$$\begin{aligned} F_{21} &= \frac{\partial F_2}{\partial n_1} = (W_{ab} + \eta_p P - t_2); \\ F_{22} &= \frac{\partial F_2}{\partial n_2} = -(1 + t_2 + P + W_{em}) - 4C_{up}n_2. \end{aligned} \tag{3.64}$$

4. Initial values: it is worth noting that in solving the NREs we need trial values as initial values in the iteration. In general, if the trial values are close to the exact solution then the solutions can be found very fast. However, if the trial values are very different with the exact solutions, not only is computing time much longer, but numerical errors can be higher and sometimes there are nonphysical solutions. One of the common mistakes in solving NREs of EDFAs (and many other systems as well) is that we tend to think of exact solutions of populations in extreme situations. For example, in an Er-doped system we would think the population of the ground state $^4I_{15/2}$ and excited one $^4I_{13/2}$ are close to '0' and '1' or 100%, respectively, in the extremely high pumping conditions. And, therefore, it is convenient to choose the trial values of the ground and excited states that are close to these values, i.e., $n_1^0 = 0$, and $n_2^0 = 1$. However, in practice there are two problems with that assumption. First, even in the conditions of extremely high pump power, the populations are quite different with '0' and '1'. It is quite simple to estimate these values in a 1480 nm pumping scheme using equations (3.48)–(3.49).

For example, the excited population in the case without up-conversion with normalization can be re-written as

$$n_2^0 = \frac{N_2}{N_{Er}} = \frac{1 + (\eta_P P_P + \eta_S P_S)}{1 + (P_P + P_S + \eta_P P_P + \eta_S P_S)} \tag{3.65}$$

Here, we ignore the ASE and using normalizations defined in equations (3.39) and (3.40) earlier. In the situation of extremely high pump power, in which $P_p \gg 1 P_P \gg P_S$, we have

$$n_2^0 \simeq \frac{\eta_P P_P}{(1 + \eta_P) P_P} = \frac{\eta_P}{(1 + \eta_P)} = \frac{\sigma_{abs}/\sigma_{ems}}{1 + \sigma_{abs}/\sigma_{ems}} = \frac{\sigma_{abs}}{\sigma_{ems} + \sigma_{abs}}. \tag{3.66}$$

$$n_1^0 \simeq \frac{P_P}{(1 + \eta_P) P_P} = \frac{1}{(1 + \eta_P)} = \frac{1}{1 + \sigma_{abs}/\sigma_{ems}} = \frac{\sigma_{ems}}{\sigma_{ems} + \sigma_{abs}}. \tag{3.67}$$

From equations (3.65)–(3.67) we can see that in extreme populations n_1, n_2 is just determined by the ratio $\sigma_{abs}/(\sigma_{ems} + \sigma_{abs})$ and $\sigma_{ems}/(\sigma_{ems} + \sigma_{abs})$, respectively. We can easily see the values are quite different with '0' and '1' as assumed earlier. From the plot of absorption and emission CRs of Er-doped in phosphate glass, i.e., figure 3.9, we can easily estimate the population n_2 and n_1 in high power condition pumping at 1480 nm. For example, $\sigma_{abs}(1840\text{ nm}) \sim 0.12$ and $\sigma_{abs}(1840\text{ nm}) \sim 0.38$ then $n_2 \sim 76\%$ and $n_1 \sim 24\%$. The second mistake in choosing the trial value of up-state population $n_2 \sim 1$, which is a wrong assumption as presented above, for the whole fiber length, is that the pump power is not uniform in the whole fiber. In the situation when pump power is not extremely high conditions, which is the reality, the up-level population is much lower than 100%. Furthermore, as the pump and signal power are not uniform along the fiber, the values of populations are also varied.

From the discussion above, we can see that choosing the trial values for Newton iteration is not as simple as first thought. However, if we understand the physics of the process, we have quite a simple solution for that. The nonlinear term $\sim C_{22} N_2^2$ of up-conversion process in EDFA is in fact quite small as compared to other terms in NREs, and its effects on populations are not very big. That statement is even more correct because in reality no EDFAs with very high Er concentration are used. Of course, even small changes of populations in the whole fiber length could make important differences in fiber amplifier operation. Not only gain, but NF is quite sensitive to the changes of populations. Therefore, it is very good to assume that the exact solutions of populations from function `NewtonSys(pp, ps)` are quite close to the solutions without nonlinear terms. These are the analytical expressions (3.48) and (3.49) and can be estimated very fast with input parameters *pp* and *ps* which are normalized power of pump and signal. By choosing the trial values as the analytical expressions (3.48) and (3.49) we can effectively deal with some problems: first, the trial values are quite close to the exact solutions, and second the trial values are determined by the pump and signal power at each step in propagation, and finally, estimates of the analytical expression should be very fast.

In the following section, we will introduce a more complicated scheme—the so called 980 nm-pump Er doped glass fibers. After understanding the main processes in Er-ion only doped in glasses, we will study by example with MATLAB® in MM cladding-pumped Yb–Er co-doped fibers.

3.4.2 Modeling 980 nm-pumped Er-doped fiber amplifiers

In this subsection we will introduce another pumping scheme in Er-doped fiber amplifiers. In this scheme, 980 nm diodes are used to pump an EDFA instead of a 1480 nm pump, as described in subsection 3.4.1. Figure 3.12 presents an energy diagram of Er-ions doped in glasses and the main processes in the amplification process with a 980 nm pump. We follow the conventional way of description in rare-earth doped fiber amplifiers [3, 16, 27, 28]. As mentioned earlier, depending on host glass, Er concentration doped in glasses is quite different. For example, Er concentration in silica glass should be low, in the range less than 0.1–0.2 wt%, to

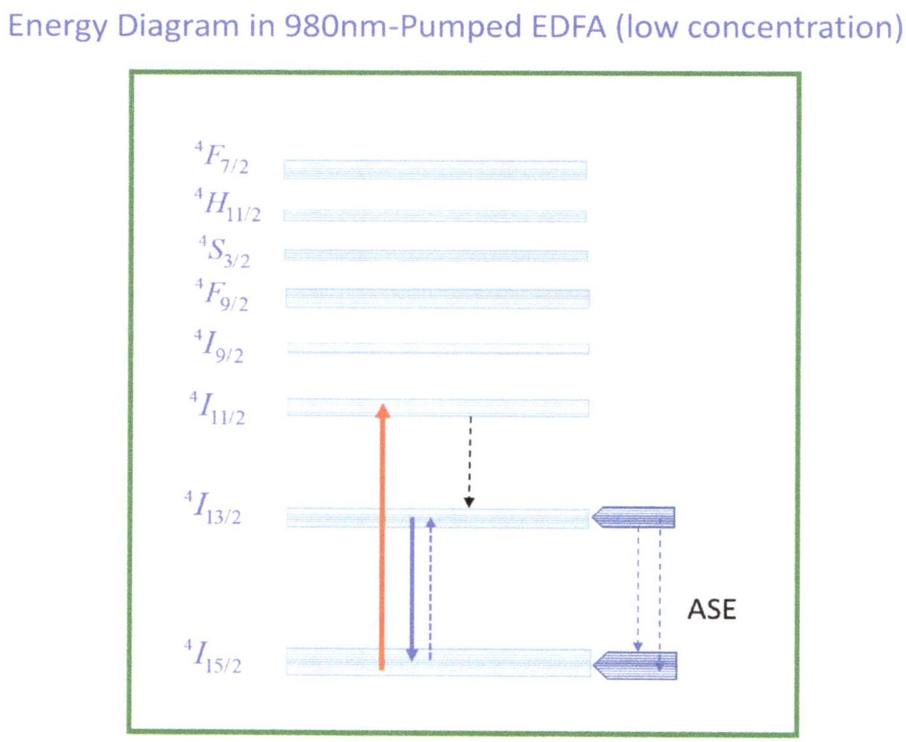

Figure 3.12. Diagram of energy of Er-ions doped glass under a 980 nm pumping scheme. Red solid arrow: pump photons, blue solid and dashed arrows: stimulated emission and absorption of signal photons, respectively. Black dashed arrows: non-radiative relaxation from high-level to lower-level states. Dashed-thin arrows: ASEs.

avoid clustering and strong up-conversion effects in amplification processes. Meanwhile, Er_2O can be doped much higher in phosphate glass, 2–3 wt% or even higher but still have reasonably low up-conversion and other negative effects. In any case, the common feature is that in low concentration we can ignore the effects that originate from the transfer energy among Er-ions that are very close. In a 980 nm pump scheme, this situation is similar to the 'ideal' three-level system, and is described in figure 3.12. However, when Er concentration increases, the Er-ions become more and more close, and at some points the excited ions will transfer energy from one to others causing up-conversion or other effects in which pump photons are lost.

First, let us consider a 980 nm pump Er-doped fiber amplifier in low concentration regime, as described in figure 3.12.

Let us describe the processes presented in the diagram of figure 3.12. First, pump photons around 980 nm are absorbed by Er-ions in ground state (level 1) $^4I_{15/2}$ exciting the ions into $^4I_{11/2}$ state (level 3) of Er^{3+}-ions. The process is presented by a red solid arrow in the diagram of figure 3.12. In low concentration regime Er-ions are considered as completely separated. In such situations, the interaction among Er-ions is negligibly small, and 980 nm pumped Er-ions behave like the three-level system or quasi-three levels system. The Er-ions in excited state $^4I_{11/2}$ (level 3) can be radiative and nonradiative relaxing to state $^4I_{13/2}$ (level 2) with relaxation time τ_{32}. Er-ions in state $^4I_{13/2}$ (level 2) or upper-level state of amplification are characterized by relaxation time τ_{21}, and these Er-ions can be stimulated by signal photons to emit photons with the same wavelength and amplify the signal. The amplification of signal in the 1.55 μm band (more exactly is from ~1520 to ~1570 nm in Er-doped glass) is presented by the solid blue arrow in figure 3.12. The Er-ions in the upper-level state ($^4I_{13/2}$-state) can also spontaneously emit photons and therefore generate the noise in the system. The spontaneously emitted photons can also get amplified or ASE by excited Er-ions. The ASEs are presented by dashed thin arrows. Note that, the Er-ions in ground state can also absorb amplified signal photons and ASE, and the process is characterized by the blue dashed arrow in the diagram. From the description, the rate equations for the 'ideal' three-level system in 980 nm pumped EDFAs can be written as follows:

$$\frac{dN_1}{dt} = -(W_{12} + R_{13})\, N_1 + (W_{21} + 1/\tau_{21})\, N_2, \tag{3.68}$$

$$\frac{dN_2}{dt} = W_{12}\, N_1 - \left(W_{21} + \frac{1}{\tau_{21}}\right) N_2 + \frac{N_3}{\tau_{32}}, \tag{3.69}$$

$$\frac{dN_3}{dt} = R_{13}\, N_1 - \frac{N_3}{\tau_{32}}. \tag{3.70}$$

Here, in equations (3.68)–(3.70) the rates of absorption W_{12} and emission W_{21} of signal and ASEs are defined as

$$W_{12} = \frac{\Gamma_S \sigma_{abs}(\lambda_S)}{h\nu_S A_{core}} P_S(z) + \sum_i \frac{\Gamma_i \sigma_{abs}(\lambda_i)}{h\nu_i A_{core}} \left\{ P_i^{(+)}(\lambda_i, z) + P_i^{(-)}(\lambda_i, z) \right\}, \tag{3.71}$$

$$W_{21} = \frac{\Gamma_S \sigma_{ems}(\lambda_S)}{h\nu_S A_{core}} P_S(z) + \sum_i \frac{\Gamma_i \sigma_{ems}(\lambda_i)}{h\nu_i A_{core}} \left\{ P_i^{(+)}(\lambda_i, z) + P_i^{(-)}(\lambda_i, z) \right\}. \tag{3.72}$$

The pumping absorption rates are different in core-pump and cladding-pump schemes and can be expressed as below:

$$R_{13} = \begin{cases} \dfrac{\Gamma_P \sigma_{abs}(\lambda_P)}{h\nu_P A_{core}} P_P(z), & \text{for core–pumped scheme} \\ \dfrac{\sigma_{abs}(\lambda_P)}{h\nu_P A_{core}} P_{P,core}(z), & \text{for cladding–pump scheme} \end{cases} \tag{3.73}$$

It should be noted that in the 980 nm core-pump scheme, because 980 nm pump wavelength is much shorter than signal wavelength ~1550 nm, the SM pump beam could still propagate with few modes in the fiber core, which is designed as SM for signal 1550 nm. In that situation, an effective overlap factor Γ_P can be used to characterize the few-mode beam that is mostly confined in the fiber core. In that situation $\Gamma_P P_P$ represents the pump power coupled to the fiber core. However, as discussed earlier in sections 3.1–3.2 in the MM cladding-pump scheme there is no good approximation of using effective overlap factor so that we should use the pump power coupled to the core $P_{P,core}$ that is calculated by the BPM.

The steady-state solutions of rate equations (3.68)–(3.70) in the low concentration regime can be easily found, and normalized populations can be expressed as

$$n_1^0 = \frac{N_1^0}{N_{Er}} = \frac{1 + W_{21}}{W_{12} + P_P + (1 + W_{21})(1 + t_1 P_P)}, \tag{3.74}$$

$$n_2^0 = \frac{N_2^0}{N_{Er}} = \frac{1 + W_{12} + P_P}{W_{12} + P_P + (1 + W_{21})(1 + t_1 P_P)}, \tag{3.75}$$

$$n_3^0 = \frac{N_3^0}{N_{Er}} = \frac{t_1 P_P (1 + W_{21})}{W_{12} + P_P + (1 + W_{21})(1 + t_1 P_P)}. \tag{3.76}$$

Here, again the super-index '0' in the populations N_i^0, $i = 1, 2, 3$ is for the 'ideal' three-level system which is applicable for low concentration regime. These expressions (3.74)–(3.76) can be used to calculate signal gain and pump absorption in a low concentration regime. They can be useful as 'trial values' in calculation of NREs of an EDFA in a high concentration regime that we will study in the next subsection.

At this point we want to point out that a 980 nm pumping scheme for EDFAs has some advantages of a three-level system as compared with the two-level system in a 1480 nm scheme. The greatest benefit of 980 nm pumping is that population inversion (N_2–N_1) can be very high and in the 'ideal' cases discussed above the upper-level population N_2 can be close to 100% in a high pump power regime. The

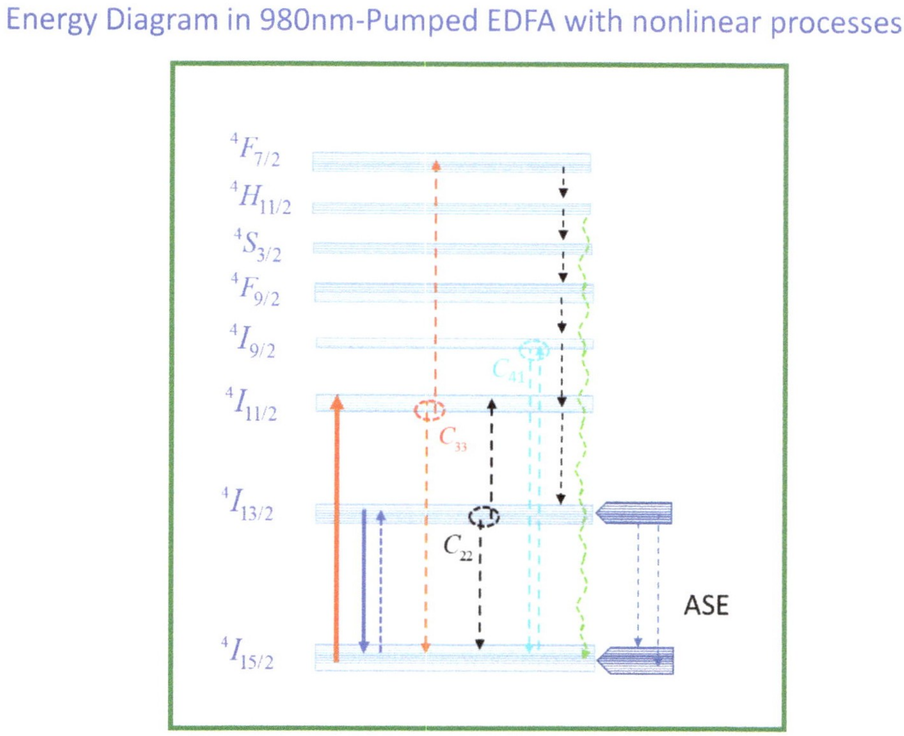

Figure 3.13. Diagram of energy of Er-ions doped fiber under a 980 nm pumping scheme. Red solid arrow: pump photons, blue solid and dashed arrow: stimulated emission and absorption of signal photons, respectively. Black dashed arrows: non-radiative relaxation from high-level to lower-level states. Dashed-thin arrows: ASEs. There are different processes: up-conversion characterized by C_{22} and C_{33} coefficients, and cross-relaxation processes characterized by C_{14} coefficient. The green curled arrow presented green light emitted from high level to the ground level of the system.

high population inversion is very important for amplification not only for high gain but also low NF, which are the two most important factors of the EDFA. However, 980 nm pumping has some negative impact for the operation of EDFA too. First, absorption cross section at and around 980 nm in Er-doped glasses is relatively low compared with the 1480 nm band of absorption (see, CRs in figure 3.13). That means a 980 nm pump is not efficient, and because of low pump absorption the pumping scheme for an Er-doped only system is not suitable for an EDFA in a high power regime. Second, the quantum defect in a 980 nm pump scheme is quite high $\eta = (1 - \lambda_P/\lambda_S) \sim 35\%$, which means that at least ~35% of pump energy will not be converted to signal energy, and mostly the waste energy generates heat by non-radiative relaxation. Therefore, thermal management in a 980 nm pumping scheme is an important issue, especially in high power operation.

Let us now turn to the more complicated systems in which higher Er-concentration glasses are used in making fiber cores. Figure 3.13 describes a 980 nm pumped EDFA with nonlinear processes which are originated from the interaction among Er-ions staying close in the host materials.

As can be seen, besides the fundamental processes such as pump absorption, signal amplification and ASE as in the 'ideal' three-level system, there are several processes characterized by C_{22}, C_{33} and C_{14} as described in figure 3.13. In general, when spacings among Er-ions decrease from the 'ideal' situation, at some point the Er-ions become close enough and they can exchange and transfer energy from one to others through different mechanisms. The closer they are, the stronger the interactions among Er-ions. In other words, Er concentrations increase from the 'ideal' situation, spacing among Er-ions decreases causing different exchanges and transfers of energy among the Er-ions. These interactions lead to several processes such as up-conversions characterized by C_{22} and C_{33} coefficients, and the cross-relaxation C_{14}. These are nonlinear processes characterized by second-order terms of the populations $\sim C_{22}\, N_2^2$, $\sim C_{33}\, N_3^2$ and $\sim C_{14}\, N_1 N_4$. Finally, once Er-ions are excited to higher-level states, they can radiatively emit photons in the green wavelength ~530 nm. The green light can be strongly visualized in highly doped Er-concentration EDFA in reality.

Let us analyze the up-conversion process described by black dashed arrows in figure 3.13. First, the 980 pump photons are absorbed by Er-ions in level 1 (ground state or $^4I_{15/2}$ state) and excite the ions to level 3 ($^4I_{11/2}$ state)—solid red arrow. From there the excited Er-ions can be non-radiative relaxed to the second state ($^4I_{13/2}$ state)—black dashed arrow. The non-radiative relaxation from level 3 to level 2 is very fast and is characterized by τ_{32}, that is, in the order of 10^{-5} to 10^{-6} s. The up-conversion process C_{22} starts when the close Er-ions are in level-2 state and exchange the energy among them. For example, if two Er-ions in level-2 state are close, one ion can transfer its energy to the other. The energy-transferred Er-ion drops to level-1 state and the energy-accepted Er-ion will be excited to higher-level state, in this case the level-3 state ($^4I_{11/2}$) and quickly relax to level 2. The process involves two Er-ions starting in level-2 state, meaning their strength is proportional the number of pairs of the two Er-ions in level-2 state, and therefore can be described by a term $\sim C_{22}\, N_2^2$. Similarly, the up-conversion process $\sim\ C_{33}\, N_3^2$ involves a pair of Er-ions that start from level-3 state. Meanwhile, the cross-relaxation involves a pair of ions, one starting from level 4 and the other start from level 1. In other words, the process is proportional to populations N_4 and N_1 of Er-ions in level-4 state and level-1 state, respectively. Physically, there are other high-order processes, but they are much weaker than the three that have just been described and can be ignored in the model.

The rate of changes of populations or simply the nonlinear rate equations corresponding to the processes in figure 3.16 can be written as below.

$$\frac{dN_1}{dt} = -(W_{12} + R_{13})\, N_1 + \left(W_{21} + \frac{1}{\tau_{21}}\right) N_2 + C_{22}\, N_2^2 - C_{14}\, N_1 N_4 + C_{33}\, N_3^2, \quad (3.77)$$

$$\frac{dN_2}{dt} = W_{12}\, N_1 - \left(W_{21} + \frac{1}{\tau_{21}}\right) N_2 + \frac{N_3}{\tau_{32}} - 2C_{22}\, N_2^2 + 2C_{14}\, N_1 N_4, \quad (3.78)$$

$$\frac{dN_3}{dt} = R_{13}\, N_1 - \frac{N_3}{\tau_{32}} + \frac{N_4}{\tau_{43}} - 2C_{33}\, N_3^2, \quad (3.79)$$

$$N_{Er} = N_1 + N_2 + N_3 + N_4. \tag{3.80}$$

Here, the system now can be approximated as a 4-level system where level-4 state $^4I_{11/2}$ has relaxation time τ_{43}, which is much shorter than τ_{21} and τ_{32}, but all these three relaxation times are still much longer than other higher-level states. Therefore, those higher level states can be considered as virtual states with no real populations in the processes.

It worth noting that the steady state solutions of NREs can be obtained numerically by solving the NREs as presented earlier with the trial values (3.74)–(3.76) of the 'ideal' system without nonlinear terms. From equation (3.80), $N_4 = N_{Er} - (N_1 + N_2 + N_3)$ and it can be used to replace N_4 in (3.77)–(3.78). As a result, the NRE becomes a system of three equations for N_1, N_2 and N_3 and can be solved by the Newton iteration method as presented above.

Our study of 980 nm pump Er-doped only fiber serves as a fundamental understanding of the EDFAs in the pumping scheme. However, because Er-absorption CR at 980 nm is very low, this system has not been used in practice. However, as can be seen in the following section, when Er and Yb are co-doped in host glasses, the co-dopant system becomes very good for the amplification in the 1.5 μm band. In an Yb–Er co-doped system, we can utilize the very high absorption of Yb-ions and the high population in 980 nm pumping Er-ions for good amplification processes.

3.5 Modeling multimode cladding-pumped Yb–Er co-doped fiber amplifiers

3.5.1 The model of 980 nm-pumped Yb–Er Co-doped fiber amplifiers

The first important feature in an Yb–Er co-doped system is very high absorption in the 980 nm band, as shown in the CR of absorption in the upper panel of figure 3.14. The main physical processes in an Yb–Er co-doped system can be described in the lower panel of figure 3.14.

In order to understand the advantages of the co-dopant system let us analyze the processes described in the diagram in figure 3.14. We follow the conventional description similar to the ones presented earlier. In this case, in the 980 nm pumping Yb^{3+}–Er^{3+} co-doped system, the cooperative energy transfer from Yb^{3+}-ions to Er^{3+}-ions doped in host glass is the reason for the concept of co-doping.

First, for simplicity the indexes of the states of Er^{3+}-ions are labelled from 1 to 4, and Yb^{3+} states from 5 to 6, respectively, as indicated in figure 3.14. Note that, we do not need to consider the higher levels in Er^{3+}-ions because the lifetime of those levels is much shorter than the lower states 1–4, therefore, ions in such states quickly relax to state 4 and lower ones.

Let us start our analysis by considering the absorption mechanism in a co-doped Yb/Er system. Pump photons around 975 nm are first strongly absorbed by Yb^{3+}-ions in ground state—level-5 state ($^4F_{7/2}$), and excite the ions into level-6 state ($^4F_{5/2}$) of Yb^{3+}-ions. From there a cooperative energy transfer process occurs between Yb^{3+} ions in excited level-6 state ($^4F_{5/2}$) and Er^{3+}-ions in the ground state—level-1 state ($^4I_{15/2}$). The transferred energy excites the Er^{3+} ion to the level-1

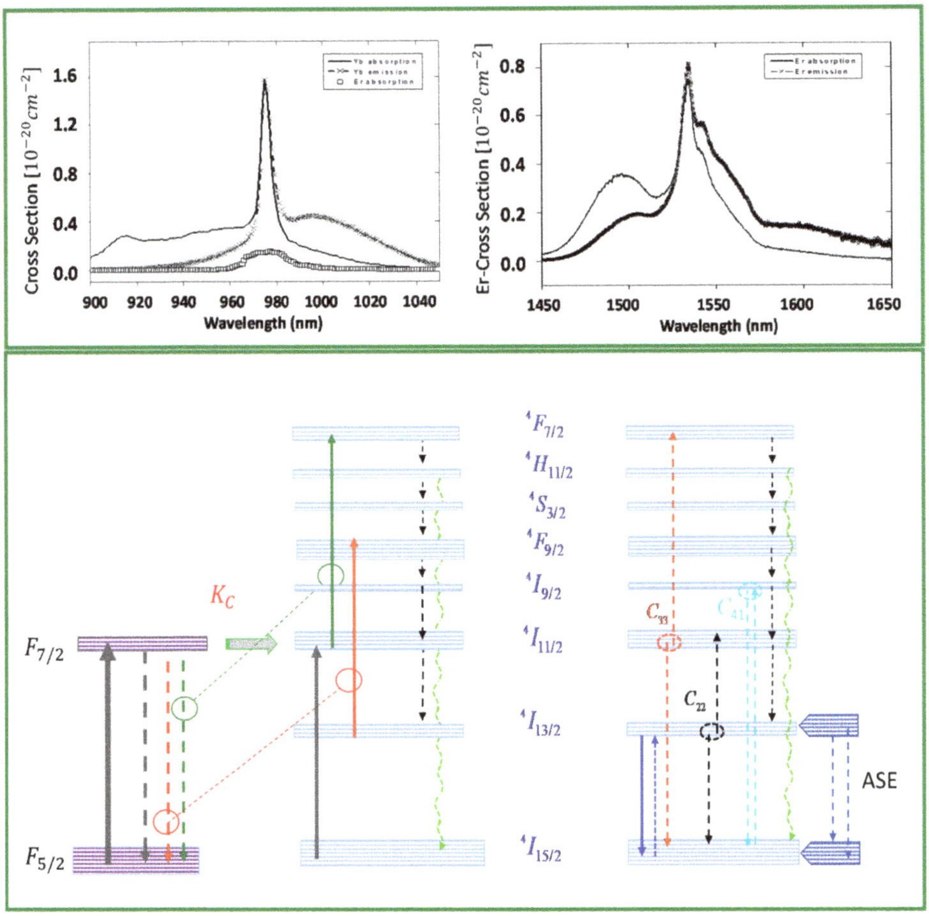

Figure 3.14. Upper: CRs of Yb- and Er-ions in 980 nm band in phosphate glass (left) and CR of Er-ion doped in phosphate glass (left). Lower: dcaiagram of energy of Yb and Er-ions doped fiber under a 980 nm pumping scheme. The main processes of absorption and transfer energy from Yb- to Er-ions (left), up-conversion and cross relaxation processes in Er-system (right). Details of all processes are described in the text.

state ($^4I_{15/2}$) while dropping back Yb^{3+} to its ground state—level-5 ($^4F_{7/2}$) state of Yb^{3+}. The radiative lifetime of the excited state of Yb is about $\tau_{56} \sim 1$ ms, while radiative lifetimes of level-3 and -2 states in Er-ions are $\tau_{32} \sim$ ns and $\tau_{21} \sim 10$ ms, respectively. The cooperative energy transfer process is described by the energy transfer coefficient K_F. Note that, Er^{3+} ions in level-1 state ($^4I_{15/2}$) can also absorb photons around 975 nm, but the absorption efficiency is much lower than absorption in Yb^{3+}-ions. It can be seen clearly from absorption CRs in the upper panel of figure 3.14, $\sigma_{abs}^{Yb} \gg \sigma_{abs}^{Er}$ in the 975 nm absorption band. As shown in the above energy diagram, due to the resonant effect, the energy levels in Er^{3+}-ions can also support two other energy transfer processes from Yb^{3+} to Er^{3+} ions described by K_D (double-energy transfer), and K_C (cumulative transfer) coefficients. These processes, as discussed earlier, can be presented by nonlinear terms of multiplication of populations of Yb^{3+}-ions states N_6

and Er^{3+}-ions states, i.e., $\sim K_F\ N_1N_6$, $\sim K_C\ N_2N_6$ and $\sim K_D\ N_3N_6$. It should be noted that in a highly co-doped Er/Yb system the K_D and K_C processes are mainly responsible for the emission of green and red light, respectively, which are always observed experimentally. Note that the green light is visible in highly doped glass fibers. It is worth noting that these processes of energy transfer are originated from interactions between the Yb-ions and Er-ions, which strongly depend on the spacing between the ions. In general, the higher the concentration of both dopants, the stronger the K-coefficients. There are some investigations, both in theory and experiment, about the strengths of the *K*-coefficients related to the ratio between the Yb- and Er-concentrations. In general, the higher the Yb concentration the stronger the *K*-coefficients of energy transfer processes. Therefore, the Yb–Er co-doped systems are suitable for high power regime EDFAs. The right part of the diagram also illustrates the up-conversion (C_{22}, C_{33}) and cross-relaxation processes (C_{14}) taking place between the two neighboring Er ions. These are nonlinear processes characterized by second-order terms of the populations $\sim C_{22}\ N_2^2$, $\sim C_{33}\ N_3^2$ and $\sim C_{14}\ N_1N_4$.

The standard rate and power equations model was modified in the present work to account for the highly elevated doping concentrations by including certain higher order terms, which are associated with inter-ion interactions and their corresponding up-conversion and pair induced quenching effects. The rate equations of the highly co-doped Yb/Er system can be written as

$$\frac{dN_1}{dt} = -\left(W_{12} + R_{13}\right) N_1 + \left(W_{21} + \frac{1}{\tau_{21}}\right) N_2 + C_{22}N_2^2 - C_{14}N_1N_4 \\ + C_{33}N_3^2, - K_FN_1N_6 + K_CN_2N_6 + K_DN_3N_6, \tag{3.81a}$$

$$\frac{dN_2}{dt} = W_{12}\ N_1 - \left(W_{21} + \frac{1}{\tau_{21}}\right)N_2 + \frac{N_3}{\tau_{32}} - 2C_{22}N_2^2 + 2C_{14}N_1N_4 - K_CN_2N_6, \tag{3.81b}$$

$$\frac{dN_3}{dt} = W_{13}N_1 - \frac{N_3}{\tau_{32}} + \frac{N_4}{\tau_{43}} - 2C_{33}N_3^2 + K_FN_1N_6 - K_DN_3N_6, \tag{3.81c}$$

$$\frac{dN_4}{dt} = -\frac{N_4}{\tau_{43}} + C_{22}N_2^2 - C_{14}N_1N_4 + C_{33}N_3^2, \tag{3.81d}$$

$$\frac{dN_6}{dt} = R_{56}N_5 - \left(R_{65} + \frac{1}{\tau_{65}}\right)N_6 - K_FN_1N_6 - K_CN_2N_6 - K_DN_3N_6. \tag{3.81e}$$

Here, the populations in two sub-systems Er- and Yb-ions are satisfied by conditions:

$$N_{Er} = N_1 + N_2 + N_3 + N_4, \quad \text{and} \quad N_{Yb} = N_5 + N_6. \tag{3.82}$$

The rate of absorption and emission in the signal and ASE band of the Er system have the form

$$W_{12(21)} = \left\{\Gamma_S \frac{\sigma_{12(21)}(\lambda_S)P_S}{h\nu_S A_{core}} + \sum_i \Gamma_i \sigma_{12(21)}(\lambda_i) \frac{P_{ASE}^{(+)}(\lambda_i) + P_{ASE}^{(-)}(\lambda_i)}{h\nu_i A_{core}}\right\}, \tag{3.83}$$

And the rate of pump absorption (absorption) in Yb-ions is

$$R_{56(65)} = \begin{cases} \dfrac{\Gamma_P \sigma_{56(65)}^{Yb}(\lambda_P)}{h\nu_P A_{core}} P_P(z), & \text{for core–pumped scheme} \\ \dfrac{\sigma_{56(65)}^{Yb}(\lambda_P)}{h\nu_P A_{core}} P_{P,core}(z), & \text{for cladding–pump scheme} \end{cases} \tag{3.84}$$

And the rate of pump absorption in 980 nm pump Er-ions is

$$R_{13} = \begin{cases} \dfrac{\Gamma_P \sigma_{13}^{Er}(\lambda_P)}{h\nu_P A_{core}} P_P(z), & \text{for core–pumped scheme} \\ \dfrac{\sigma_{13}^{Er}(\lambda_P)}{h\nu_P A_{core}} P_{P,core}(z), & \text{for cladding–pump scheme} \end{cases} \tag{3.85}$$

Note that in a 980 nm pumped Yb–Er co-doped system, the pump photons are absorbed by both Yb- and Er-ions. However, only excited Yb-ions in $^4F_{5/2}$ state emit the 980 nm photons with $\tau_{65} \sim 10$ ms. Meanwhile, 980 nm excited Er-ions in $^4I_{11/2}$ state quickly radiatively and non-radiatively relax to $^4I_{13/2}$ state, and the relaxation probability to the ground state $^4I_{15/2}$ state is negligibly small.

3.5.2 Normalization and dimensionless notations

3.5.2.1 Normalization

For simplicity of numerical calculation, we use dimensionless notations in our programs. In the following we will present the modeling equations that are normalized and are in dimensionless notations which are in our programs. Although the description below is quite boring and may not be necessary for readers with good experience, we believe it would be useful for many others, especially for beginners. Moreover, the detailed description makes the programs for studying examples much clearer to follow.

First, we start with the rate of absorption and emission W_{21} and W_{12}, respectively. We have

$$W_{21}(\nu_S) = \Gamma_S \left\{\frac{\sigma_{21}(\nu_S)P_S}{h\nu_S A_{core}} + \sum_i \sigma_{21}(\nu_i) \frac{P_{ASE}^{(+)}(\nu_i) + P_{ASE}^{(-)}(\nu_i)}{h\nu_i A_{core}}\right\}, \tag{3.86}$$

$$W_{12}(\lambda_S) = \Gamma_S \left\{\frac{\sigma_{12}(\lambda_S)P_S}{h\nu_S A_{core}} + \sum_i \sigma_{12}(\lambda_i) \frac{P_{ASE}^{(+)}(\lambda_i) + P_{ASE}^{(-)}(\lambda_i)}{h\nu_i A_{core}}\right\}, \tag{3.87}$$

We can do a simple process as below

$$\begin{aligned}\tau_{21}W_{21}(\nu_S) &= \tau_{21}\Gamma_S\left\{\frac{\sigma_{21}(\nu_S)P_S}{h\nu_S A_{core}} + \sum_i \sigma_{21}(\nu_i)\frac{P_{ASE}^{(+)}(\nu_i) + P_{ASE}^{(-)}(\nu_i)}{h\nu_i A_{core}}\right\}\\ &= \left\{\frac{P_S}{Q_S} + \sum_i \frac{P_{ASE}^{(+)}(\nu_i) + P_{ASE}^{(-)}(\nu_i)}{Q_i}\right\}\\ &= S(\lambda_S) + \sum_i \{F_i(\lambda_i) + B_i(\lambda_i)\} = W_{ems}.\end{aligned} \tag{3.88}$$

Here, $Q_{S,i} = \frac{hcA_{core}}{\tau_{21}\Gamma_S\lambda_{S,i}\sigma_{21}(\lambda_{S,i})}$ is the normalized power coefficients for signal and ASE powers.

In equation (3.88), we have defined new dimensionless powers S, F_i and B_i of signal, forward and backward ASEs, respectively:

$$S = \frac{P_S(\lambda_S)}{Q_S}, \quad F_i = \frac{P_{ASE}^{(+)}(\lambda_i)}{Q_i}, \quad \text{and} \quad B_i = \frac{P_{ASE}^{(-)}(\lambda_i)}{Q_i}. \tag{3.89}$$

Similarly, we have

$$\begin{aligned}\tau_{21}W_{12}(\nu_S) &= \left\{\frac{\sigma_{12}}{\sigma_{21}}\frac{P_S}{Q_S} + \sum_i \frac{\sigma_{12}}{\sigma_{21}}\frac{P_{ASE}^{(+)}(\nu_i) + P_{ASE}^{(-)}(\nu_i)}{Q_i}\right\}\\ &= \eta_S S + \sum_i \eta_i\{F_i(\lambda_i) + B_i(\lambda_i)\} = W_{abs},\end{aligned} \tag{3.90}$$

Because we will focus only on a cladding pumping scheme in this part, from (3.86)–(3.87) we have

$$\tau_{65}R_{65}(\lambda_P) = \tau_{65}\frac{\sigma_{65}(\lambda_P)}{h\nu_P A_{core}}P_{P,core} = \frac{P_{P,core}}{Q_P} = P, \tag{3.91}$$

$$\tau_{65}R_{56}(\lambda_P) = \frac{\sigma_{56}(\lambda_P)}{\sigma_{65}(\lambda_P)}\frac{P_{P,core}}{Q_P} = \eta_{56}P, \tag{3.92}$$

$$\tau_{32(21)}R_{13}(\lambda_P) = \tau_{32(21)}\frac{\sigma_{13}(\lambda_P)P_{P,core}}{h\nu_P A_{core}} = \frac{\sigma_{13}(\lambda_P)}{\sigma_{65}(\lambda_P)}\frac{\tau_{32(21)}}{\tau_{65}}\frac{P_{P,core}}{Q_P} = \eta_{13}s_{36(26)}P, \tag{3.93}$$

Here, in (3.86)–(3.91) new notations are defined as

$$\eta_{S,i} = \frac{\sigma_{12}(\lambda_{S,i})}{\sigma_{21}(\lambda_{S,i})}; \quad \eta_{56} = \frac{\sigma_{56}(\lambda_P)}{\sigma_{65}(\lambda_P)}; \quad \eta_{13} = \frac{\sigma_{13}(\lambda_P)}{\sigma_{65}(\lambda_P)}; \quad s_{36(26)} = \frac{\tau_{32(21)}}{\tau_{65}}. \tag{3.94}$$

Next, in the following we present the processes of normalizing NREs (3.81)–(3.82) in steady state. The normalized NREs are used in our MATLAB® program in the example we will study later in the section. We will use the Newton method for numerically solving the NREs in the program. The NREs (3.81)–(3.82) in steady-state can be rewritten as

$$0 = -(W_{12} + R_{13})N_1 + \left(W_{21} + \frac{1}{\tau_{21}}\right)N_2 + C_{22}N_2^2 - C_{14}N_1N_4 + C_{33}N_3^2 - K_FN_1N_6 + K_CN_2N_6 + K_DN_3N_6 \quad (3.95a)$$

$$0 = W_{12}N_1 - \left(W_{21} + \frac{1}{\tau_{21}}\right)N_2 + \frac{N_3}{\tau_{32}} - 2C_{22}N_2^2 + 2C_{14}N_1N_4 - K_CN_2N_6, \quad (3.95b)$$

$$0 = W_{13}N_1 - \frac{N_3}{\tau_{32}} + \frac{N_4}{\tau_{43}} - 2C_{33}N_3^2 + K_FN_1N_6 - K_DN_3N_6, \quad (3.95c)$$

$$0 = -\frac{N_4}{\tau_{43}} + C_{22}N_2^2 - C_{14}N_1N_4 + C_{33}N_3^2, \quad (3.95d)$$

$$0 = R_{56}N_5 - \left(R_{65} + \frac{1}{\tau_{65}}\right)N_6 - K_FN_1N_6 - K_CN_2N_6 - K_DN_3N_6. \quad (3.95e)$$

The system of nonlinear equation (3.93) can be numerically solved by Newton iteration method, in which good trial values of solutions are needed to ensure fast iterations and highly accurate solutions. As described in section 3.3, using solutions of the *linear* rate equations for the trial values of the iteration methods is a good approach, in which the trial values are quite close to the exact solutions. More importantly, by doing that we can always find good trial values quickly along the whole fiber because they can be estimated from analytical expressions.

In the linear regime, all nonlinear terms in equation (3.93) are neglected. In that case we can re-write the equations for normalized populations as

$$-(W_{12} + R_{13})n_1^0 + \left(W_{21} + \frac{1}{\tau_{21}}\right)n_2^0 = 0, \quad (3.96a)$$

$$W_{12}n_1^0 - \left(W_{21} + \frac{1}{\tau_{21}}\right)n_2^0 + \frac{n_3^0}{\tau_{32}} = 0, \quad (3.96b)$$

$$R_{13}n_1^0 - \frac{1}{\tau_{32}}n_3^0 = 0, \quad (3.96c)$$

$$R_{56}n_5^0 - \left(R_{65} + \frac{1}{\tau_{65}}\right)n_6^0 = 0. \quad (3.96d)$$

Here, $n_i^0 = N_i^0/N_{Er}$, $i = 1 \div 4$ and $n_{5,6}^0 = N_{5,6}^0/N_{Yb}$. The normalized linear populations satisfy the following conditions:

$$n_1^0 + n_2^0 + n_3^0 = 1, \quad \text{and} \quad n_5^0 + n_6^0 = 1. \quad (3.97)$$

The solutions of equations (3.96) can be easily found, and their expressions can be written as

$$n_1^0 = \frac{(W_{21} + 1/\tau_{21})}{(W_{12} + R_{13}) + (1 + \tau_{32}R_{13})(W_{21}1/\tau_{21})}, \tag{3.98a}$$

$$\begin{aligned} n_2^0 &= 1 - \frac{(W_{21} + 1/\tau_{21})(1 + \tau_{32}R_{13})}{(W_{12} + R_{13}) + (W_{21} + 1/\tau_{21})(1 + \tau_{32}R_{13})} \\ &= \frac{(W_{12} + R_{13})}{(W_{12} + R_{13}) + (W_{21} + 1/\tau_{21})(1 + \tau_{32}R_{13})}, \end{aligned} \tag{3.98b}$$

And

$$n_6^0 = \frac{\tau_{65}R_{56}}{1 + \tau_{65}(R_{65} + R_{56})}, \quad \text{and} \quad n_5^0 = 1 - n_6^0 \tag{3.98c}$$

Using the normalized notations (3.89)–(3.94) we can now re-write equations (3.98) as:

$$\begin{aligned} n_1^0 &= \frac{1 + \tau_{21}W_{21}}{\tau_{21}(W_{12} + R_{13}) + (1 + \tau_{32}R_{13})(1 + \tau_{21}W_{21})} \\ &= \frac{1 + W_{ems}}{(W_{abs} + \ \eta_{13}s_{26}P) + (1 + \eta_{13}s_{36}P)(1 + W_{ems})}, \end{aligned} \tag{3.99a}$$

$$\begin{aligned} n_2^0 &= \frac{\tau_{21}(W_{12} + R_{13})}{\tau_{21}(W_{12} + R_{13}) + (1 + \tau_{32}R_{13})(1 + \tau_{21}W_{21})} \\ &= \frac{(W_{abs} + \eta_{13}s_{26}P)}{(W_{abs} + \eta_{13}s_{26}P) + (1 + \eta_{13}s_{36}P)(1 + W_{ems})}, \end{aligned} \tag{3.99b}$$

$$n_6^0 = \frac{\tau_{65}R_{56}}{(1 + \tau_{65}R_{65}) + \tau_{65}R_{56}} = \frac{\eta_{56}P}{1 + (1 + \eta_{56})P}, \tag{3.99c}$$

And

$$n_3^0 = 1 - n_2^0 - n_1^0, \ n_5^0 = 1 - n_6^0 \tag{3.99d}$$

As described earlier in this chapter, in our BPM modeling method for MM pump propagation we will need the pump absorption at each step of increment in propagation. In a 980 nm pumping scheme, the pump is absorbed by both Yb-ion and Er-ions, and the total attenuation including absorption and background loss α_P can be written as

$$\begin{aligned} \alpha(z) &= \{[\sigma_{56}N_5(z) - \sigma_{65}N_6(z) + \sigma_{13}N_1(z)] + \alpha_P\} \\ \tilde{\alpha} &= \{\beta_Y[\eta_{56}(1 - n_6) - n_6] + \beta_E n_1 + \alpha_P\}. \end{aligned} \tag{3.100}$$

Here, in equation (3.100), $\tilde{\alpha} = L_0\alpha$, $\tilde{\alpha}_P = L_0\alpha_P$, $\beta_E = L_0 N_{Er}\sigma_{13}$, $\beta_Y = L_0 N_{Yb}\sigma_{65}$, where $L_0 = 1$ cm is length unit in our programs.

Similarly, the propagation of the SM signal beam can be written in dimensionless notations as

$$\frac{dS(x)}{dx} = \{\gamma_S[n_2(x) - \eta_S n_1(x)] - \tilde{\alpha}_S\} S(x), \tag{3.101}$$

In equation (3.101), the normalized coefficients are $\gamma_S = L_0 N_{Er}\sigma_{21}$, $\tilde{\alpha}_S = L_0\alpha_S$, $x = z/L_0$.

3.5.2.2 Newton iteration solving nonlinear rate equations

After some simple calculations using the normalization process presented earlier in subsection 3.5.2.1, we can rewrite these NREs in normalized notations as below

$$-(W_{abs} + \eta_{13}s_{26}P_1)n_1 + (1 + W_{ems})n_2 + s_{24}n_4 - k_{fy}n_1n_6 + k_{cy}n_2n_6 + k_{dy}n_3n_6 = 0, \tag{3.102a}$$

$$W_{abs}n_1 - (1 + W_{ems})n_2 + s_{23}n_3 - c_{up}(n_2^2 - n_1n_4) - k_{cy}n_2n_6 = 0, \tag{3.102b}$$

$$\eta_{13}s_{26}P\, n_1 - s_{23}n_3 + s_{24}n_4 - c_{up}n_3^2 + k_{fy}n_1n_6 - k_{dy}n_3n_6 = 0, \tag{3.102c}$$

$$\eta_{56}P_1\, n_5 - (P_1 + 1)n_6 - (k_{fe}n_1n_6 + k_{ce}n_2n_6 + k_{de}n_3n_6) = 0. \tag{3.102d}$$

The Newton method for solving NREs (3.102) that are coded in the MATLAB® function is described next. For simplicity, let us use simpler notations $n_{1,2,3} = x_{1,2,3}$, $n_6 = x_4$.. Then the system of NREs (3.102) can be simplified by using the relations $\sum_{i=1}^{4} n_i = 1$ and $\sum_{i=5}^{6} n_i = 1$ as

$$\begin{aligned} &-(W_{abs} + \eta_{13}s_{26}P_1)x_1 + (1 + W_{ems})x_2 + s_{24}(1 - x_1 - x_2 - x_3) \\ &\quad - k_{fy}x_1x_4 + k_{cy}x_2x_4 + k_{dy}x_3x_4 = 0. \end{aligned} \tag{3.103a}$$

$$W_{abs}x_1 - (1 + W_{ems})x_2 + s_{23}x_3 - c_{up}[x_2^2 - x_1(1 - x_1 - x_2 - x_3)] - k_{cy}x_2x_4 = 0, \tag{3.103b}$$

$$\eta_{13}s_{26}P\, x_1 - s_{23}x_3 + s_{24}(1 - x_1 - x_2 - x_3) - c_{up}x_3^2 + k_{fy}x_1x_4 - k_{dy}x_3x_4 = 0, \tag{3.103c}$$

$$n_{56}P_1(1 - x_4) - (1 + P_1)x_4 - k_{fe}x_1x_4 - k_{ce}x_2x_4 - k_{de}x_3x_4 = 0. \tag{3.103d}$$

We define dimensionless notations as:

$$s_{23} = \frac{\tau_{21}}{\tau_{32}},\quad s_{26} = \frac{\tau_{21}}{\tau_{65}},\quad s_{24} = \frac{\tau_{21}}{\tau_{43}},\quad s_{32} = \frac{\tau_{32}}{\tau_{21}},\quad s_{34} = \frac{\tau_{32}}{\tau_{43}},\quad s_{36} = \frac{\tau_{32}}{\tau_{65}}, \tag{3.104a}$$

$$c_{up} = 2C_{22}\tau_{21}N_{Er},\quad k_{f(c,\,d)y} = K_{F,C,D}\tau_{65}N_{Yb},\quad k_{f(c,\,d)e} = K_{F,C,D}\tau_{65}N_{Er}. \tag{3.104b}$$

In Newton's method, we need to find the Jacobian of the equations (3.103). We have

$$\begin{aligned} F_1(x) = s_{24} - (s_{24} + W_{abs} + \eta_{12}s_{26}P)x_1 + (1 - s_{24} + W_{ems})x_2 - s_{24}x_3 \\ - (k_{fy}x_1x_4 - k_{cy}x_2x_4 - k_{dy}x_3x_4), \end{aligned} \tag{3.105a}$$

$$\begin{aligned} F_2(x) = (W_{abs} + c_{up})x_1 - (1 + W_{ems})x_2 + s_{23}x_3 \\ - c_{up}(x_1^2 + x_2^2 + x_1x_2 + x_1x_3) - k_{cy}x_2x_4, \end{aligned} \tag{3.105b}$$

$$\begin{aligned} F_3(x) = s_{24} + (\eta_{12}s_{26}P - s_{24})x_1 - s_{24}x_2 - (s_{24} + s_{24})x_3 \\ - c_{up}x_3^2 - k_{fy}x_1x_4 - k_{dy}x_3x_4, \end{aligned} \tag{3.105c}$$

$$F_4(x) = \eta_{56}P - [1 + P(1 + \eta_{56})]x_4 - (k_{fe}x_1x_4 + k_{ce}x_2x_4 + k_{de}x_3x_4). \tag{3.105d}$$

Then we need to solve the Jacobian equation:

$$J[\hat{x}] \times \hat{x} = -\hat{F}[\hat{x}]. \tag{3.106}$$

The matrix elements are:

$$\begin{aligned} F_{11} = \frac{\partial F_1}{\partial x_1} = -(s_{24} + W_{abs} + \eta_{13}s_{26}P) - k_{fy}x_4, \quad F_{12} = \frac{\partial F_1}{\partial x_2} = (1 - s_{24} + W_{ems}) + k_{cy}x_4, \\ F_{13} = \frac{\partial F_1}{\partial x_3} = -s_{24} + k_{dy}x_4, \quad F_{14} = \frac{\partial F_1}{\partial x_4} = -k_{fy}x_1 + k_{cy}x_2 + k_{dy}x_3. \end{aligned} \tag{3.107a}$$

$$\begin{aligned} F_{21} = (W_{abs} + c_{up}) - c_{up}(2x_1 + x_2 + x_3), \quad F_{22} = -(1 + W_{ems}) - c_{up}(2x_2 + x_1) \\ - k_{cy}x_4, \\ F_{22} = -s_{23} - c_{up}x_1, \quad F_{24} = -k_{cy}x_2. \end{aligned} \tag{3.107b}$$

$$\begin{aligned} F_{31} = (-s_{24} + \eta_{13}s_{26}P) + k_{fy}x_4, \quad F_{32} = -s_{24}, \\ F_{33} = -(s_{23} + s_{24}) - 2c_{up}x_3 - k_{dy}x_4, \quad F_{34} = k_{fy}x_1 - k_{dy}x_3. \end{aligned} \tag{3.107c}$$

$$\begin{aligned} F_{41} = -k_{fy}x_4, \quad F_{42} = -k_{ce}x_4, \\ F_{43} = -k_{de}x_4, \quad F_{44} = [1 + P(1 + \eta_{56})] - (k_{fe}x_1 + k_{ce}x_2 + k_{de}x_3). \end{aligned} \tag{3.107d}$$

3.6 Modeling example: MM cladding-pumped Yb–Er Co-doped fiber amplifiers

In this section we will study examples of MM cladding-pumped Yb–Er co-doped phosphate glass fiber amplifiers. The reason we use the phosphate glass fiber is because that can be doped with very high concentrations, which is important for high power operations, especially in high energy pulse amplifications. In such situations, not only high power of signal but also low *B*-number is very important for a high quality beam. Note that both thermal and mechanics properties of phosphate glass fibers are not as good as silica glass fibers, in particular they are quite fragile. Therefore, handling them requires more care. In many cases, the thermal and mechanical properties are the main obstacle for good performance of the fiber amplifiers.

Before studying the examples by computing programs, we want to stress that modeling real fiber amplifiers requires a lot of experimental parameters of the materials. As can be seen earlier in this section, the modeling equations require tens of experimental parameters, some depend on concentration and temperature etc. For example, pump and signal CRs, refractive indexes can be considered as independent of concentrations, but they are temperature dependent. On the other hand, lifetimes of excited states in Yb-ions and Er-ions, and all nonlinear processes *K*- and *C*-coefficients strongly depend on concentrations. Determining all these parameters is not easy, and sometimes published data is conflicting—usually because of different measurement conditions, but sometimes due to mistakes. Therefore, it is critical that modelers work closely with material and characterization scientists to obtain as reliable as possible parameters. After having good model and computation methods, these accurate parameters are most important for obtaining reliable modeling and simulation results. At the same time, we should understand that experiments are never perfect. There are always tolerance and measurement errors. Therefore, on one hand we must always respect the experimental data, on the other hand do not trust them 100% before analyzing data and completely understanding it and the experiment conditions for obtaining the data. It is not rare for modelers to point out some mistakes in experiments from analyzing the data. In order to have a good model we must test the simulation results with all available experiment data that are related to the model, and that work requires not only a good model and computing program but also patience. Sometimes it takes years to build a reliable modeling and simulation method that can be used for a broad range of problems. In figures 3.15 and 3.16 we show the simulation results as compared with experiments of Yb–Er co-doped fiber amplifiers. Details of the results can be found in [16].

Finally, we want to stress that the parameters used in the program as stated earlier, should be determined by a process in which experimental data is fed to the numerical model to simulate the systems. Because the models are always based on some approximations and experiments do not always measure all parameters in the models, there is the need to adjust the model's equations and parameters. For example, physically among the nonlinear processes in the Er-ions system, C_{22} is

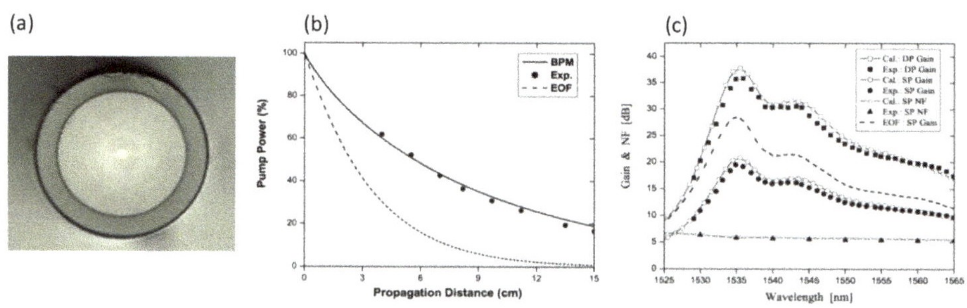

Figure 3.15. Simulation results of a double clad Yb–Er co-doped fiber amplifier [16]. Details of the simulations and experiments can be found in the reference.

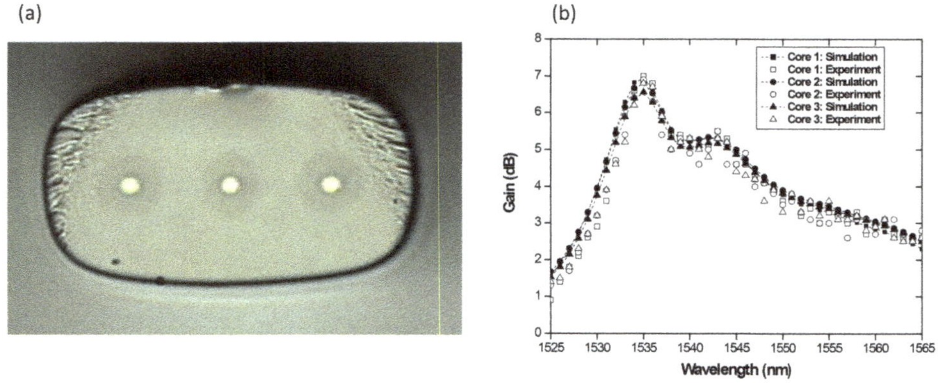

Figure 3.16. Simulation results of a three-core Yb–Er co-doped fiber amplifier. Reprinted with permission from [16], copyright (2007) IEEE. Details of the simulations and experiments can be found in the reference.

strongest. The process to determine C_{22} is much easier than for the other coefficients C_{33} and C_{14}. Therefore, although we want to have all experiment data of C_{ij} we mostly end up with having only experimental C_{22}. In such situations, we can adjust the model equations with approximations $C_{up} = C_{22} \sim C_{33} \sim C_{14}$. Obviously, these approximations have some effects on the simulation results because the higher the values of the nonlinear coefficients C_{ij} the more negative the impacts on the performance of the fiber amplifiers. Therefore, we may consider the simulation results in approximations $C_{14} \sim C_{22}$ and $C_{33} \sim C_{22}$ as the lower limit of performance. In the case of the cooperative transfers from Yb-ions to Er-ions, the K_F is much stronger than K_C and K_D and measurements of K_C and K_D are difficult. The three cooperative transfers play very different roles in the amplification process. The stronger the K_F the more efficient the transfer energy from pump absorption in Yb-ions to the population inversion in Er-ions and therefore better amplification processes. On the other hand, stronger K_C and K_D is less efficient because these two processes excite Er-ions to the high-level states and most of the energy is wasted in radiative and non-radiative relaxations. Similar to the situation of C_{ij}, we may only have the experiment data of K_F, and we need to determine the other two K-coefficients numerically. Physically, we know K_F is much stronger than K_C and K_D, and the relationships are concentration dependent. By varying those parameters in our model and comparing the simulations with experiments of signal gain, NF and also the pump residue—the pump power measured at the end of the fiber—we would be able to come up with numerical values, and their dependence on concentrations that are good for the model in a broader range of concentrations. As described, the work requires patience and very good collaborations with experimentalists.

Below are examples of simulation results for two very different cases of MM cladding-pumped Yb–Er co-doped fiber amplifiers. It is clear that the BPM results not only agree qualitatively but also quantitatively with experiment results.

Let us now study an example for MATLAB® program 3.6. In the example, we will simulate highly doped Yb–Er in phosphate glass fibers. We will see the simulation can help optimizing concentration and fiber length.

3.6.1 Programming example

```
%xxxxxxxxxxxxxxxxxxxxxxxxxxxxxxxxxxxxxxxxxxxxxxxxxxxxxxxxxxxxxxxxxxxxxx
%  PM_EY01.m - Propagation in PM Er-Yb codoped fiber Amplifier        x
%  Multi-Mode Propagation of Pump, SM signal                          x
%                                                                     x
%                 d=16, NA=0.072, Gs = 0.82, clad=125                 x
%                    Er, Yb concentration changes                     x
%                                                                     x
%                 Modified from Rayth3.m  (09/29/20143)               x
%                    by Dan Nguyen  02/08/2020                        x
%xxxxxxxxxxxxxxxxxxxxxxxxxxxxxxxxxxxxxxxxxxxxxxxxxxxxxxxxxxxxxxxxxxxxxx

    function [zar,rval,Pz,Pabs,Pc,Psig,Irz] = PM_EY01;

      global t1 t2a t2b t3 t4 alfa;
      global gama eta_s eta_p;
      global cup kfy kcy kdy kfe kce kde;

    hbar = 6.626e-34;          % planck constant, JxS
      c0 = 3e+14;              % light velocity in um/S
% ******* FIBER STRUCTURE *****************************************
%
       d  = 16.0;              % core diameter in um
       A  = pi*d^2/4.0;        % Core Area, in um^2
      MFD = 17.70;
     Aeff = pi*MFD^2/4.0;
   xcore  = 0.00;              % Core position
   ycore  = 0.00;              %
                               % Indices
    ncore = 1.5675;            % core index
    nclin = 1.5658;            % inner clad index
    nclou = 1.5500;            % jacket index
     deln = 000e-3;            % refractive-index fluctuations
      alf = 0.03;              % propagation loss in dB/cm

%
   Rclin  = 62;               % inner clad radius
   Rclou  = 100;              % jacket radius

   xymax   = 256.0;           % grid size in micron
    nxy    = 512;             % number of transverse points
    L      = 40;
    zmax   = L*1e+4;           % progation distance in micron
    nz     = L*1e+4;           % number of propagation points
    ncycle = L*1e+1;           % points between outputs

 %**** end of fiber structute ************************************

 %********** PUMPING CONDITIONS **********************************

   lamP = 0.975;            % Pump wavelength in um
   lamS = 1.535;            % Signal wavelength in um

   w0x     = 62;          % Pump spot size in um
   w0y     = 62;
   nxsup   = 12;             % supper-gaussian order
   nysup   = 12;
   Rpump   = 62;          % Pumping Area Radius
```

```
   ycen     = 00.0;            % beam displacement from center
   inrand   = -1;              % type of input
%                          = 0   Gaussian or modal input
%                          = +1  Random amplitude fluctuations on a top hat
%                                that fills the cladding
%                          = -1  Random amplitude fluctuations on a top hat
%                                of radius rtop

   noiseA   = 0.2;              % amplitude noise level
   noiseP   = 0.2;              % phase noise level
   wcorX    = 4.0;              % spatial correlation length in X, in micron
   wcorY    = 4.0;              % spatial correlation length in Y, in micron

   tau      = 500e-12;          % pulse width 500 ps
   fre      = 300e3;            % Rep Rate 200 kHz
   Trep     = 1/fre;            % T=1/f
   NonI     = 1.0e-8;           % nonlinear index of phosphate glass
                                % n2 = 2 10^(-16) cm2/W
   coef     = 0.88*Trep/tau;

%**** end of pumping condition *********

%**** Input Material and Normalized Coefficients*****************
     ner = 3.450e+08;             % x=3% Er-concen (ions/um^3), = x*1.15
     nyb = 6.60e+08;              % y=6% Yb-concen (ions/um^3) =  y*1.10
      GS = 0.82;                  % Overlap factor of signal


     s13  = 0.160e-12;            % Er-pump abs CS @975, um^2
     s56  = 1.495e-12;            % 974-Yb-pump abs CS, in um^2
     s65  = 1.495e-12;            % 974-Yb-pump emission CS

     %s12 = 0.484e-12;            % 1530-absorption CR, in um^2
     %s21 = 0.450e-12;            % 1530- **  emission

     s12 = 0.647e-12;            % 1535-absorption CR, in um^2
     s21 = 0.664e-12;            % 1535- **  emission

     %s12 = 0.302e-12;            % 1550-absorption CR, in um^2
     %s21 = 0.409e-12;            % 1550- **  emission

     gama = GS*ner*s21;
     eta_s = s12/s21;


      t21 = 6.0e-3;               % Lifetime t21, in s
      t32 = 2.0e-6;               % Lifetime t32, in s
      t43 = 1.0e-7;               % Lifetime t43, in s
      t65 = 1.0e-3;               % Lifetime t65, in s
      c22 = 4.50e-6;              % C22, um^3/s (2%: 3.0, 3%: 4.5)
      kf = 2.80e-4;               % Kf, um^3/s (4%:2.6; 6%:2.8; 8%:3.0,15% 4.0)

      kc = 0.10*kf;                % Kc, um^3/s
      kd = 0.10*kf;                % Kd, um^3/s

    %**** --- Normalized dimensionless parameters ------***********
```

```
     t1 = t32/t21;
    t2a = t65*s56/(t21*s13);
    t2b = t65*s65/(t21*s13);
    t3 = t43/t21;
    t4 = t43/t32;
    alfa = alf*1e-4/4.343;         % clad intensity loss per micron
    eta_p = s56/s65;
    betay = nyb*s65;
    betae = ner*s13;
    cup = 2*t32*c22*ner;
    kfy = t43 * kf * nyb;
    kcy = t32 * kc * nyb;
    kdy = t43 * kd * nyb;
    kfe = t65 * kf * ner;
    kce = t65 * kc * ner;
    kde = t65 * kd * ner;

% Normalized power
    qs   = hbar*c0*A/(t21*s21*lamS*GS);  % SM Signal in W !!!!!
    qp   = hbar*c0*A/(t21*s13*lamP);     % MM Pump   in W !!!!!

%****** SET UP INPUT FOR BPM *******************************************
%
    nout    = nz/ncycle;
    dz      = zmax/nz;
    dx      = xymax/nxy;
    n1      = round(d/dx);
    dx      = d/n1;
    xymax   = nxy*dx;
     kmax    = 2*pi/dx;
     dk      = kmax/nxy;
    nmid    = floor(nxy/2);

%***Set Up Input Beam
%
    efield      = zeros(nxy,nxy);
    pcore       = zeros(nxy,nxy);
    pclin       = zeros(nxy,nxy);
    pclou       = zeros(nxy,nxy);
    ppump       = zeros(nxy,nxy);

                % Grid in x-y Space
     v          = [0:nxy-1];
     rval       = v*dx - xymax/2;
     [x,y]      = meshgrid(v,v);
     x          = x*dx - xymax/2;
     y          = y*dx - xymax/2;

                % Grid in K-Space
     p          = find(v > nmid);
     v(p)       = nxy-v(p);
     v          = v*dk;
     [k1,k2]    = meshgrid(v,v);

                % Fiber Core Profile
    rcore       = sqrt((x-xcore).^2+(y-ycore).^2);
    p           = find(rcore <= d/2);
    pcore(p)    = 1;
```

```
                    % inner clad profile
   r                = sqrt(x.^2+y.^2);
   p                = find(r <= Rclin);
   pclin(p)         = 1;
   pclin            = pclin - pcore; %

                    % Pumping Input Area
   p                = find(r <= Rpump);
   ppump(p)         = 1;

                    % Fiber Jacket Profile
   p                = find((sqrt(x.^2+y.^2)<=Rclou));
   pclou(p)         = 1;
   pclou            = pclou - pclin - pcore;


                      % Normalized Noisy Input Beam
   GausC            = exp(-x.^2/wcorX^2-y.^2/wcorY^2);
   normC            = dx^2*sum(sum(GausC));
   GausC            = GausC/normC;

                    % Amplitude Noise
   stocA            = ifft2( fft2(rand(nxy)-0.5).*fft2(GausC) );
   stocA            = noiseA*stocA/max(max(abs(stocA)));
                    % Phase Noise
   stocP            = ifft2( fft2(rand(nxy)-0.5).*fft2(GausC) );
   stocP            = noiseP*stocP/max(max(abs(stocP)));

   efield           = exp( -(x/w0x).^(2*nxsup)-((y-ycen)/w0y).^(2*nysup) ).*...
                      (1+stocA).*exp(2*pi*sqrt(-1)*stocP);

     if inrand == +1
       efield = (1-pclou).*(1+stocA).*exp(2*pi*sqrt(-1)*stocP);
     end

     if inrand == -1
       efield = ppump.*(1+stocA).*exp(2*pi*sqrt(-1)*stocP);
     end

                    % Normalized initial field
   norm             = dx*dx*sum(sum(abs(efield).^2));
   efield           = sqrt(1/norm)*efield;

                    % Transfer Function
   arg              = -dz*(k1.^2 + k2.^2)/(4*pi*nclin);
    freq            = exp(sqrt(-1)*arg);
   absdz            = dz*alfa.*(pcore + pclin);

                    % absorbing boundary
    fil             = exp(-(2.0*r/xymax).^32);

%**** END OF SET UP FOR BPM ******
```

```
%*************************************************************************
%                      START BEAM PROPAGATION                           *
%*************************************************************************

      Irz        = zeros(nxy,nz/ncycle+1);
      Irz(:,1)   = abs(efield(nmid,:)).^2';
      zout       = zeros(1,nz/ncycle+1)';
      zout(1)    = 0;
      BInt       = 0;
      Psum       = 0;
      Pblos      = 0;
      count      = 1;
         z       = 0;


      dphi0      = 2*pi*dz*((ncore^2 - nclin^2)/(2*nclin))*pcore...
                 + 2*pi*dz*((nclou^2 - nclin^2)/(2*nclin))*pclou;

               % Input Pump and Signal Power
        Pin      = 100.0;                  % Input power in W
          pp     = (d/Rpump)^2*Pin/qp;     % Normalized to qp


         Ps0    = 1.00;                    % Signal Power in W
          ps    = Ps0/qs;                  % normalized to qs


    for k = 1:nz
         dphi   = dphi0 + 2*pi*dz*deln*rand(nxy);

            sj = ps;


             x = NewtonSys(pp, ps);

       % Core absorption per micron & dz
       coreabs = (betay*(eta_p - (1 + eta_p)*x(4)) + betae*x(1));
           adz = dz*coreabs*pcore;


         step = (gama*(x(2) - eta_s*x(1))- alfa)*ps;
           ps = sj + dz*step;


      % B-Intergral
           BInt = BInt + 2*pi*NonI*dz*ps*qs/(lamS*Aeff);
           B0(k)= BInt;


                % Norm Pfield w diffraction & absorption in dz
      efield   = efield.*exp(sqrt(-1)*dphi-adz/2-absdz/2);


               % FFT after K-Transfomation
      efield   = fft2(ifft2(efield).*freq).*fil;

  %*********end of transform procedure *********************
```

```
%***** Calculation of Pump and Signal Propagation*********

      zar(k)  = z*1e-4;
              % Total Pump Power [%]
      Pz(k)   = 100*dx^2*sum(sum(abs(efield).^2));
              % Pump Power in Core Pcpt [%] Pcore [W]
      Pcpt    = 100*dx^2*sum(sum(abs(efield).^2.*pcore));
                % Pump Power in Core Pc [W]
       Pc(k)  = Pcpt*Pin/100;
              % Absorption Pump Power [%]
       Psum   = Psum + dz*coreabs*100*dx^2*sum(sum(abs(efield).^2.*pcore));
      Pabs(k)= Psum;               % pump absorption

                % Signal Power in W
      Psig(k)= ps*qs;

%***** OUTPUT DATA *************************************
      if mod(k+1,ncycle)==0,
         count = count + 1
         Irz(:,count)   = abs(efield(nmid,:)).^2';
         zout(count)    = z*1e-4;
         pump = Pz(k)
         absorp = Pabs(k)
            sig = Psig(k)
        bnumber = coef*B0(k)

       end

       z   = z + dz;
       pp = Pc(k)/qp;
       ps= Psig(k)/qs;

 end
 save('Pze3y6.dat', 'Pz', '-ASCII')
 save('Pae3y6.dat', 'Pabs', '-ASCII')
 save('Pse3y6.dat', 'Psig', '-ASCII')
 save('Ipe3y6.dat', 'Irz', '-ASCII')
 save('B2e3y6.dat', 'B0', '-ASCII')
 save('z.dat', 'zar', '-ASCII')
 save('rval.dat', 'rval', '-ASCII')
  % ****  END OF PROPAGASTION ******************

%
       figure(1)
       plot(zar, Pz,'b', zar,Pabs,'g', zar,Psig,'r','LineWidth',2);
       set(gca,'FontSize',20);
       axis([ 0 L 0 100 ]);
       set(gca,'YTick',[0:10:100])
       grid on
       legend('P_p', 'P_{abs}', 'P_{sig}');
       xlabel('Propagation distance (cm)');
       ylabel('Power [%]')

       figure(2)
       plot(zar,coef*B0,'r','LineWidth',2);
       set(gca,'FontSize',20);
       axis([ 0 L 0 Inf]);
       grid on
       xlabel('Propagation distance (cm)');
       ylabel('B-Integral [rad]')
```

```
       figure(3)
       colormap(jet);
       imagesc(zout,rval,Irz);
       axis([ 0 L -80 80 ]);
       set(gca,'FontSize',20);
       set(gca,'YTick',[-62 -8 8 62])
       xlabel('Propagation distance (cm)');
       ylabel('Lateral distance (microns)');

%      End of Main Program
%===========================================================
% function:
% NewtonSys Newton method for system of nonlinear equations
%===========================================================

function x = NewtonSys(pp, ps)
  global t1 t2a t2b t3 t4;
  global eta_s;
  global cup kfy kcy kdy kfe kce kde;

 tol = 1e-5;
 kmax = 5;
 %initial-value
    wa  = eta_s*ps;
    n10 = (1 + ps)/( wa + pp + (1 + ps)*(1 + t1*pp) );
    n20 = (wa + pp)/(wa + pp + (1 + ps)*(1 + t1*pp));
    n30 = t1*pp*(1 + ps)/(wa + pp +(1 + ps)*(1 + t1*pp));
    n60 = t2a*pp/(1 + (t2a + t2b)*pp );
    x0 = [n10, n20, n30, n60];
  xold = x0; iter = 1;


  while (iter <= kmax)
% system of nonlinear equations
f1 = 1 - (1+t3*(wa + pp))*xold(1) + (t3*(1+ps)-1)*xold(2) - xold(3) ...
    - kfy*xold(1)*xold(4) + t4*kcy*xold(2)*xold(4) + kdy*xold(3)*xold(4);

f2 = (t1*wa + cup)*xold(1)-t1*(1+ps)*xold(2)+xold(3)-
kcy*xold(2)*xold(4)...


     - cup*(xold(1)^2 + xold(2)^2 + xold(1)*xold(2)+xold(1)*xold(3));

 f3 = 1+(t3*pp-1)*xold(1)-xold(2)-(1+t4)*xold(3)-t4*cup*xold(3)^2...
     + kfy*xold(1)*xold(4) - kdy*xold(3)*xold(4);

 f4 = t2a*pp -(1 + pp*(t2a + t2b))*xold(4)...
     - kfe*xold(1)*xold(4) - kce*xold(2)*xold(4) - kde*xold(3)*xold(4);

  % Jacobian
   df11 = -(1 + t3*(wa + pp)) - kfy*xold(4);
   df12 = t3*(1 + ps) - 1 + t4*kcy*xold(4);
   df13 = - 1 + kdy*xold(4);
   df14 = - kfy*xold(1) + t4*kcy*xold(2) + kdy*xold(3);

   df21 = t1*wa + cup - cup*(2*xold(1) + xold(2) + xold(3));
   df22 = - t1*(1 + ps) - cup*(2*xold(2) + xold(1)) - kcy*xold(4);
   df23 = 1 - cup*xold(1);
   df24 = - kcy*xold(2);
```

```
df31 = t3*pp - 1 + kfy*xold(4);
df32 = -1;
df33 = - (1 + t4)- 2*t4*cup*xold(3) - kdy*xold(4);
df34 = kfy*xold(1) - kdy*xold(3);

df41 = - kfe*xold(4);
df42 = - kce*xold(4);
df43 = - kde*xold(4);
df44 = - (1 + (t2a + t2b)*pp) - kfe*xold(1) - kce*xold(2) - kde*xold(3);

F = [f1; f2; f3; f4];

J = [df11, df12, df13, df14;
     df21, df22, df23, df24;
     df31, df32, df33, df34;
     df41, df42, df43, df44];

        y = - J\F;
       xnew = xold + y';
       dif = norm(xnew - xold);
       if dif <= tol
           x = xnew;  %disp([iter   xnew   dif]);
           return;
       else
           xold = xnew;
       end
       iter = iter + 1;
   end
   disp('Newton method did not converge')
   x = xnew;

%end

========
```

Results

After running program 3.6, we obtain the results for high power fiber amplifiers as shown in figure 3.17 for different concentrations of Yb and Er co-doped phosphate glass fibers. The fibers in simulations have the same structures with core diameter $d = 16$ μm and inner cladding diameter $\Phi = 125$ μm. Pump beam is top-hat and couples to the inner cladding. For simple discussion and analysis we use pump power $P_P = 100$ W and input signal $P_S = 1$ W. The background loss of the fiber is assumed as $\alpha = 3$ dB m^{-1}. The figures on the left are for fiber with 2% and 6wt% (in weight) for Er and Yb dopants, respectively. The results in the center figures are for fiber with 3% Er and 6% Yb concentrations, and the right figures are for fiber with 3% Er and 8% Yb. The upper figures are the 2D images of beam propagation, and the lower figures are the plots of pump power (blue curve), pump absorption (green curve) and signal (red) powers along the fibers. The vertical dashed lines indicate the peak of signal power for each case.

As usual, let us analyze the results in figure 3.17. In simulation we assumed the background loss $\alpha = 3$ dB m^{-1} or total background loss in 40 cm is 1.2 dB or ~25% that is included in the total pump attenuation P_{abs} in the core and inner cladding. The results for the first fiber on the left (E2Y6) P_{abs} ~ 83% and pump power residue

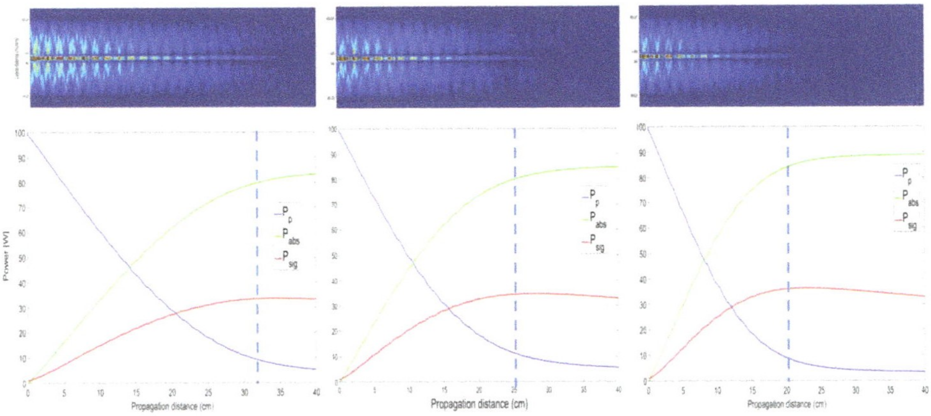

Figure 3.17. Simulation results of 980 nm cladding-pumped highly doped Er–Yb fiber amplifiers. Fiber cores 16 mm, pump power 100 W and input signal power 1 W at 1535 nm. Pump beam is top-hat filled inner cladding of diameter 125 μm. Background loss α = 3 dB m^{-1}. 2% Er and 6% Yb (left), 3% Er and 6%Yb (center) and 3% Er and 8% Yb. Upper: image of MM pump propagation in the fiber. Lower: powers of pump beam (blue), pump absorption (P_{abs}) and amplified signal along the YDFA. Reprinted with permission from [16], copyright (2007) IEEE.

is ~7%, meaning the power loss in the outer cladding and air is about 10%. Stronger absorption in the core will reduce the loss in the outer cladding. For example, results of the fiber (E3Y8) in the right figures has P_{abs} = 90% and pump residue ~5%, or loss in outer cladding and air is about 5%.

At first glance, we may think that the differences in output signal powers and fiber are not significant. For example, the peak signal powers are ~32, 34 and ~36 W for fiber with (2E,6Y), (3E,6Y) and (3E,8Y), respectively. Here, (qE,pY) stand for $q$wt% Er and $p$wt% Yb. The difference between maximum signal powers output from fiber (E2Y6) and (E3,Y8) is about ~12.5%. The fiber lengths corresponding to the peak power of signal are 32, 25 and 20 cm, respectively. However, with other factors of the fiber performance, the difference could be quite important. For example, in many applications with pulse amplification we need to consider the B-integral factor which is discussed earlier. The B value is the phase change accumulated during the pulse propagation due to the nonlinear refractive index induced in the propagation. Although the simulation results in this example are for CW signal, the amplified power of signal in this case can be considered approximately as average power of the pulse amplification [30, 31]. Form this calculation of average power along the fiber we can estimate the peak intensity and integrate the values for the whole fiber length. It is well established that the smaller the B values, the higher the beam quality, and B-values <= 3 are reasonably accepted for many applications. In figure 3.18 we plot the signal powers and B-values for two cases of laser repetition rates 200 kHz and 400 kHz (the program of the plots is too simple to list it here). The figure on the left shows signal powers and B-values in three fibers (E2,Y6), (E3,Y6) and (E3,Y8) for the pulse repetition rate 200 kHz. The horizontal dashed line is for B = 3 limit. The vertical lines are values of signal powers corresponding to the B = 3 limit. It is similar for the figure on the right but with repetition rate 400 kHz. In the case of 200

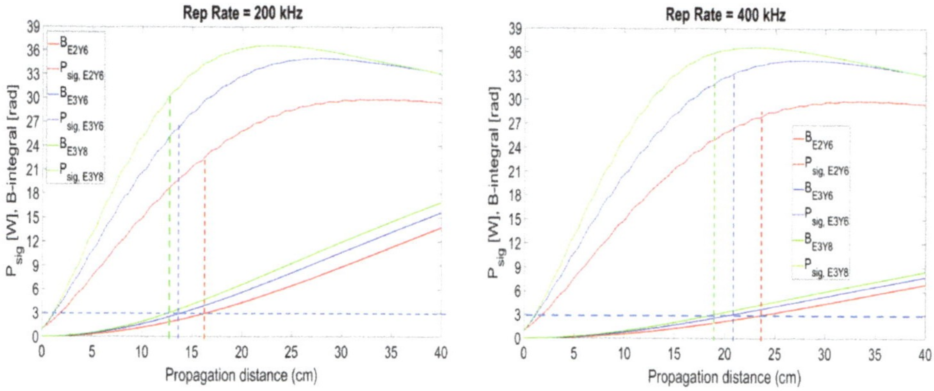

Figure 3.18. Simulation results of 980 nm cladding-pumped highly doped Er–Yb fiber amplifiers. Fiber cores 16 mm, pump power 100 W and input signal power 1 W at 1535 nm. Pump beam is top-hat filled inner cladding of diameter 125 μm. Background loss $\alpha = 3$ dB m^{-1}. 2% Er and 6% Yb (left), 3% Er and 6%Yb (center) and 3% Er and 8% Yb. Upper: image of MM pump propagation in the fiber. Lower: powers of pump beam (blue), pump absorption (P_{abs}) and amplified signal along the YDFA. Reprinted with permission from [16], copyright (2007) IEEE.

kHz, the signal powers for fiber (E3,Y8) is ~30 W, fiber (E3,Y6) ~ 27 W and fiber (E2,Y6) ~ 22 W. Now the difference is ~36%. For the case of repetition rate 400 kHz, we have powers of ~36, ~33 and 28 W, and the difference is ~29%. In these estimates we still ignore the ASE effects. When ASE effects are taken into account in the simulations as will be presented in next section, we can see the ASE effect is more negative for the gain and NF in longer fibers. As a result, the differences become even more significant.

The results presented above are good examples for designing EDFAs. A good design of EDFA, especially in the high power regime must take into account all factors that impact on the final performance of the device. Depending on the requirements of the targets, some emphasize beam quality, others can relax the restriction of *B*-values but prioritize total output power and fiber lengths etc. Pump power budget is also another restriction that modeling cannot ignore in the calculation.

One important point that may affect the performance of fiber amplifiers is the non-uniform distribution of the MM pump beam, as shown in figure 3.19. As discussed earlier, due to the mode couplings, the spatial distribution of the MM beam is changing, and the modes may converge at some positions along the fiber. As a result, the intensity in the core can spike much higher than the core intensity at the input position. In the high power regime, when pump power is very high the spike of intensity in the core can create hot-spots, which is not easy to understand from conventional understanding based on effective overlap factor (EOF) approximation, in which beam distribution is assumed uniform along the fiber and is represented by an EOF.

Next, we will study the simulation problem of cladding-pump Er/Yb co-doped fiber amplifiers with ASE. The problem is more complicated since we cannot apply the conventional approach of treating ASE, which is developed for the SMP scheme for SM pumping EDFAs or EOF method for cladding-pumped EDFAs.

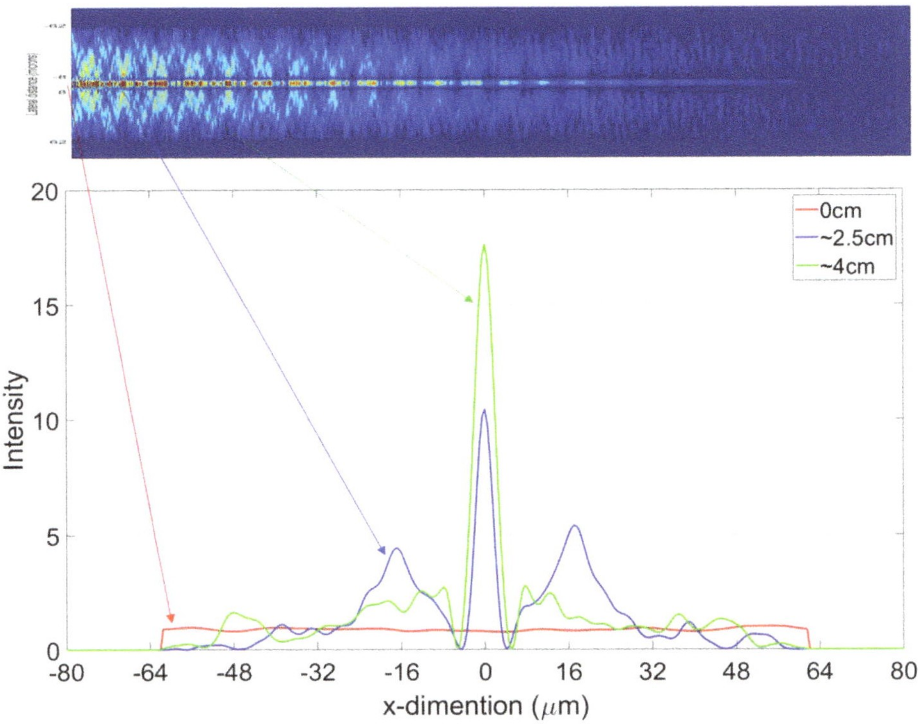

Figure 3.19. Simulation results of spatial distribution of MM pump in cladding-pumped highly doped Er–Yb fiber amplifiers. Fiber cores 16 mm, pump beam is top-hat with amplitude and phase noise (red curve) filled inner cladding of diameter 125 μm. 2% Er and 6% Yb (left), Upper: image of MM pump propagation in the fiber. Lower: spatial distribution at three positions in the cores: input (red), $z \sim 2.5$ cm (blue) and ~ 4 cm (green). Note that the distribution in x- and y-dimensions can be different during the propagation. Reprinted with permission from [16], copyright (2007) IEEE.

3.7 Modeling MM cladding-pumped fiber amplifiers with ASEs

So far we have ignored the ASEs in our modeling examples. The reason for this is that the problem of modeling cladding-pumped fiber amplifiers is extremely complicated if we follow the conventional method for single mode pump (SMP) or EOF MMP schemes. As described earlier, both forward and backward ASEs in the emission band of Er-ions can be treated as multiple channels of noise counter-propagating in the fiber. Each channel is characterized by a bandwidth $\Delta\lambda_i$ and a wavelength λ_i at the center of $\Delta\lambda_i$. In general the smaller bandwidth, the more accurate the simulation results that can be achieved. However, it takes much longer computing time and computer memory if too narrow bandwidth of ASE is used in simulations because it means too many propagation equations for all ASE channels. Desurvire showed that $\Delta\lambda_i = 0.5$ nm is small enough in balancing good simulation results and computing resources including computing time and hardware [1].

If we follow the conventional method for ASE, in our effective BPM method we have to solve the following equations

$$P_P(x, y, z + \Delta z) \doteq BMP\{\alpha\{n_i(P_P(z), P_S(z))\}\Delta n(x, y, z), P_P(x, y, z)\}, \quad (3.108a)$$

$$\frac{dP_S(z)}{dz} = \{\Gamma_S N_{Yb}[\sigma_{ems}(\lambda_S)n_2(z) - \sigma_{abs}(\lambda_S)n_1(z)] - \alpha_S\}P_S(z), \tag{3.108b}$$

$$\frac{dP_i^{(+)}(z)}{dz} = \{\Gamma_i[\sigma_{ems}(\lambda_i)N_2(z) - \sigma_{abs}(\lambda_i)N_1(z)] - \alpha_i\}P_i^{(+)}(z) + h\nu_i\Delta\nu_i\Gamma_i\sigma_{ems}(\lambda_i)N_2, \tag{3.108c}$$

$$\frac{dP_i^{(-)}(z)}{dz} = -\{\Gamma_i[\sigma_{ems}(\lambda_i)N_2(z) - \sigma_{abs}(\lambda_i)N_1(z)] + \alpha_i\}P_i^{(-)}(z) - h\nu_i\Delta\nu_i\Gamma_i\sigma_{ems}(\lambda_i)N_2. \tag{3.108d}$$

Note that in solving equations (3.108) we should consider the ASE band from 1450 to 1650 nm as can be seen from the CR of Er-doped glasses in figure 3.9. However, if we use a bandwidth $\Delta\lambda = 0.5$ nm for each ASE channel, then equations (3.108c) and (3.108d) become 200 equations for forward ASEs and 200 for backward ASEs. There are problems with that. First, if we propagate the system of equations (3.108) including the BPM propagation for an MM pump beam, one SM propagation for signal and 400 for ASEs, it would require a lot of computer memory. Second, because the problem of ASEs propagation has two-boundary conditions, it requires multiple back and forth propagations with trial values of the backward ASEs. That involves multiples of BMP propagations which consume huge computing time. Therefore, we should find a different and better way for simulating cladding-pumped EDFA with ASE. First, we notice that the BPM propagation in Yb–Er co-doped fiber strongly depends on absorption of the Yb subsystem, but weakly on absorption of the Er subsystem. This characterization is even more clear in a highly Yb/Er co-doped system, which is typical for high power fiber amplifiers. Furthermore, as discussed earlier in the high power regime the fiber lengths are usually short to avoid nonlinear effects in the fibers. These nonlinear effects include stimulated Raman scattering (SRS) and stimulated Brillouin scattering (SBR), and also the nonlinear phase change with high value of *B*-integral. In such situations, ASE powers are quite low as compared to pump and signal powers which dominate the pump absorption. Therefore, we can approximately solve the system of equation (3.108) by steps, as in figure 3.20.

The whole process can be described as follows.

I. Solving equations (3.108a) and (3.108b) as presented earlier in this section. Saving the obtained values of pump power $P_P(z_i)$ with $z_i = 0, \Delta z, 2\Delta z, \cdots N\Delta z = L$, L is fiber length. The most important solution of equation (3.108a) is the pump power in the core.

$$P_P(z) = P_{P,core}(z_i), \tag{3.109}$$

where $P_{P,core}(z_i)$ is function of $z_i = 0, \Delta z, 2\Delta z, \cdots N\Delta z = L$ in the core obtained by the BPM running earlier. Because populations are determined by $P_{P,core}(z)$, $P_S(z)$ and $P^{(+)}_{ASE}(z)$, $P^{(-)}_{ASE}(z)$ therefore we now can use

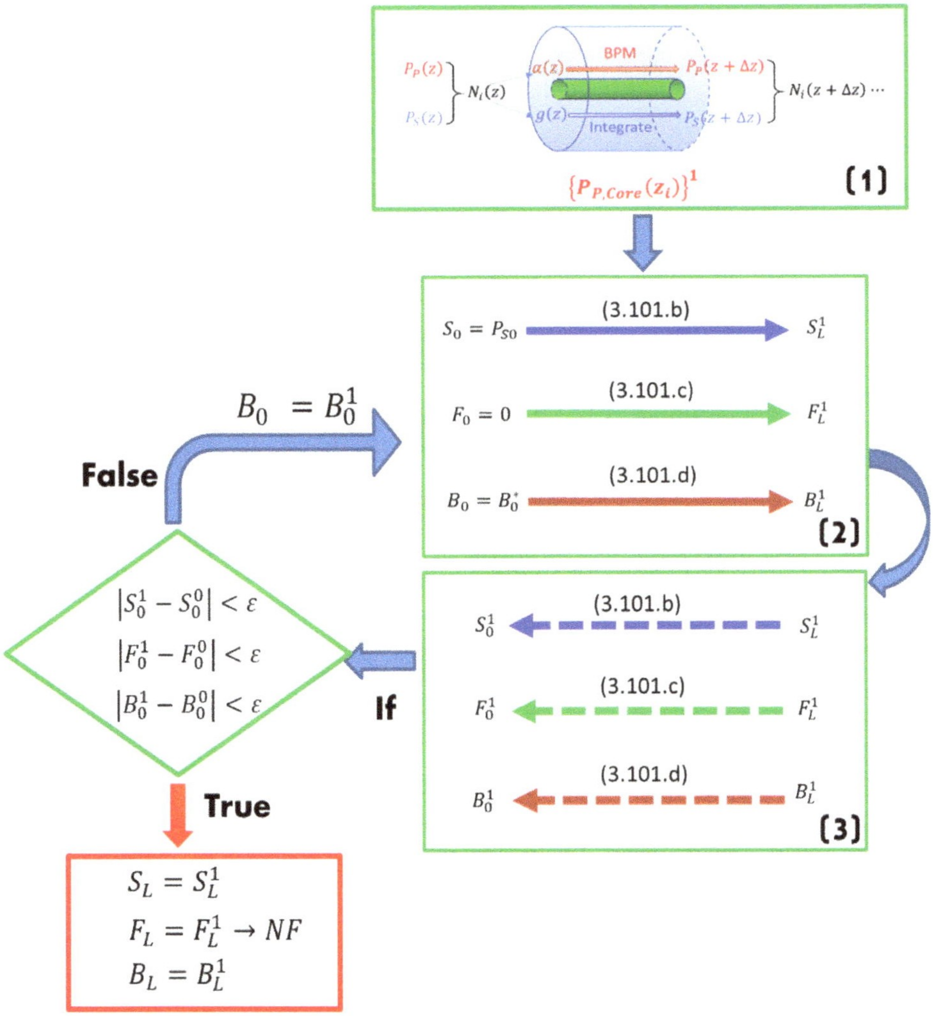

Figure 3.20. Schematic simulations of cladding-pumped fiber amplifiers with ASE.

$P_{P,core}(z_i)$ not the power of the whole pumping beam to solve the rate equations in the next steps.

II. Solving propagation equations (3.108b), (3.108c) and (3.108d) in forward direction $z_i = 0, \Delta z, 2\Delta z, \cdots N\Delta z = L$ $z = 0$ with pump power in the core $P_P(z) = P_{P,core}(z_i)$.

In this step, the initial conditions for signal and forward ASE at $z = 0$ are known:

$$S_0 = P_S(0) \quad \text{and} \quad F_0 = \sum_{i=1}^{M} P_{ASE}^{(+)}(\lambda_i, z = 0) \equiv 0 \tag{3.110a}$$

But the initial condition for backward ASE is unknown. Therefore, a trial power $P_{ASE}^{(-)*}(\lambda_i, z = 0)$ for each backward ASE channel is chosen

$$B_0 = B_0^* = \sum_{i=1}^{M} P_{ASE}^{(-)*}(\lambda_i,\, z = 0) \qquad (3.110b)$$

III. Using the values of signal power $S_L^1 = P_S(z = L)$ and forward ASE $F_L^1 = \sum_{i=1}^{M} P_{ASE}^{(+)}(\lambda_i,\, z = L)$ and backward ASE $B_L^1 = \sum_{i=1}^{M} P_{ASE}^{(-)}(\lambda_i,\, z = L)$ obtained by step (2) above as initial conditions to integrate equations (3.108b)–(3.108d) in backward direction $z_i = N\Delta z,\, (N - 1)\Delta z,\, \dots \Delta z,\, 0$, with the pump power in the core now in the backward direction $P_P(z) = P_{P,core}(z_i = N\Delta z,\, \dots \Delta z,\, 0)$.

IV. The results obtained from step (3) will be compared with the initial conditions in step (2):
   - If conditions $|S_0^1 - S_0| \leqslant \varepsilon$, $|F_0^1 - F_0| < \varepsilon$ and $|B_0^1 - B_0^*| \leqslant \varepsilon$ are not satisfied then repeat step (2) with new initial condition for backward ASE $B_0^* = B_0^1$;
   - If the conditions are satisfied, the output results are recorded, and the forward ASE spectrum will be used to calculate NF of the amplifier.

Exercise

Below are the details of the instruction of the program following figure 3.20. In general, this program is suitable for simulating the ASE effects in cladding-pumped fiber amplifiers under conditions of low ASE power in comparison with signal and pump powers. Under that condition, we can assume the impact of ASE power on pump absorption is negligibly weak. The ASE impacts on pump absorption through its role in populations, see equation (3.96) for example.

I. The program of stage I of the process is program 3.6. Near the end of the program, adding a line: `save('Pc.dat', 'Pcore', '-ASCII')`

II.

   1. First, we need loading/input of CRs for description of ASE spectrum in Er-ions. Below is an example of the absorption and emission CRs (in c++)

```
float lamb[120] = {1520, 1520.5, 1521, 1521.5, 1522, 1522.5, 1523, 1523.5, 1524, 1524.5,
    1525, 1525.5, 1526, 1526.5, 1527, 1527.5, 1528, 1528.5, 1529, 1529.5,
    1530, 1530.5, 1531, 1531.5, 1532, 1532.5, 1533, 1533.5, 1534, 1534.5,
    1535, 1535.5, 1536, 1536.5, 1537, 1537.5, 1538, 1538.5, 1539, 1539.5,
    1540, 1540.5, 1541, 1541.5, 1542, 1542.5, 1543, 1543.5, 1544, 1544.5,
    1545, 1545.5, 1546, 1546.5, 1547, 1547.5, 1548, 1548.5, 1549, 1549.5,
    1550, 1550.5, 1551, 1551.5, 1552, 1552.5, 1553, 1553.5, 1554, 1554.5,
    1555, 1555.5, 1556, 1556.5, 1557, 1557.5, 1558, 1558.5, 1559, 1559.5,
    1560, 1560.5, 1561, 1561.5, 1562, 1562.5, 1563, 1563.5, 1564, 1564.5,
    1565, 1565.5, 1566, 1566.5, 1567, 1567.5, 1568, 1568.5, 1569, 1569.5,
    1570, 1570.5, 1571, 1571.5, 1572, 1572.5, 1573, 1573.5, 1574, 1574.5,
    1575, 1575.5, 1576, 1576.5, 1577, 1577.5, 1578, 1578.5, 1579, 1579.5};
```

float es[120] = {0.184, 0.188, 0.195, 0.200, 0.208, 0.213, 0.223, 0.229, 0.240, 0.248, 0.261, 0.270, 0.285, 0.297, 0.316, 0.330, 0.353, 0.370, 0.398, 0.418, 0.450, 0.473, 0.509, 0.534, 0.570, 0.593, 0.625, 0.642, 0.661, 0.666, 0.664, 0.656, 0.635, 0.617, 0.588, 0.569, 0.545, 0.532, 0.519, 0.513, 0.509, 0.508, 0.508, 0.507, 0.506, 0.504, 0.500, 0.495, 0.487, 0.479, 0.467, 0.458, 0.445, 0.436, 0.422, 0.413, 0.401, 0.393, 0.383, 0.377, 0.368, 0.363, 0.356, 0.352, 0.347, 0.343, 0.339, 0.336, 0.332, 0.329, 0.324, 0.321, 0.316, 0.313, 0.308, 0.305, 0.301, 0.298, 0.293, 0.290, 0.285, 0.282, 0.277, 0.274, 0.269, 0.266, 0.261, 0.258, 0.252, 0.249, 0.242, 0.238, 0.232, 0.228, 0.221, 0.216, 0.208, 0.203, 0.195, 0.189, 0.181, 0.175, 0.166, 0.160, 0.153, 0.148, 0.141, 0.136, 0.130, 0.126, 0.122, 0.120, 0.116, 0.114, 0.112, 0.110, 0.107, 0.106, 0.105, 0.104};

float as[120] = {0.243, 0.247, 0.253, 0.256, 0.263, 0.268, 0.276, 0.282, 0.291, 0.299, 0.310, 0.318, 0.332, 0.343, 0.360, 0.373, 0.394, 0.410, 0.435, 0.454, 0.484, 0.504, 0.536, 0.556, 0.588, 0.608, 0.633, 0.645, 0.657, 0.657, 0.647, 0.634, 0.603, 0.581, 0.548, 0.524, 0.495, 0.480, 0.462, 0.454, 0.445, 0.441, 0.436, 0.432, 0.425, 0.421, 0.412, 0.405, 0.393, 0.384, 0.369, 0.360, 0.345, 0.335, 0.321, 0.311, 0.298, 0.290, 0.279, 0.273, 0.263, 0.258, 0.249, 0.245, 0.238, 0.234, 0.229, 0.224, 0.219, 0.215, 0.209, 0.206, 0.201, 0.197, 0.191, 0.188, 0.183, 0.180, 0.175, 0.171, 0.167, 0.164, 0.159, 0.156, 0.152, 0.149, 0.144, 0.141, 0.137, 0.133, 0.129, 0.125, 0.121, 0.117, 0.113, 0.109, 0.104, 0.101, 0.096, 0.092, 0.087, 0.084, 0.078, 0.075, 0.070, 0.068, 0.064, 0.061, 0.058, 0.055, 0.053, 0.052, 0.049, 0.048, 0.047, 0.045, 0.044, 0.043, 0.043, 0.042};

here we only consider the spectra in the region $\lambda_i = \{1520, \cdots 1560\}$ nm where bandwidth of each ASE channel is $\Delta\lambda_i = 0.5$ nm as input array lamb[200] above. The emission and absorption CRs are described by arrays es[200] and as[200].

2. Loading/input Pc.dat from step 1.
3. Initial conditions (again, this is in c++, you need to change to MATLAB®)

```
z = 0;
ps = ps0/qs;
pp = pz[0]/qp;

f = new double [size]; // initial for forward ASE F0
for( int nf = 0; nf < size; nf++) f[nf] =0;

b = new double [size];
for( int nb = 0; nb < size; nb++) b[nb] =0; // trial
values for backward ASE
```

4. Propagate signal and forward ASE(+) and backward ASE(-) from $z = 0, \Delta z, \cdots N\Delta z = L$ with step Δz used in the BPM propagation in

stage I and recorded in $P_C(z = \Delta z, \cdots N\Delta z)$. In this propagation, we may have two functions for propagation of ASE(+) and ASE(−) as below

```
void ASEF( double Ff[], double f[], double n1, double
n2, int sizeOfArray){double g[size], pn[size];
    for( int j = 0; j < sizeOfArray; j++) {
              g[j] = L0*GS*ne*es[j]*crs;
              pn[j]=GS*c*t21*es[j]*crs/(pow(lamb[j],2)
              *delta*A);
              Ff[j] = (g[j] * (n2 - as[j]/es[j] * n1) - alfa)
              *f[j] + g[j]*pn[j]*n2;
        }
}
void ASEB( double Fb[], double b[], double n1, double
n2, int sizeOfArray){double g[size], pn[size];
    for( int j = 0; j < sizeOfArray; j++){
              g[j] = L0*GS*ne*es[j]*crs;
              pn[j]=GS*c*t21*es[j]*crs/(pow(lamb[j],2)
              *delta*A);
              Fb[j] = - (g[j] * (n2 - as[j]/es[j] * n1) -
              alfa)*b[j] - g[j]*pn[j]*n2;
        }
}
```

Remember, the ASE(+) and ASE(−) are described by equations (3.108c) and (3.108d), respectively, and the equation for the signal is (3.108b).

At the end of stage II, we obtained powers of signal, ASE(+) and ASE(-) $S_L^1 = P_S(L)$, $F_L^1 = \sum P_i^{(+)}(z = L)$ and $B_L^1 = \sum P_i^{(-)}(z = L)$, respectively.

III. In this stage, using pump power in reverse order $P_C(N\Delta z, \cdots\Delta z, 0)$ and propagating signal, ASE(+) and ASE(−) backward from $z = L$ to 0. That means equations (3.108b)–(3.108d) now have (−) sign before the original equations. For functions of ASEs above, now change the sign in the last lines.

```
Ff[j] =-((g[j] * (n2 - as[j]/es[j] * n1) - alfa)*f[j] +g[j] *
pn[j]*n2);     // ASE(+) backward propagation
```

and

```
Fb[j] = (g[j] * (n2 - as[j]/es[j] * n1) - alfa)*b[j] + g[j]
*pn[j]*n2;     // ASE(-) backward propagation
```

At the end of this stage III, we obtained powers of signal, ASE(+) and ASE(−) at $z = 0$: $S_0^1 = P_S(z = 0)_{\text{calculated}}$, $F_0^1 = \sum P_i^{(+)}(z = 0)_{\text{calculated}}$ and $B_0^1 = \sum P_i^{(-)}$

$(z = L)_{\text{calculated}}$, respectively. Note that, at $z = 0$, we know initial values of signal power $S_0^0 = P_{S0}$ and forward ASE(+) power $F_0^0 = \sum P_i^{(+)}(z = 0) \equiv 0$. Therefore, at the end of this stage we can compare the values of calculated and known values. If the differences between these values are larger than a tolerance we want for the results, we repeat the steps of stage II until we get the convergence of the results in both stages II and III.

Once we get the convergence we can calculate the noise figure of the amplifier by

```
gain = 10*log10((ps - f[j0])*qs/ps0);
fase = f[j0]*qs;
fz = planck*c*c*0.5/pow(lams, 3);
NF = 10*log10(1 + fase/fz) - gain;
```

Here, j0 = 1 to 200 is index of spectra of wavelength from 1520 to 1560 nm defined in the arrays of wavelength lamb[200] and es[200] and as[200].

Figure 3.15 is an example of results of gain and NF calculated.

Finally, in the cases in which ASE powers are comparable with signal power, the situation becomes much more complicated and requires much more computing resource including computing memory and time. In such situations, at least we have to add the ASEs in propagation in stage I as described above. However, with so many equations of ASEs the required computer memory becomes a big issue knowing that the BPM requires a very small propagation step, in the order of microns (in our examples, we use $\Delta z = 1$ μm). In practice, we may not need to use 200 equations for forward and backward ASEs because the power of ASEs is only high around the peak gain spectrum. Instead, 20 equations for forward and 20 equations for backward ASEs are enough. Even with reduced number of equations for ASE, the simulation still consumes much time as compared with the simple cases described above. Therefore, in principle, although this effective BPM method can simulate the cases with high ASE power, we do not recommend using this method for these situations.

References

[1] Desurvire D 1994 *Erbium-Doped Fiber Amplifiers, Principles and Applications* (New York: Wiley)

[2] Becker P C, Olson N A and Simpson J R 1999 *Erbium-Doped Fiber Amplifiers, Fundamentals and Technology* (New York: Academic)

[3] Digonnet M J F (ed) 1993 *Rare Earth Doped Fiber Lasers and Amplifiers* (New York: Marcel Dekker, Inc.)

[4] Giles C R and Desurvire D 1991 Modeling erbium-doped fiber amplifiers *J. Light. Technol.* **9** 271–83

[5] Paschotta R, Nilsson J, Tropper A C and Hanna D C 1997 Ytterbium-doped fiber amplifiers *IEEE J. Quantum Electron.* **33** 1049–56

[6] Hardy A and Oron R 1997 Signal amplification in strongly pumped fiber amplifiers *IEEE J. Quantum Electron.* **33** 307–13

[7] Sorbello G, Taccheo S and Laporta P 2001 Numerical modeling and experimental investigation of double-cladding erbium-ytterbium-doped fibre amplifiers *Opt. Quantum Electron* **33** 599–619

[8] Valley G C 2001 Modeling cladding-pumped ER/YB fiber amplifiers *Opt. Fiber Tech.* **7** 21–44
[9] Kouznetsov D, Moloney J V and Wright E M 2001 Efficiency of pump absorption in double-clad fiber amplifiers. I. Fiber with circular symmetry *J. Opt. Soc. Amer.* B **18** 743–9
[10] Bedö S, Lüthy W and Weber H P 1993 The effective absorption coefficient in double-clad fibres *Opt. Commun.* **99** 331–5
[11] Liu A and Ueda K 1996 The absorption characteristics of circular, offset, and rectangular double-clad fibers *Opt. Commun.* **132** 511–8
[12] Kouznetsov D, Moloney J V and Wright E M 2001 Efficiency of pump absorption in double-clad fiber amplifiers. II. Fiber with circular symmetry *J. Opt. Soc. Amer.* B **18** 743–9
[13] Feit M D and Fleck J A 1978 Light propagation in graded-index optical fibers *Appl. Opt.* **17** 3990–8
[14] Feit M D and Fleck J A 1979 Calculation of dispersion in graded-index multimode fibers by a propagating-beam method *Appl. Opt.* **18** 2843–51
[15] Feit M D and Fleck J A 1980 Computation of mode properties in optical fiber waveguides by a propagating beam method *Appl. Opt.* **19** 1154–64
[16] Nguyen D T *et al* 2007 A novel approach of modelling cladding-pump highly doped Er/Yb fiber amplifiers *IEEE J. Quantum Electron.* **43** 1018–27
[17] Babaeian M, Nguyen D T, Demir V, Akbulut M, Blanche P-A, Kaneda K, Neifeld M and Peyghambarian N 2019 A single shot coherent Ising machine based on a network of injection-locked multicore fiber lasers *Nat. Commun.* **10** 3516
[18] Demir V, Akbulut M, Nguyen D T, Kaneda K, Neifeld M and Peyghambarian N 2019 Injectionlocked, single frequency, multi-core Yb-doped phosphate fiber laser *Sci. Rep.* **9** 356
[19] Marcuse D 1991 *Theory of Dielectric Optical Waveguides* (New York: Academic)
[20] Synder A W and Love J D 1983 *Optical Waveguide Theory* (New York: Chapman and Hall)
[21] Snyder W 1972 Coupled-mode theory for optical fibers *J. Opt. Soc. Am.* **62** 1267–77
[22] Yariv A 1973 Coupled-mode theory for guided-wave optics *IEEE J. Quantum Electron.* **9** 919–33
[23] Haus H, Huang W, Kawakami S and Whitaker N 1987 Coupled-mode theory of optical waveguides *J. Lightwave Technol.* **5** 16–23
[24] Hwang B-C, Jiang S, Luo T, Watson J, Sorbello G and Peyghambarian N 2000 Cooperative upconversion and energy transfer of new high Er- and Yb-doped phosphate glasses *J. Opt. Soc. Amer.* B **17** 833–9
[25] Page R H, Schaffers K I, Waide P A, Tassano J B, Payne S A, Krupke W F and Bischel W K 1998 Upconversion-pumped luminescence efficiency of rare-earth doped hosts sensitized with trivalent ytterbium *J. Opt. Soc. Amer.* B **15** 996–1008
[26] Lester C, Bjarklev A, Rasmussen T and Dinesen P G 1995 Modeling of Yb-sensitized Er-doped silica waveguide amplifiers *J. Lightw. Technol.* **13** 740
[27] Karasek M 1997 Optimum design of Er–Yb codoped fiber for largesignal high pump power applications *IEEE J. Quantum Electron.* **33** 1699–705
[28] Karasek M and Kanka J 1998 Numerical analysis of Yb-sensitized Er-doped fiber ring laser *Proc. Inst. Elect. Eng. Optoelectron.* **145** 133–7
[29] Jiang S *et al* 2003 Multi-mode pumped amplifier using a newly developed 8-cm long erbium-doped phosphate glass fiber *Opt. Eng.* **42** 2817–20
[30] Peng X *et al* 2013 High efficiency, monolithic fiber chirped pulse amplification system for high energy femtosecond pulse generation *Opt. Express.* **21** 25440–51
[31] Peng X *et al* 2014 Monolithic fiber chirped pulse amplification system for millijoule femtosecond pulse generation at 1.55 μm *Opt. Express.* **22** 2459–64

Dan T Nguyen

Chapter 4

Modeling ultrafast mode-locked fiber lasers

Ultrashort lasers have become an invaluable tool for both fundamental research and numerous applications in different areas of science and technology. For example, ultrafast lasers have been used for time resolved studies in chemistry and optical coherence tomography. Importantly, ultrashort lasers can provide lasers with high peak intensities and a broad optical spectra, which are critical for numerous applications such as micro-machining and marking, material processing, nonlinear optics, metrology etc. Related medical applications such as laser surgery are also widely used nowadays.

It is important to know that most ultrafast solid-state lasers currently use passive mode-locking technique using saturable absorption (SA). And the most commercialized mode-locked lasers currently are semiconductor saturable absorber mirrors (SESAMs) mode-locked lasers, after the invention of the SESAM in the early 1990s [1, 2]. On the other hand, for the last decade or so mode-locked fiber lasers have received much attention due to their low cost, low power consumption, long term robustness, and ease for coupling to optical fibers for long distance transmission and much else. Research and developments of mode-locked lasers including new saturable absorption materials, laser configurations, dispersion regimes etc, have become a huge field of research and development.

Mode-locked lasers have been investigated thoroughly and comprehensively for more than 30 years, and excellent reviews of the field can be found in [3–7]. However, details of theoretical modeling and simulation of the whole mode-locked laser system is still lacking, at least in the literature. To the best of our knowledge there are few publications on details of modeling works of mode-locked lasers, although all components and configurations of the lasers can be characterized very well experimentally and theoretically. Therefore, our main subject in this chapter is to provide a detailed description of modeling and simulation of whole systems of mode-locked laser, especially the mode-locked fiber lasers include nonlinearity, dispersion, saturable absorption, gain and losses in the laser cavity. The presentation

doi:10.1088/978-0-7503-2272-0ch4

of this chapter starts with a brief introduction to mode-locked lasers. Readers are encouraged to study comprehensively the concept and working principles of mode-locked lasers in the above references. Descriptions of several key components that interplay to produce the mode-locking in laser cavity such as SAs, dispersion, nonlinear index are then given in the simplest way but clear enough for even beginners to understand the basic concepts and to be able to work in later examples. Then, the nonlinear Schrödinger equation (NLSE) and generalized NLSE (GNLSE) are introduced as a mathematical tool for numerical modeling of the system [8]. In particular, the numerical method, the fourth-order Runge–Kutta in the interaction picture (RK4IP) [9] for numerically solving the NLSE and GNLSE is presented. It is important to note that there are several numerical methods of solving the NLSE and GNLSE. However, the RK4IP has been proved to be the best in numerical errors and efficient algorithm as compared to other methods [8–10], and that becomes very important for numerically solving light that circulates multiple rounds with gain in highly complicated structures including nonlinear absorption by SA, nonlinear index, dispersion etc. After understanding all these components and their mathematical descriptions as well as the numerical method of solving the NLSE we then study some examples using MATLAB® programs. The modeling programs can describe in detail the laser beam behavior depending on these important parameters.

4.1 A brief introduction to mode-locked lasers

In this section, we will give a brief introduction to mode-locked lasers. The aim of this simple introduction is to prepare some basic knowledge of the systems for beginners to study the problem of modeling mode-locked fiber lasers by examples with MATLAB® programs.

First, let us recall the results in chapter 1 in which we have studied resonators or laser cavities with multiple longitudinal modes spanning a broad region of frequency/wavelength. These cavities under proper conditions $G > L$ (G: gain, and L: loss) can produce a laser beam with many different frequency modes. In the cases of mode-locked lasers, the mode-locking term refers to a concept of phase locking of many neighboring longitudinal modes of the laser cavity. As can be seen in the following, this phase locking, if done properly, can produce a laser of a continuous train of extremely short pulses. In general, if the number of longitudinal modes that are locked is sufficiently large, a laser with a broad spectrum or a short pulse can be generated. Once the relative phases of a large number of modes in the cavity are locked, the laser cavity produces a periodic laser output whose periodicity is determined by the round-trip time of signal in the cavity. In fact, mode-locked lasing is a very complicated process, and it is difficult to describe the whole process even just only qualitatively. Although basic understandings of the mode-locking lasing processes has developed since the 1970–80s [4–7], quantitative modeling of the lasers is still lacking. As we can see later through the examples, it turns out that modeling and simulation are extremely useful for understanding the process of mode-locked lasers. In particular, in these examples the numerical solutions of light that is built up in gain medium and is locked by saturable absorption, and is interplayed by dispersion and nonlinearity in the cavity can be visualized. The simulations are very effective for theoretical description of the mode-locked lasers. Readers can observe for themselves the whole process visually through modeling and simulation works in several examples presented later in this chapter.

Let us start from the better understood problem of laser cavities and laser operations without mode locking. Recall the descriptions of resonators and resonant modes of laser cavities in chapter 2. For each mode to be lased it has to satisfy the lasing condition, which states that total gain is higher than the total loss in the cavity for the lasing mode. In general, a cavity may have multiple resonant modes. However, depending on the gain and loss of the cavity, both are wavelength-dependent functions, the laser may lase in single mode, few modes or multiple modes regimes. For example, figures 4.1(a) and (b) show a laser cavity with two mirrors M_1 and M_2 and reflections R_1 and R_2. By changing the cavity length the cavity can have three resonant modes (figure 4.1(c)) and multiple modes (figure 4.1(d)). Assume the gain spectrum is given by green curves in figures 4.1(c) and (d). As examples, by changing the loss (black lines) a laser that can operate with single mode (loss is the dashed line—figure 4.1(c)), few modes (loss is the solid line—figure 4.1(c)), or multi modes (figure 4.1(d)) can be realized. Mathematically, the simplest case to understand and describe is a single mode laser in steady state condition or continuous wave (cw) regime. Let us first look at a cw regime of a laser cavity satisfied lasing condition for a single mode with frequency ω of the cavity. Figure 4.1(c) shows lasing condition for single mode with total loss (dashed black line) and the gain (green curve). In such a situation, the output laser beam is single mode with narrow

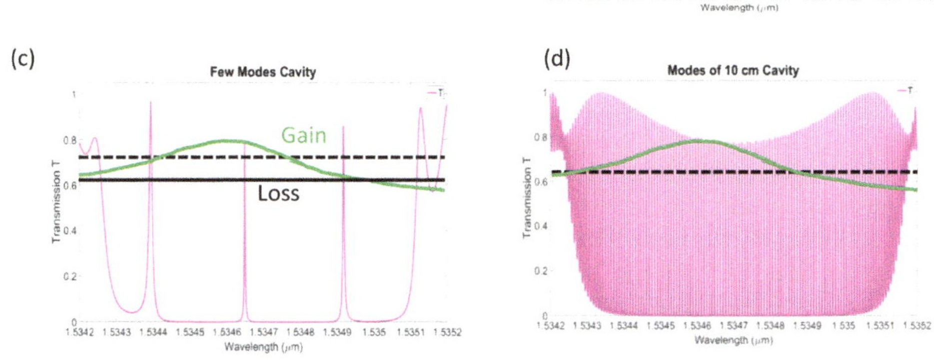

Figure 4.1. (a) Schematic of a laser cavity, (b) calculated reflection R_1 and R_2 of two mirrors M_1 and M_2. (c) and (d) resonant modes of cavity with $L = 1$ mm and $L = 10$ cm, respectively. The green curves and black lines are the cavity gains and losses, respectively.

frequency at ω. In time domain the electrical field of the laser beam can be expressed as $E(t) = A_1 e^{i(\omega t+\varphi)}$, where φ is the phase of the lasing mode. In general, light of propagating mode accumulates in a phase $\varphi = \beta z = knz + \varphi_0$, where β is propagation constant of the mode in a medium of index n, k is wave number and φ_0 is the initial value of the mode phase.

In general, however, the gain medium could overlap with several or even a large number of modes and is greater than the losses of these modes, as shown in figure 4.1(d). In such situations the output of the laser in cw regime and in the time domain can be expressed as:

$$E_{\text{Total}}(t) = \sum_{i=1}^{N} A_i e^{i(\omega_i t+\varphi_i)}. \tag{4.1}$$

Here the sum is over all of the lasing cavity modes, A_i is the amplitude of the ith mode, ω_i is the angular frequency of the ith mode, and φ_i is the phase of the ith mode. As presented in chapter 2, in the cavity of multiple longitudinal modes, the frequency difference between consecutive longitudinal modes $\Delta\nu$ can be determined by $\Delta\nu = c_0/2nL$, where c_0 is light speed in vacuum, n is refractive index of medium and L is cavity length. In the multimode case, each mode has a phase $\varphi_i = \beta_i z = k_i n_i z + \varphi_{0i}$, where β_i is propagation constant of the ith mode, n_i,and k_i are refractive index and wave number for the mode, respectively. For each mode, φ_{0i} is the initial value of mode phase, which can have different values. As we can see from a following simple example, the output of such a laser depends critically on the phase relationship between these modes. If these modes have no phase correlation or these lasing modes have random phases, readers can easily verify for themselves by taking summation of these waves in equation (4.1), and obtaining that the output intensity is a cw beam with a large amount of intensity noise, as shown in figure 4.2. In the example shown in figure 4.1, we assume 20 lasing modes with equal amplitude $A_0 = A_i$ from a cavity having free spectral range FSR = 100 MHz (for more details of laser description, see reference [11]). At first glance the output in figure 4.2 is somewhat random with time, but we can see it has certain features—the waveform of the whole beam is periodic with a period $\tau_P = 1/\Delta\nu$, while a frequency domain detector would show us that the energy was contained in narrow spikes spaced evenly by FSR of the cavity. This calculation is simple and readers are encouraged to check for themselves the results.

Now let us see what will happen if we assume the relative phases of some lasing modes are fixed to a set value, i.e., the phases of these modes are correlated or locked when these modes circulate in the cavity. At this moment, we do not pay attention to how to lock phases of these modes. From equation (4.1) after some simple calculations with assumption that all the phases are fixed $\varphi_i - \varphi_j = \delta\varphi$, $i, j \in \{N\}_{\text{lock}}$, where $\{N\}_{\text{lock}}$ is for N locking modes. The result as an example is shown in figure 4.3. We can see these phase-locked modes interfere constructively at multiples of round-trip time in the cavity, while they destructively interfere for other frequencies. This process creates shorter pulses as the number of phase-locked modes increases.

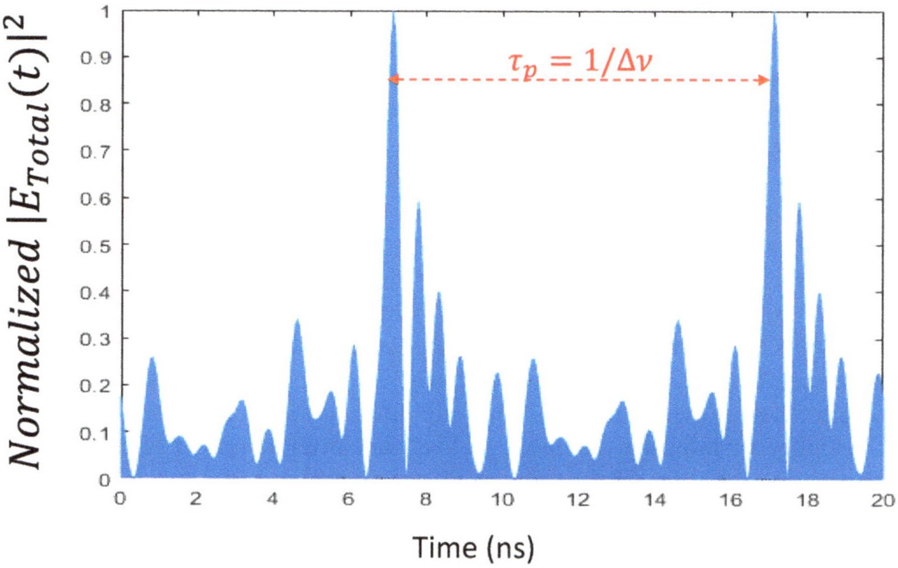

Figure 4.2. Time behavior of the squared amplitude of the total electric field with random phases or no phase coherence between 20 lasing modes of a cavity with FRS = 100 MHz.

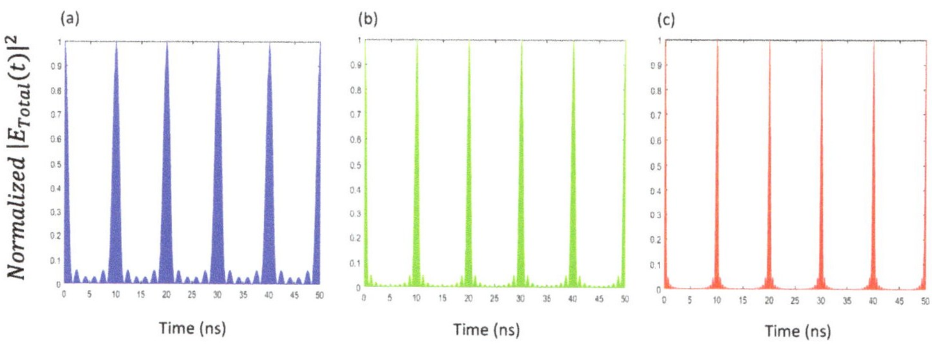

Figure 4.3. Time behavior of the squared amplitude of the total electric field where a number N of lasing modes are phase-locked $\Delta\varphi_{n,n+1} = \varphi_{n+1} - \varphi_n = 0$, with $n \in \{1, \cdots N\}$ lasing modes of the same cavity with $\tau_p = 10$ ns (FSR = 100 MHz). (a) $N = 5$. (b) $N = 10$ and (c) $N = 20$.

Note that the results from simple calculations in figure 4.3 assumed all lasing modes have the same amplitudes $A_i = A_0$, which mostly is not the case in reality. However, the results explain very well mathematically how locking phases of lasing modes can create short and even ultrashort pulse lasers. Figure 4.3 also shows that in the same cavity (resulting in the same FRS) the results where the larger the number of phase-locked lasing modes which span over broader frequency band, the shorter the laser pulses that can be generated.

From the results in figure 4.3 of the above simple calculations, we can now understand at least qualitatively that short pulse lasers can be generated by locking phases of lasing modes. Note that a number of different mechanisms (both active and passive) have been used for mode-locking lasers, which is a very big research area that we do not intend to go deeply into. A review of the mode-locked laser and its historical research and development can be found in reference [3]. Furthermore, a review of mode-locked fiber lasers which are more related to our examples in this chapter can be found in [12, 13].

Let us now turn to the description of some mechanisms for achieving phase locking of longitudinal modes of a resonator. First, it is worth noting from the examples in figure 4.3 that the mode-locked lasers produce ultrashort pulses at a rate τ_p, that is, one round-trip propagation time in the cavity. This suggests that the phase locking can be achieved by some parts or elements of the laser cavity that operate repeatedly for every circulation of light in the cavity. In general, an element/part that can introduce temporally periodic loss of the cavity can essentially do the job. In addition to that, if the element can provide intensity-dependent loss in such a way that the higher the intensity the lower the loss, then after a number of circulations the modes with low gain are likely suppressed due to higher loss, and only modes with highest gain would be able to get higher and higher energy and intensity. By doing that, the mode locking of longitudinal modes with high intensity (low loss) can be achieved. This loss-controlled element can be operated actively—corresponding to the so-called actively mode-locked lasers, and can be realized passively by an element/part known as an SA. The working principles of an SA can be described as simply as that its absorption is high/low absorption (loss) of low/high intensity light. In other words, the main functionality of SA is to decrease loss of light with increasing intensity. Figure 4.4 illustrates an example of a mode-locked laser operated with a nonlinear loss induced by an SA.

In the example shown in figure 4.4, a layer of SA is inserted inside a laser cavity whose gain (G) is determined by laser medium under pumping conditions. The total loss (L) of the cavity is a combination of typical losses such as $R1$, $R2$, background loss etc, which are constant with light intensity, and a loss by SA absorption, which is nonlinearly dependent on light intensity. Inside the cavity, all the modes with low intensity will suffer higher loss than gain, and are suppressed by the loss. Only the modes with highest intensity and satisfying condition $G > L$ can accumulate more energy and increase intensity when circulating inside the cavity. Light corresponding to these modes circulates in the cavity with round-trip time τ_p, and for each circulation, when the light hits the SA, the low intensity modes are suppressed and the high intensity modes get more gain. The process is repeated and provides a mechanism to lock all the lasing modes. As a result, the lasing modes can be locked with repetition rate τ_p, to produce output laser pulses. Note that dispersion also plays an important role in mode locking and will be discussed later in this part.

4.2 General model of mode-locked fiber lasers

Before studying examples of mode-locked fiber lasers with MATLAB® programs, let us first describe qualitatively a general configuration of a mode-locked fiber laser. In figure 4.5, a mode-locked fiber ring cavity that includes several important components such as SA, gain fiber, dispersion compensated fiber (DCF), output

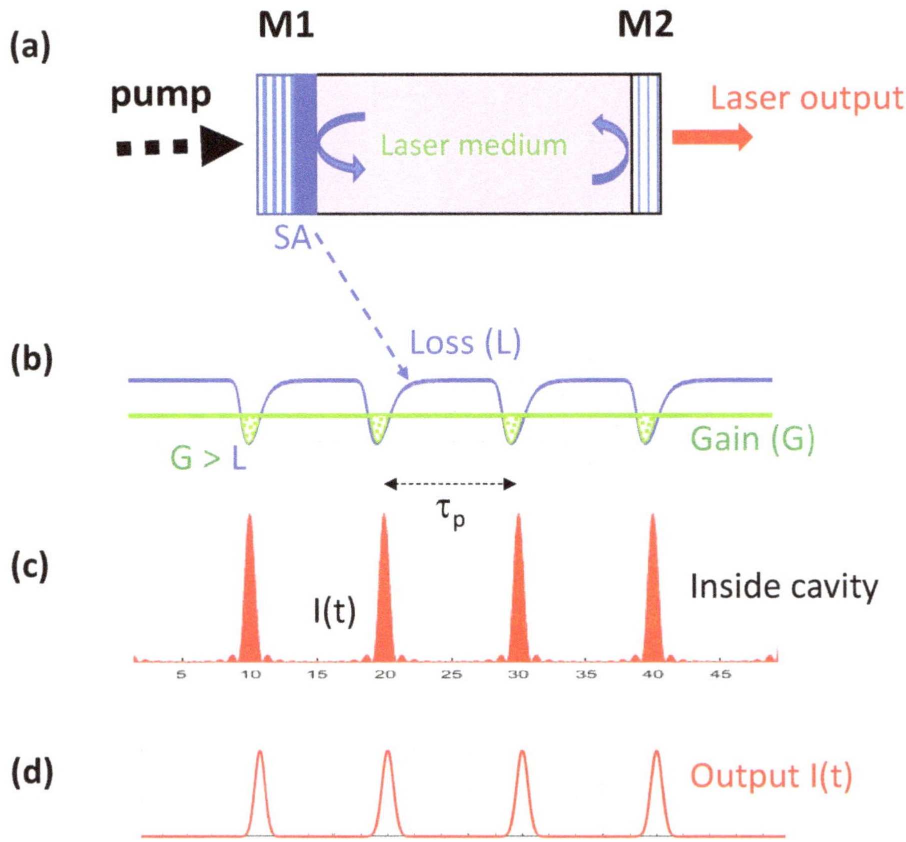

Figure 4.4. Illustration of mode-locked laser pulses generated with nonlinear loss by SA. (a) A layer of SA is inserted inside a conventional cavity. (b) Time behavior of the total loss includes nonlinear saturable absorption (blue) of SA, and gain (green). Modes with low intensity will be suppressed due to loss being higher than gain. Absorption reduces for high intensity modes, and only modes satisfying $G > L$ can get more gain to circulate inside the cavity. (c) The lasing modes are locked in phase as they circulate and hit SA repeatedly. (d) The mode locking inside the cavity produces the output lasers pulses.

coupler (OC), polarization controller (PC) etc. Note that the fiber ring configuration has been widely used in practice but linear cavities are very popular too, and we will study examples with both configurations. As shown in figure 4.5, the active fiber (red) plays a role for gain medium, an integrated device of the SA providing saturable absorption, the SMF (black) is the single-mode fiber for integrating all parts of the cavity, DCF (blue) is a dispersion compensated fiber providing dispersion compensation for other parts of the cavity. Other parts of the cavity include WDM (wavelength-division multiplexer), PC (polarization controller), ISO (isolator) and OC (output coupler).

Modeling and simulation of the mode-locked ring fiber laser (figure 4.5) essentially is numerical solving a propagation equation of light in complicated systems. The equation must describe well the nonlinear absorption by SA, dispersion, gain effects, and nonlinear propagation. One of the most important properties of laser

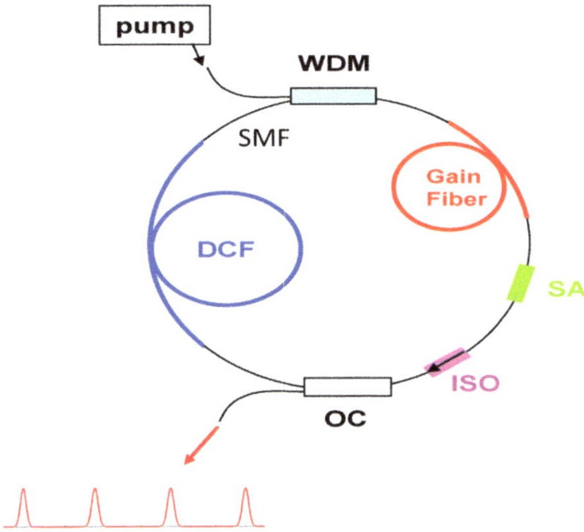

Figure 4.5. Illustration of mode-locked fiber ring cavity including gain fiber (red), dispersion compensated fiber (DCF—blue) and passive single-mode fiber (SMF—black). Other parts of the cavity include PC: polarization control, ISO: isolation, OC: output coupler, SA: saturable absorber.

pulses is that the laser intensity can reach to very high or even extremely high. As a result, the nonlinear index term $n_2 I$ becomes important and cannot be ignored in the theoretical model. As will be seen in the following, we will use the NLSE or more generally, the GNLSE for describing light propagation in a mode-locked laser system. Figure 4.6 shows numerical solutions of the GNLSE as an example of the behavior of a mode-locked ring fiber laser.

In figure 4.6, the plot on the top shows the evolution of saturable absorption versus round-trip number N of light circulation. Meanwhile, the middle plot shows the build-up intensity inside the cavity, and the bottom plot shows the intensity in wavelength domain. Let us take a first look at the results shown in figure 4.6. In all three plots, the green curves are assumed as the initial values. For example, the initial value of absorption is the linear absorption assumed for low intensity. In the above example the linear absorption is assumed as $\alpha_0 = 60\%$ (green curve) and the saturated absorption is 52% (or modulation $\Delta\alpha = 8\%$). The initial values of light intensity are assumed to be noise or spontaneous emission from gain medium as shown by green curves in the middle and bottom plots for time and wavelength domain, respectively. At first, when light intensity is very weak the absorption roughly is close to the linear absorption α_0 as shown for the first 10 round trips in the plot of absorption. As the light gets higher amplification with increasing number of circulations in the cavity (round trips), the light intensity increases as shown in the middle plot (in time domain) and bottom plot (frequency domain), and the absorption behavior change dramatically. In the time domain, the absorption reaches the saturated values when intensity reaches a certain value and in return the loss by the intensity-dependent absorption suppresses the low-intensity light and

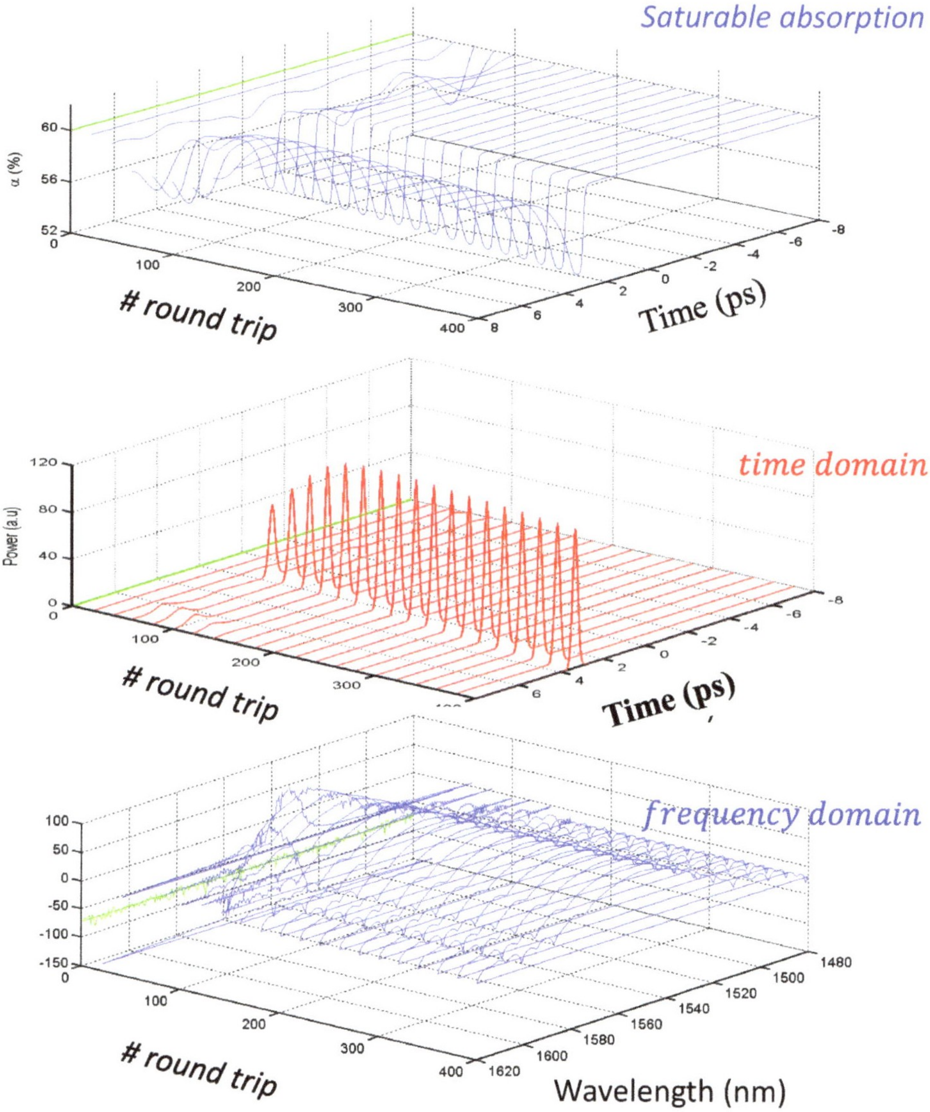

Figure 4.6. Examples of numerical results of modeling and simulation of a mode-locked fiber laser. Top: evolution of saturable absorption at different round-trip number of light build-up inside the cavity. Middle: light intensity in time domain is built up at different round-trip number at the OC of the cavity; and bottom: light intensity in frequency (wavelength) domain.

locks the high-intensity light. As a results, the light is transformed from noisy to pulses. As shown in the bottom plot, in this example the spectrum of the light pulse is not Gaussian but has some features. As will be seen later, those special features of the pulse spectrum are related to the total dispersion of the cavity. It turns out that cavity dispersion plays an important role for pulse duration, pulse shapes in

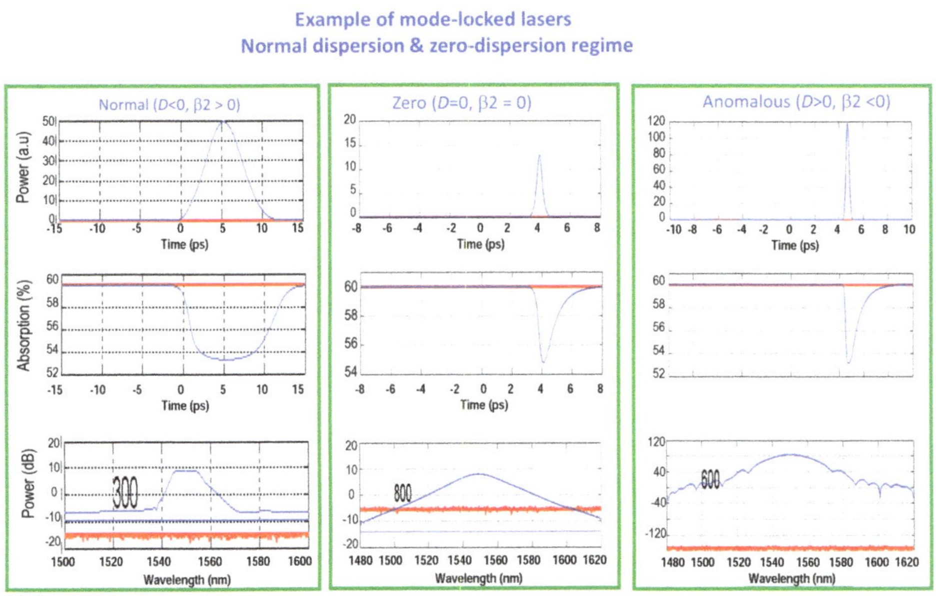

Figure 4.7. Examples of numerical results of modeling and simulation of mode-lock fiber lasers in three dispersion regimes. All panels: laser pulse inside cavity in time domain (top), absorption (middle) and laser spectrum (bottom). Left panel: normal dispersion, and results are calculated at number of circulations inside cavity $N = 300$. Middle panel: zero-dispersion with $N = 800$, and Right panel: anomalous dispersion with $N = 600$. In all figures, red lines are the starting values of light as noise in cavity in time domain (top) and frequency domain (bottom), and linear value of absorption, in these examples $\alpha_0 = 60\%$.

mode-locked lasers. Other parameters such as gain, nonlinear index n_2, saturated energy of SA etc, also play very important roles in laser operation such as for pulse duration, laser stability etc.

It is worth noting that the gain fiber in the configuration in figure 4.5 can be different doped-material fibers, depending on laser operation wavelengths; for example, Er-doped fiber (EDF) and Yb/Er co-doped fiber for $\lambda \sim 1.55$ μm, Yb-doped fiber ($\lambda \sim 1$ μm) and Tm-doped or Tm/Ho co-doped fibers (~2 μm) etc. When modeling and designing the laser cavity, it is important to know the gain fibers and also all components that work for the relevant wavelengths. These components can have very different losses, therefore, it is necessary to take these differences seriously in designs.

For the total dispersion, it is necessary to know the dispersion of all components in the cavity. Usually, the dispersion of fibers can be found from specifications by manufacturers. However, in many cases, the designer should know how to calculate the dispersion of the fiber. Once we know dispersion of each fiber, we can adjust the lengths of fibers to have a desired total dispersion of the cavity. The total dispersion of cavity is an important factor for laser operation, as shown in figure 4.7. There are three different regimes of laser operation depending the total dispersion:

1. $D > 0$ ($\beta_2 < 0$) anomalous dispersion;
2. $D = 0$ ($\beta_2 = 0$) zero-dispersion;
3. $D < 0$ ($\beta_2 > 0$) normal dispersion.

Figure 4.7 shows examples of the simulation results of mode-locked fiber lasers in three different dispersion regimes. In each panel of figure 4.7, the top figure shows the laser pulse in time domain, the middle one shows absorption and the bottom figure shows the spectrum. The number in the bottom figure is the number of round trips of light in the cavity. The red curves are initial values of intensity and absorption.

Let us first examine the most important parts of the cavity that we will describe in our model and simulation works.

4.2.1 Saturable absorption

In general, the temporal behavior of loss due to a SA $\alpha(t)$ can be described by the following equation [4, 5]

$$\frac{d\alpha(t)}{dt} = -\frac{\alpha(t) - \alpha_0}{\tau_S} - \frac{\alpha(t)P(t)}{E_{sat}}. \tag{4.2}$$

Here, α_0 is unsaturated loss—or linear absorption, τ_S is recovery time of the SA, $P(t)$ is light power and E_{sat} is saturated energy of the SA. It is clear that the saturable absorption/loss is intensity-dependent.

In slow SA regime, pulse duration is much slower than recovery time $\tau \ll \tau_S$. Within a pulse duration we have

$$\frac{d\alpha(t)}{dt} \approx -\frac{\alpha(t)P(t)}{E_{sat}}, \tag{4.3}$$

For integration equation (4.2), we have the analytical solution

$$\alpha(t) = \alpha_0 \exp\left[-\frac{E_P}{E_{sat}}\int_0^t f(t')dt'\right]. \tag{4.4}$$

Here, for laser pulses $P(t) = E_P f(t)$, where $\int_0^{T_R} f(t')dt' = 1$.

Meanwhile, in the opposite regime— the fast SA, where $\tau \gg \tau_S$ equation (1) can be written approximately as

$$0 = \frac{\alpha(t) - \alpha_0}{\tau_S} + \frac{\alpha(t)P(t)}{E_{sat}}, \tag{4.5}$$

and the analytical solution can also be obtained as

$$\alpha(t) = \frac{\alpha_0}{1 + I(t)/I_{sat}}. \tag{4.6}$$

Here $P_{sat} = E_{sat}/\tau_S$ is saturable power and $P(t)/P_{sat} = I(t)/I_{sat}$ with I_{sat} being saturated intensity.

The analytical solutions above are very useful for quick estimates and have been usually used, especially the solution of fast SA regime, e.g., equation (4.6). However, in reality the SA may be in between the two opposite regimes, and numerical solution is necessary. In figure 4.8 we show numerical solutions of equation (4.2) for

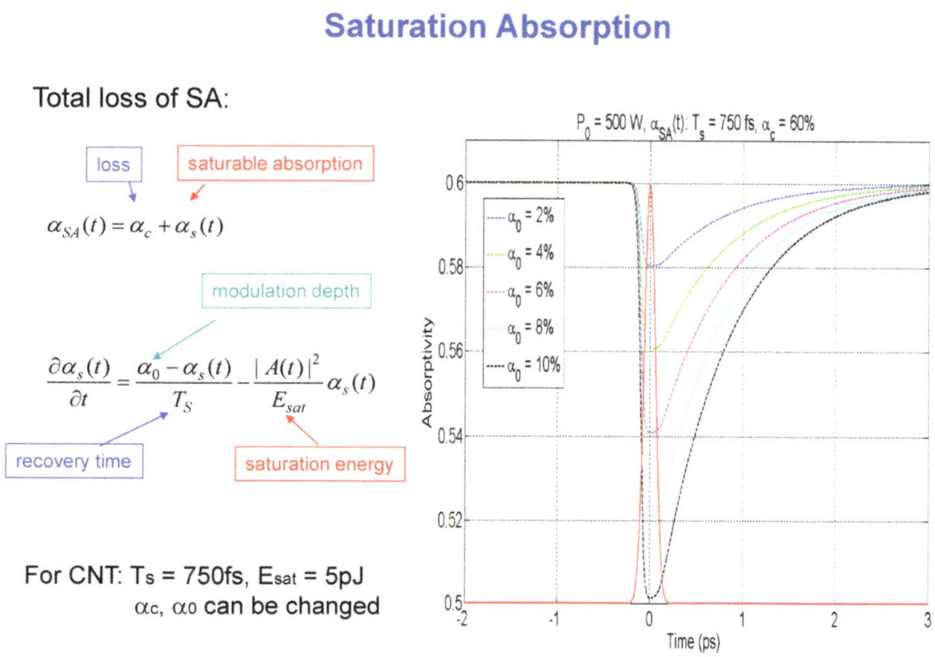

Figure 4.8. Numerical calculation of saturable absorption of a SA versus modulation depth $\Delta\alpha$ with recovery time τ_S = 750 fs, E_{sat} = 5 pJ, α_0 = 60% under excitation by a laser pulse with τ = 500 fs, P_0 = 500 W.

an SA of recovery time τ_S = 750 fs under excitation of a laser with pulse duration τ = 500 fs with peak power of P0 = 1 and 10 kW on the left and right figures, respectively. It is clear that this situation is neither slow nor fast SA. The temporal behavior of the saturable absorption with different modulation depth $\Delta\alpha$, is the difference between losses in the limits of low and high intensity. Therefore, while the analytical solution in fast SA is more often used in description of saturable absorption, the numerical one shown in figure 4.8 is very useful in simulation of the whole mode-locked laser system. We will see later in our examples that $\Delta\alpha$ is an important parameter for mode-locked laser operation, and it usually is determined by experiment.

In solving equation (4.2) we will need other parameters—the saturated energy E_{sat} and modulation depth $\Delta\alpha$—which are material parameters and are determined by experiments. Note that, both modulation depth $\Delta\alpha$ and the saturated energy of SA are also important parameters for laser operation. A good modeling and simulation method of mode-locked lasers should take into account these parameters in describing the mode-locked laser system. Moreover, the saturated energy is important to determine the point when the laser becomes unstable. Qualitatively, it can be understood as follows. In the high gain regime, the light intensity inside cavity would reach a high value that the energy of the pulse can be higher than the saturated energy E_{sat} of SA. In such a situation, the mode-locking function of SA will not operate normally. As a result, the mode-locked laser pulses become unstable. Because of that, in mode-locked lasers, the gain and saturated energy much be

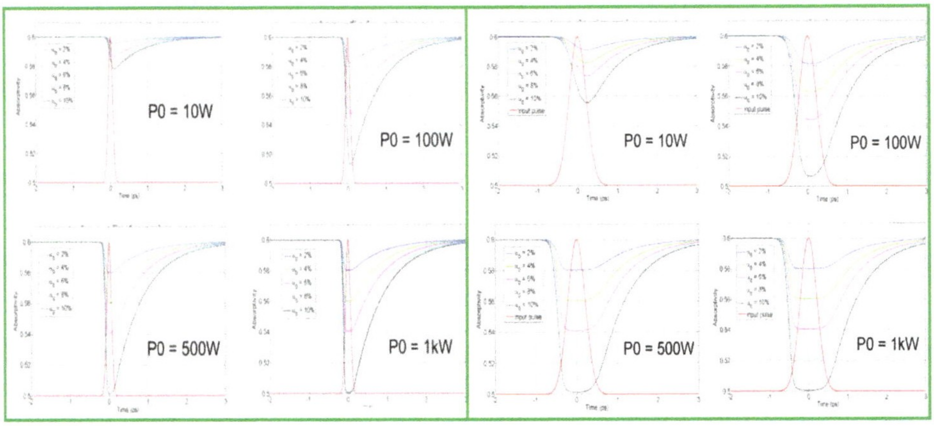

Figure 4.9. Numerical calculation of saturable absorption of a SA versus modulation depth $\Delta\alpha$ with recovery time τ_S = 750 fs, E_{sat} = 5 pJ, α_0 = 60% under excitation by a laser pulse with τ = 200 fs (left panel) and 500 fs (right panel), with different peak powers.

chosen carefully. We now know numerical solving SA incorporated in solving GNLSE is necessary for modeling and simulating the mode-locked laser. We will learn more of these numerical solutions in Exercise 1 later in this chapter.

The results in figure 4.8 not only show the temporal behavior of the SA absorption driven by temporal shape of the laser pulse, but also the saturation of absorption in high peak power. For example, the absorption $\alpha = \alpha_0$ = 60% in zero-power limit (non-saturated value) and α decreases with increasing power. In this case with P_0 = 500 W, the absorption almost reached the saturated values which equal α_0—$\Delta\alpha$. Note that, the saturated values of absorption strongly depend on the saturated energy of the SA, and this plays an important role for stability of the mode-locked lasers, as we will see in the following examples. For example, figure 4.9 shows the saturable absorption of the same SA with saturated energy E_{sat} = 5 pJ, recovery time T_s = 750 fs, α_0 = 60% under excitation by a laser pulse with τ = 200 fs (left panel) and τ = 500 fs (right panel) with different peak powers and modulation depths $\Delta\alpha$. The behavior of SA under interaction with the light circulating in the cavity will also determine the characterization of the laser pulse at the output.

At this point we want to stress that the topics of saturable absorption and dispersion are very broad areas of optical material sciences, and we do not try to cover it in this chapter. Readers can easily find many references and textbooks for comprehensive reviews of theory, experiment and application aspects of these topics. Here, we only want to introduce the concept of saturable absorption to readers unfamiliar with the concept. Although only basic characteristics of SA are presented, we believe they are enough for readers to understand and make use of it in the modeling and simulation works later in this chapter.

As stated earlier, modeling and simulation of the whole mode-locked laser systems, especially the mode-locked fiber lasers, require some basic understanding of SA and others such as dispersion, nonlinear propagation of light pulses that is described by the NLSE. In the following, we will describe briefly the physics and mathematics models of these problems. Our description of these topics will be simple and introductorily brief aiming to prepare for beginners some basics for understanding the examples and the MATLAB® programs. We strongly encourage readers to study these topics in detail in several outstanding references [4, 5].

4.2.2 Material and waveguide dispersions

So far, our attention has been on saturable absorption in making mode-locking. However, it turns out another effect of the cavity also plays a crucial role in locking multiple longitudinal modes of the cavity. From earlier discussion and description, especially in simple examples of the calculations shown in figures 4.3–4.4, we are now able to understand at least qualitatively that locking phases of lasing modes, e.g., the modes satisfy lasing condition (Gain > Loss) can produce short pulse lasers. In the examples, when fixing the relative phases of the lasing modes, we have implicitly assumed that all these modes circulate inside the cavity with the same speed $v_i = c/n = const$, where v_i is speed of the ith mode in the cavity, c is speed of light in vacuum and n is refractive index (RI) of the material filling the cavity. From fundamentals of physics, only RI of vacuum is const $n = 1$ and in such condition (cavity filled with vacuum) all resonant modes have $v_i = c = const$. Otherwise, all materials would have *chromatic dispersion* or material dispersion, which is the dependence of RI on wavelength (frequency) of light $n(\lambda)$ (or $n(\omega)$) as shown by the blue curve in figure 4.10 for RI $n(\lambda)$ of a phosphate glass.

As can be seen from figure 4.1, the longitudinal modes of a cavity can span in a band of wavelength (frequency) or a region of wavelength (frequency), in which each mode is resonant in a very narrow bandwidth $\Delta\lambda$ around the resonant wavelength λ_i. Due to the material dispersion each mode has different speed $v_i = v(\lambda_i) = c/n(\lambda_i)$ where material dispersion $n_i = n(\lambda_i)$ is wavelength-dependence. When these modes

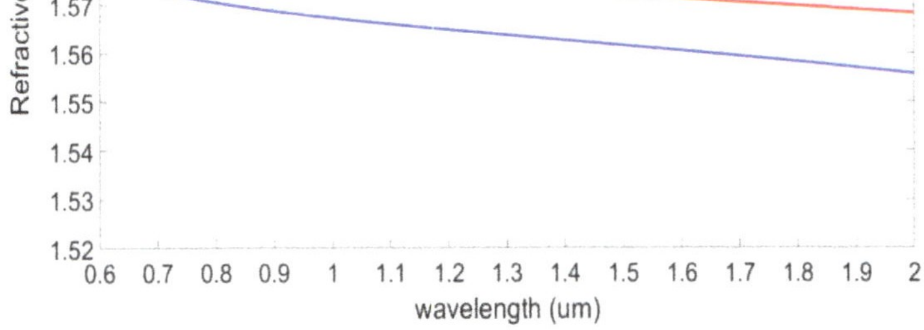

Figure 4.10. Example of refractive index $n(\lambda)$ and group index of a phosphate glass fiber.

propagate in the same length distance L inside the cavity their phases can be written as (ignoring the initial values)

$$\varphi_i = k_i n_i L = \frac{2\pi}{\lambda_i} n_i L. \tag{4.7}$$

Although the differences of RI are quite small for resonant modes which usually span in narrow band of wavelength (up to few tens of nm), these differences become significant when these resonant modes circulate multiple round trips inside the cavity without any mode-locking mechanism. Even with locking for each circulation, the phases are not completely locked with all $\Delta\varphi_{ij} = \varphi_i - \varphi_j \equiv 0$ for all i, j due to the dispersion, as indicated by equation (4.7). Because of that, the dispersion of a cavity strongly impacts the locking phases of these modes. We will study the dispersion effects in mode-locked fiber lasers in the examples later in this chapter.

Designing photonics devices, especially any devices that deal with ultra-short lasers requires a good knowledge of dispersion of the system. We will not always obtain dispersion from manufacturers and publications or even by measurements. Sometimes designers must calculate dispersions, especially the waveguide dispersions, which are essential tasks in designing. In the following, we will present a simple way to calculate the waveguide dispersion in optical fibers. As we will study the mode-locked fiber lasers whose cavities consist of many components made from glasses such as rare-earth doped glass fibers (active fibers), and guiding and dispersive fibers (passive fibers), we should pay attention to some fundamentals of the dispersion of glasses. From figure 4.7, we can see the characterizations of mode-locked lasers can be very different depending on the total dispersion of the cavity. Therefore, mastering dispersion relations in cavities is very useful not only for modeling but also optimizing mode-locked lasers. The most well-known formula for calculating material dispersion is the Sellmeier formula, found in 1871 by Wolfgang Sellmeier [11, 14], which showed that the refractive index dependence on wavelength is very similar in many transparent and non-transparent media in the visible and near infrared range expressed as

$$n^2(\lambda) = 1 + \sum_{i=1}^{3} A_i \frac{\lambda_i^2}{\lambda^2 - \lambda_i^2}. \tag{4.8}$$

Equation (4.8) is the well known Sellmeier equation, where λ is the wavelength, and A_i and λ_i are experimentally determined Sellmeier coefficients. These coefficients are usually quoted for λ in micrometers. Today the Sellmeier formula is widely used in optical science and optical industry to describe and characterize the dispersion in glasses and crystals. Many companies who offer optical materials deliver their products together with the corresponding Sellmeier coefficients. In many cases, especially for working on modeling and design of mode-locked lasers, it is often necessary to know dispersion $n(\lambda)$, therefore, the Sellmeier equation can be very useful.

For practical use it is often useful to use derivatives of refractive index, i.e. group index

$$n_g = n(\lambda) - \frac{dn(\lambda)}{d\lambda}, \tag{4.9}$$

and group velocity dispersion (GVD) resulting in the material dispersion parameter D_{Material}.

$$D_{\text{Material}}(\lambda) = \frac{d}{d\lambda}\left(\frac{n_g(\lambda)}{c}\right) = -\frac{\lambda}{c}\frac{d^2n(\lambda)}{d\lambda^2}. \tag{4.10}$$

In many cases, especially in fiber lasers, cavities are made up of active and passive waveguides whose dispersion is a total sum both material dispersion, e.g., equation (4.10), and waveguide dispersion, defined as

$$D_{\text{Waveguide}} = -\frac{n_{\text{clad}}\Delta n}{\lambda c} \times \frac{Vd^2(bV)}{dV^2}. \tag{4.11}$$

Here $V = 2\pi a\sqrt{n_{\text{core}}^2 - n_{\text{clad}}^2}/\lambda$ is V-number, a is core radius of the waveguide. For simplicity we only use the formula for single-mode and step index fiber. A more comprehensive theory of dispersion of waveguides, readers can find in many textbooks, for example [8, 11, 14]. We will see in this chapter and the next chapter the importance of dispersion in dealing with pulse lasers. In figure 4.10, an example of material and waveguide dispersions of an SM phosphate glass fiber with $a = 3$ μm, $n_{\text{clad}} = 1.5600$, $n_{\text{core}} = 1.5660$ at 1.552 μm.

One important parameter of dispersion is so-called zero-dispersion wavelength, which is the point where the dispersion curve is zero $D(\lambda_0) = 0$. In figure 4.11 the zero-dispersion of material and total dispersion is $\lambda_{0D} \sim 1.42$ μm and is $\lambda_0 \sim 1.556$ μm, respectively. Note that in SM fiber with single cladding the waveguide dispersion is usually negative (see the green curve in figure 4.10). Therefore, the total dispersion (red curve) $D_{\text{Total}} = D_{\text{Material}} + D_{\text{Waveguide}}$ of a waveguide is usually lower than the material

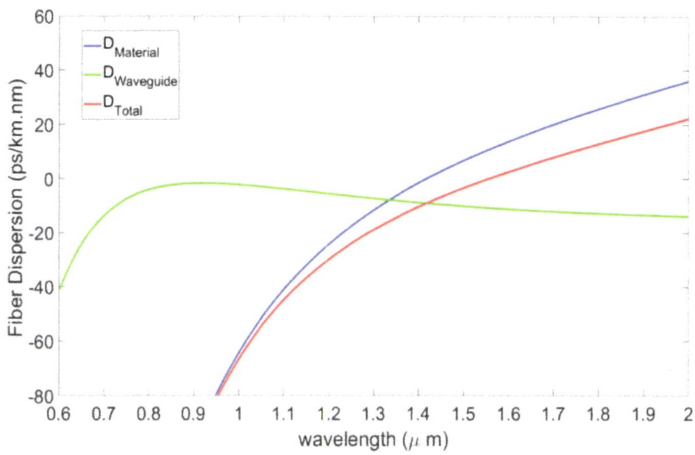

Figure 4.11. Example of dispersion of a phosphate glass fiber $a = 3$ μm, $n_{clad} = 1.5600$, $n_{core} = 1.5660$ at 1.552 μm.

dispersion, and as a result the zero-dispersion wavelength of the total dispersion is pushed to longer than the material dispersion $\lambda_0 > \lambda_{0D}$. It is easy to see that when waveguide core is larger the total dispersion is closer to the material dispersion. This feature is important to understand because in many cases we want to manipulate the dispersion of the whole cavity and smaller core waveguide will push zero-dispersion wavelength λ_0 to the longer side. Sometimes we would like to have λ_0 changes in the opposite direction. In such situations multiple claddings or different fiber structures such as grating fibers are needed. More information on these dispersion compensate devices can be found in [8]. In Exercise 2 we will work out and obtain the results in figures 4.10 and 4.11.

Let us now present the basics of the NLSE, which is our main equation for modeling mode-locked fiber lasers.

4.2.3 Nonlinear Schrödinger equation

The NLSE has been widely used to simulate pulse propagation in photonics/optics structures including optical fibers. The theoretical foundations of the NLSE for description of nonlinear propagation have extensively been described in many textbooks, for example [8]. Readers are encouraged to read these references to understand comprehensively the foundations of the NLSE and how it can be used to modeling the pulse propagation, especially in optical fibers. In this chapter, the NSE is used for modeling the mode-locked fiber lasers, therefore, a brief description of the equation is introduced. Solving the NLSE with SA and dispersion device and other parts in the laser cavities essentially is the main job of the simulation of mode-locked fiber lasers. We will present the MATLAB® programs for studying examples later in this chapter.

In general, the NLSE has usually been used to describe the nonlinear propagation of light pulses in situations when only second-order dispersion and self-phase modulation (SPM) are taken into account. As stated earlier, fiber dispersion plays a very important role in evolution of short laser pulses because different spectral components associated with the pulse travel at different speeds given by $v = c/n(\omega)$ or $c/n(\lambda)$, where c is light speed in vacuum. It is easy to see that even when the nonlinear effects are negligibly weak and can be ignored, dispersion-induced effects can still play a critical role in the evolution laser pulses in propagation. In the nonlinear regime, where the nonlinear term in refractive index $n = n_0 + n_2 \cdot I$ is not negligible, the combination of dispersion and nonlinearity can result in dramatically different changes of pulse evolution. Those have been detailed and discussed in many well-known references, e.g., [8, 11]. In the following, we will again want to present a brief and simple introduction to the derivation of the NLSE. Readers, especially beginners in this field, can find details of the derivation in [8]. However, the simple derivation as presented below will be very useful for readers to understand the basics of the NLSE, which is our main equation in modeling mode-locked fiber lasers in the examples.

For simplicity, let us first restrict ourselves to the problem of pulse propagation in a single mode fiber having propagation constant β. From the theory of wave propagation [8, 11], the electric field of the linearly x-polarized light that propagates in the z-direction can be written as $\vec{E}(x, y, z, \omega) = \vec{x}A(\omega)f(x, y)\exp[i\beta(\omega)z]$. In the following, we want to show in the simplest way how the effects of fiber dispersion can be described mathematically in the NLSE. A more general and comprehensive treatment can be found in the above mentioned references. Mathematically, we can expand the mode-propagation constant $\beta(\omega)$ in a Taylor series about the frequency ω_0 at which the pulse spectrum is centered:

$$\beta(\omega) = n(\omega)\frac{\omega}{c} = \beta_0 + \beta_1(\omega - \omega_0) + \frac{1}{2}\beta_2(\omega - \omega_0)^2 + \frac{1}{6}\beta_3(\omega - \omega_0)^3 + \cdots, \quad (4.12)$$

where

$$\beta_m = \left(\frac{\partial^m \beta}{\partial \omega^m}\right)_{\omega=\omega_0}, \; m = 0, 1, 2 \ldots. \quad (4.13)$$

It is easy to verify that parameters β_1 and β_2 are related to the refractive index n and its derivatives through the relations

$$\beta_1 = \frac{n_g}{c} = \frac{1}{c}\left(n + \omega\frac{dn}{d\omega}\right) = \frac{1}{v_g}, \quad (4.14)$$

$$\beta_2 = \frac{1}{c}\left(2\frac{dn}{d\omega} + n\frac{d^2\beta}{d\omega^2}\right). \quad (4.15)$$

where n_g is the group index and v_g is the group velocity. As described in [8] the envelope of an optical pulse moves at the group velocity while the parameter β_2 represents dispersion of the group velocity and is responsible for pulse broadening, which is an important effect in pulse propagation. This phenomenon is the GVD, and β_2 is the GVD parameter. It is worth mentioning here that in the fiber-optics literature the dispersion parameter D plotted in figure 4.11 is commonly used instead of β_2. It is related to β_2 by the relation

$$D = \frac{d\beta_1}{d\lambda} = -\frac{2\pi c}{\lambda^2}\beta_2 \approx \frac{\lambda}{c}\frac{d^2 n}{\partial \lambda^2}. \quad (4.16)$$

Let us now derive the NLSE in the simplest way, meaning that we do not start from the beginning with Maxwell equations which we believe readers should know in more or less detail. Instead, we will start from a well known wave equation for the slowly varying amplitude $E(\vec{r}, t)$ but in the Fourier domain $\tilde{E}(r, \omega)$, which satisfies the Helmholzt equation [8]

$$\nabla^2 \widetilde{E} + \varepsilon(\omega)k_0\widetilde{E} = 0, \quad (4.17)$$

where $k_0 = \omega/c$, and

$$\varepsilon(\omega) = 1 + \tilde{\chi}^{(1)}(\omega) + \varepsilon_{NL}. \quad (4.18)$$

Using $\varepsilon(\omega) = \{n + i\alpha/2k_0\}^2$ with $n = n_0 + n_2|E|^2$ and $\alpha = \alpha_0 + \alpha_2|E|^2$ where $n_0(n_2)$ and $\alpha_0(\alpha_2)$ are linear (nonlinear) refractive index and absorption coefficient, respectively. In most of the cases of interest, the dielectric constant in equation (4.18) can be approximated by

$$\varepsilon = (n_0 + \Delta n)^2 \approx n_0^2 + 2n_0\Delta n = n_0^2 + 2n_0\left(n_2 I + \frac{i\alpha_0}{2k_0}\right), \tag{4.19}$$

where the nonlinear absorption is assumed negligibly small. The solution of equation (4.15) can be found in the form [8]

$$\tilde{E}(x, y, z, \omega - \omega_0) = F(x, y)\tilde{A}(z, \omega - \omega_0)e^{i\beta_0 z}, \tag{4.20}$$

where $\tilde{A}(z, \omega - \omega_0)$ is a slowly varying function of z and β_0 is the wave number to be determined later. Using equations (4.20) and (4.16) we have

$$\frac{\partial^2 F}{\partial x^2} + \frac{\partial^2 F}{\partial y^2} + \left\{\varepsilon(\omega)k_0^2 - \tilde{\beta}^2\right\}F(x, y) = 0, \tag{4.21}$$

$$2i\beta_0\frac{d\tilde{A}}{dz} + \left(\tilde{\beta}^2 - \beta_0^2\right)\tilde{A} \simeq 2i\frac{d\tilde{A}}{dz} + (\beta + \Delta\beta - \beta_0)\tilde{A} = 0. \tag{4.22}$$

Note that we have ignored the term $\partial^2\tilde{A}/\partial z^2$ in equation (4.22) since $\tilde{A}(z, \omega)$ is assumed to be a slowly variation function in z. The wave number $\tilde{\beta} = \beta + \Delta\beta$ is determined by solving the eigenvalue equation for the fiber modes which is not the subject of our study, and it can be found in the above references. In obtaining equation (4.22) we have also used the approximation $\tilde{\beta}^2 - \beta_0^2 \simeq 2\beta_0(\tilde{\beta} - \beta_0) = 2\beta_0(\beta + \Delta\beta - \beta_0)$, where $\Delta\beta$ is given by

$$\Delta\beta = \frac{k_0\iint_{-\infty}^{+\infty} \Delta n|F(x, y)|^2 dxdy}{\iint_{-\infty}^{+\infty} |F(x, y)|^4 dxdy}. \tag{4.23}$$

If we take Fourier transform of equation (4.22) with

$$A(z, t) = \frac{1}{2\pi}\int_{-\infty}^{+\infty} \tilde{A}(z, \omega - \omega_0)\exp[-i(\omega - \omega_0)t]d\omega, \tag{4.24}$$

and using Taylor expansion (4.12) of propagation constant $\beta(\omega)$, then we obtain

$$\frac{dA}{dz} = -\beta_1\frac{\partial A}{\partial t} - i\frac{\beta_2}{2}\frac{\partial^2 A}{\partial t^2} + i\Delta\beta A. \tag{4.25}$$

Equation (4.25) can be approximated further using (4.23)

$$\frac{dA}{dz} + \beta_1\frac{\partial A}{\partial t} + i\frac{\beta_2}{2}\frac{\partial^2 A}{\partial t^2} + \frac{\alpha_0}{2}A = i\gamma|A|^2 A. \tag{4.26}$$

Here, in equation (4.26) the nonlinear parameter γ is defined as $\gamma = \frac{n_2\omega}{cA_{eff}}$ with A_{eff} is effective core area and is defined as

$$A_{eff} = \frac{\left(\iint_{-\infty}^{+\infty} F(x, y)|dxdy\right)^2}{\iint_{-\infty}^{+\infty} |F(x, y)|^4 dxdy}. \tag{4.27}$$

Equation (2.26) can be simplified further if we write it in retarded frame $T = t - z/v_g \equiv t - \beta_1 z$, which is a frame of reference moving with the pulse at the group velocity v_g, then we have

$$\frac{\partial A(z, T)}{\partial z} = -\frac{\alpha}{2}A(z, T) - i\beta_2\frac{\partial^2 A(z, t)}{\partial T^2} + i\gamma|A(z, T)|^2 A(z, T). \tag{4.28}$$

Equation (4.28) or the NLSE governs the propagation of light in an optical fiber exhibiting second-order dispersion β_2, γ is the nonlinear coefficient (proportional with nonlinear refractive index n_2 and optical power loss α. In many cases, this equation is good enough to describe the propagation of laser pulse in optical fibers. However, in deriving this NLSE several effects originated from other nonlinear processes and higher order dispersion have been neglected.

To accurately describe the process of pulse propagation, addition of terms representing higher order dispersions $\beta_3, \beta_4 \cdots$, stimulated Raman scattering, and other processes is necessary [2, 3]. Depending on specific problems, usually we have to deal with more complicated equations, the GNLSEs [8–10, 15]. GNLSE equations can be derived from analytical simplifications of Maxwell's equations. A form of the GNLSE that is commonly employed for numerical simulations of complicated processes in nonlinear materials such as supercontinuum generation is [9, 16]

$$\begin{aligned}\frac{\partial A(z, T)}{\partial z} &= -\frac{\alpha}{2}A(z, T) - \left(\sum_{n\geqslant 2}\beta_n\frac{i^{n-1}}{n!}\frac{\partial^n A(z, t)}{\partial T^n}\right) + \\ &+ i\gamma\left(1 + \frac{1}{\omega_0}\frac{\partial}{\partial T}\right) \times \{(1 - f_R)\}A|A|^2 + f_R A\int_0^{\infty} h_R(\tau)|A(z, T - \tau)|^2 d\tau\end{aligned} \tag{4.29}$$

Here, in (4.29) β_n are the higher order dispersion coefficients obtained by a Taylor series expansion (4.11) of the propagation constant $\beta(\omega)$ around the center frequency ω_0. The first and second terms on the right-hand side of (4.29) describe fiber loss and dispersion, respectively. The third term describes the nonlinear effects. The temporal derivative in this term is responsible for self-steepening and optical shock formation, whereas the convolution integral describes the delayed Raman response, which leads to effects such as intra-pulse Raman scattering [8–10, 17]. The Raman response function $h_R(t)$ is given by [8, 17]

$$h_R(t) = \frac{\tau_1^2 + \tau_2^2}{\tau_1\tau_2^2}\exp(-t/\tau_2)\sin(t/\tau_1). \tag{4.30}$$

In (4.29) and (4.28) the fractional contribution of the delayed Raman response to nonlinear polarization is represented by f_R, and parameters τ_1 and τ_2 are two adjustable parameters that are chosen to provide a good fit to the actual Raman-gain spectrum. In silica glass fibers their appropriate values $f_R = 0.18$ and $\tau_1 = 12.2$ fs and $\tau_2 = 32$ fs are usually assumed [8].

As can be seen, the NLSE or GNLSE is the equation we need to describe all effects including absorption/loss, dispersion and nonlinear propagation of pulses that occur in mode-locked fiber lasers. Let us now study the numerical method that we will use to solve the NLSE and GNLSE in the task of modeling the mode-locked fiber lasers.

4.2.4 Fourth-order Runge–Kutta in interaction picture method

In general, the NLSE and GNLSE are nonlinear PDEs that, in most cases of practical interest, cannot be solved analytically, and numerical approaches are therefore needed. Numerical modeling of pulse propagation in different photonics systems not only helps understanding the evolution of the pulses but can also be used to guide fiber design and system architecture in order to achieve desired performances of the systems. Numerical solutions of the NLSE, especially GNLSE are often time-consuming, and sometimes costly in terms of hardware requirements. Therefore, efficient numerical schemes for solving the NLSE and GNLSE are desirable.

As stated earlier, our tasks of modeling and simulation of mode-locked fiber lasers are essentially solving the NLSE with SA and dispersion device and other parts in the laser cavities. As we can imagine, the modeling and simulation of mode-locked lasers are quite complicated for light propagation in non-uniform materials and different structures. Furthermore, the nonlinear absorption due to SA, which is the key for the mode-locking process makes the situation even more complicated. Because of that, modeling and simulation of mode-locked lasers are very challenging. One of the main reasons for the difficulty is the requirement of high accuracy in simulation of lasers, in particular mode-locked lasers. Light propagation in laser cavities has a resonant feature that makes the resonators significantly different from other systems. Multiple circulations of light inside the cavity also mean numerical error can also be accumulated in the propagation. Even worse, numerical solutions of circulated light in amplified medium will have increasing errors as compared with non-circulation propagation problems. In other words, a combination of the nonlinear dynamics due to SA and multiple circulations in the cavity with amplified medium is a very challenging problem. Therefore, modeling and simulation of mode-locked fiber lasers by numerically solving a GNLSE in the laser cavity require a very high accuracy scheme of numerical iterations of nonlinear PDEs.

The most widely-used numerical scheme for solving the GNLSE is the split-step Fourier method [8]. In the split-step scheme, dispersive and nonlinear effects are separately integrated, and the results are combined to construct the full solution. The dispersive term, which is linear, is evaluated in the frequency domain through the use of the fast Fourier transform (FFT), whereas the nonlinear term is treated in the time domain. As discussed in [9], the simple split-step schemes of global second- or third-order accuracy [18–22] have been widely used in solving the GNLSE. Note that it is also possible to construct higher order split-step schemes such as the fourth-order scheme presented by Blow and Wood [10]. However, global accuracy of the scheme cannot exceed the accuracy of the method that is used to integrate the nonlinear step. Usually, Runge–Kutta or implicit schemes have been used for the integration.

It is worth noting that a highly efficient and accurate algorithm, which is called the fourth-order Runge–Kutta in the interaction picture (RK4IP) method and is described in detail by Caradoc-Davies [23], has been developed to solve the Gross–Pitaevskii equation to describe the dynamics of Bose–Einstein condensates. Interestingly, it has a structure that is similar to the optical NLSE and GNLSE equations, but with time and space variables playing opposite roles. This RK4IP method has been applied for simulating supercontinuum generation in optical fibers solving GNLSE numerically [9]. In that work, Hult has shown a high efficiency algorithm for numerically solving NLSE and GNLSE by combining a fourth-order Runge–Kutta technique for stepping, with an appropriate choice of separation between the normal and interaction pictures. The resulting method exhibits fourth-order global accuracy, is efficient, and easy to implement in solving the NLSE or GNLSE equations in optical fibers. In this chapter, we will follow closely the RK4IP scheme developed by Hult and apply it for simulating mode-locked fiber lasers. In general, the RK4IP method is closely related to the split-step Fourier method. The advantage of the RK4IP algorithm comes from transforming the problem into an interaction picture, which is originated from quantum mechanics. This transform to interaction picture allows the use of conventional explicit techniques to step the solution forward, which make the RK4IP algorithm highly efficient. Readers are encouraged to study the method from the references, especially Hult's work in [9]. In the following, we will present briefly the method aiming to provide some basics equations that will be used in our examples of mode-locked fiber lasers with MATLAB® programs.

First, let us describe briefly the split-step method, which serves as basics of the RK4IP.

Brief description of split-step Fourier method

For numerical integration, it is useful to represent (4.28) and (4.29) in the form

$$\frac{\partial A(z, T)}{\partial z} = (\hat{D} + \hat{N})A(z, T), \tag{4.31}$$

where $\hat{D}$ is a dispersion operator, and $\hat{N}$ is a nonlinear operator. From (4.28), they are given by

$$\hat{D} = -\frac{\alpha}{2} - i\beta_2\frac{\partial^2}{\partial T^2}, \quad \text{and} \quad \hat{N} = i\gamma|A(z, T)|^2, \tag{4.32}$$

whereas for the GNLSE (4.29), they are given by

$$\hat{D} = -\frac{\alpha}{2} - \left(\sum_{n\geqslant 2}\beta_n\frac{i^{n-1}}{n!}\frac{\partial^n}{\partial T^n}\right), \tag{4.33a}$$

$$\hat{N} = i\gamma\left(1 + \frac{1}{\omega_0}\frac{\partial}{\partial T}\right) \times \{(1 - f_R)\}|A|^2 + f_R \int_0^{\infty} h_R(\tau)|A(z, T - \tau)|^2 d\tau. \quad (4.33b)$$

In the symmetric split-step Fourier method, the solution to (4.30) over a step h is approximated by

$$A(z + h, T) = \exp\left(\frac{h}{2}\hat{D}\right)\exp\left(\int_z^{z+h} \hat{N}(z')dz'\right)\exp\left(\frac{h}{2}\hat{D}\right)A(z, T). \quad (4.34)$$

where the exponential dispersion operator is conveniently evaluated in the Fourier domain through the use of the FFT. Since the dispersion and nonlinear operators do not commute, in general, the solution (4.34) is only an approximation to the exact solution, with a global error that is second-order in the step size $O(h^2)$. Many different approaches in approximating the nonlinear term, which are described by the integral in the middle exponential, have been reported. The simplest consists of approximating it with $\exp(h\hat{N})$, which will henceforward be referred to as the symmetric split-step method [22]. Second and fourth-order Runge–Kutta methods (symmetric split-step RK2 and symmetric split-step RK4, respectively) [18–22], as well as an implicit scheme (symmetric split-step) [8], have also been employed. Simpler schemes are the split-step and reduced split-step methods [18, 22], which rely on approximating the solution of (4.31) with $A(z + h, T) = \exp(h\hat{D})\exp(h\hat{N})A(z, T)$, leading to a global error that is only first-order in the step size $O(h)$. It is also possible to construct higher order split-step schemes by various forms of extrapolation [18–22]. As discussed in detail by Hult in [9], and we do not want to repeat here, none of the aforementioned Fourier split-step schemes can possess a global accuracy exceeding that of the numerical scheme RK4IP for integration of the nonlinear term.

Before presenting the basic equations of the RK4IP method we want to stress here that the term 'interaction picture' originated from quantum mechanics. In general, the 'interaction picture' is a hybrid representation that is useful in solving problems with time-dependent Hamiltonians. Readers who want to understand the concept and usefulness of the method are encouraged to take some deep reviews or studies from many well-known textbooks of quantum mechanics. Due to the limited space and scope of this chapter, we will only present the main equations of RK4IP, and then apply it in the MATLAB® programs for modeling and simulation of mode-locked fiber lasers. For simplicity, we will follow the presentation in reference [9].

Fourth-order Runge–Kutta in the interaction picture method

In the RK4IP method, the NLSE (or GNLSE) is transformed into an interaction picture in order to separate the effect of dispersion contained in $\hat{D}$ from the nondispersive terms in $\hat{N}$. This allows the use of explicit techniques to put the solution forward. Field envelope A is transformed into the interaction picture representation A_I by

$$A_I = \exp(-(z - z')\hat{D})A, \quad (4.35)$$

where z' is the separation distance between the interaction and normal pictures. Differentiating (4.35) gives the evolution of A_I.

$$\frac{\partial A_I}{\partial z} = \hat{N}_I A_I \tag{4.36}$$

where

$$\hat{N}_I = \exp(-(z - z')\hat{D})\hat{N}\exp(-(z - z')\hat{D}). \tag{4.37}$$

is the nonlinear operator in the interaction picture. The differential equation (4.36) can now be solved by using conventional explicit schemes such as Runge–Kutta methods, as the stiff linear parts of the PDE (4.31) have now been improved by moving into the interaction picture.

A straightforward solution of (4.36), which employs a fourth-order Runge–Kutta method, with the exponential operators evaluated in the frequency domain, would require 16 FFTs per step. The use of the RK4IP algorithm, however, reduces by half the required number of FFTs [9, 23]. This is achieved by choosing the step midpoint as the separation distance $z' = z + h/2$, which eliminates the dispersion exponentials in (4.37) for the two midpoint trajectories k_2 (4.38c) and k_3 (4.38d). The algorithm that advances $A(z, T))$ to $A(z + h, T)$ in a spatial step h, expressed in the normal picture A, is now written as

$$A_I = \exp\left(\frac{h}{2}\hat{D}\right)A(z, T), \tag{4.38a}$$

$$k_1 = \exp\left(\frac{h}{2}\hat{D}\right)\left\{h\hat{N}[A(z, T)]\right\}A(z, T), \tag{4.38b}$$

$$k_2 = h\hat{N}[A_I + k_1/2]\cdot(A_I + k_1/2), \tag{4.38c}$$

$$k_3 = h\hat{N}[A_I + k_2/2]\cdot(A_I + k_2/2), \tag{4.38d}$$

$$k_4 = h\hat{N}\left\{\exp\left(\frac{h}{2}\hat{D}\right)\cdot(A_I + k_3)\right\}\exp\left(\frac{h}{2}\hat{D}\right)\cdot(A_I + k_3), \tag{4.38e}$$

and the solution can be expressed as

$$A(z + h, T) = \exp\left(\frac{h}{2}\hat{D}\right)\{A_I + k_1/6 + k_2/3 + k_3/2\} + k_4/6. \tag{4.39}$$

The transformation into the normal picture (4.39) introduces an overhead of two FFTs per step; however, this overhead is eliminated by keeping the last trajectory k_4 (4.38e) in the normal picture. In total, each step thus requires four evaluations of the nonlinear operator $\hat{N}$ and four evaluations of the exponential dispersion operator $\exp(h\hat{D}/2)$, which requires eight FFTs. The RK4IP algorithm has a local error which is a fifth-order $O(h^5)$ and, thus, is a globally fourth-order accurate method $O(h^4)$. In [9], Hult has compared the efficiency and accuracy of different schemes in numerical solving problems of second-order solitons and supercontinuum generation. The results show that RK4IP is the best in computation times and with lowest errors. In particular, the average relative error of RK4IP can be lower than the split-step method, which has been widely used for simulating pulse propagation problems, at least few orders of magnitude (more than 4 orders is possible). Although we do not show a comparison of the results of modeling and simulation of the mode-locked fiber lasers using different numerical methods, our results show RK4IP is the most effective method for this work.

Let us now start studying several examples of mode-locked fiber lasers with ring cavity configurations. The examples of linear cavities will be considered later.

4.3 Example of modeling mode-locked ring fiber lasers

Let us now study the examples of mode-locked fiber laser with MATLAB® programs. We show on the left of figure 4.12 a schematic of a typical mode-locked fiber laser with ring cavity. The cavity consists of several important components (in clockwise order) such as: pump-diode 974 nm and input coupling (WDM), a section of Er/Yb doped fiber for gain medium, a polarization control (PC), an output coupler (OC), an integrated device of carbon nanotube (CNT) for the SA, an isolation (ISO), a section of dispersion-compensated fiber (DCF) for adjusting and controlling total dispersion of the cavity. Finally, all components are integrated by passive SMF. Also, in figure 4.12 is the gain curves of 2 cm highly Yb-Er co-doped phosphate glass fiber, whose gain is saturated at around 974 nm pump power of 300 mW. The dispersion parameters of the passive SMF D_{SMF} = 17 (km · nm)$^{-1}$, active gain Yb-Er co-doped fiber (EDF) D_{EDF} = 8 ps/(km · nm), and dispersion-compensated fiber (DCF) D_{DCF} = −40 ps/(km · nm) at 1550 nm are also listed in the figure.

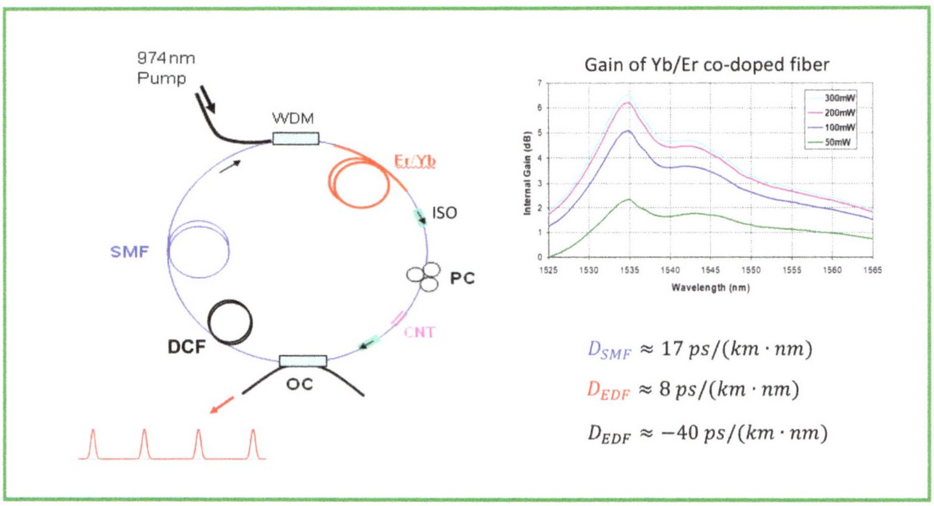

Figure 4.12. Illustration of mode-locked fiber ring cavity including Er–Yb co-doped gain fiber (red), dispersion compensated fiber (DCF—black) and passive SMF (blue). Other parts of the cavity include PC, ISO, OC. Gain curves of 2 cm EDF are plotted on the right figure.

Example 4.1

In this first example, a configuration of a laser cavity is shown in figure 4.12, and the lengths of SMF, DCF and Yb/Er co-doped fibers are assumed to have values as $L_{SMF} = 1.0$ m, $L_{EDF} = 0.2$ m and $L_{DCF} = 0.8$ m. Let us list all important parameters of the components that we will describe in the model and in the MATLAB® program. All fibers are assumed to have nonlinear index $n_2 = 2.6 \cdot 10^{-20} \mathrm{m^2\,W^{-1}}$.

1. SA CNT: the saturable absorption is assumed to have $\alpha_0 = 60\%$, absorption modulation $\Delta\alpha = 5\%$, recovery time $T_S = 750$ fs and saturated energy $E_p = 5$ pJ.
2. Passive fiber SMF: propagation loss $\alpha_{SMF} = 0.2$ db km^{-1}, mode field diameter, MFD = 10.0 μm, $D_{SMF} = 17$ ps/(km · nm). Note that in a real cavity, the passive fiber SMF is used to integrate components of the cavity so that the SMF is in fact divided by several parts, and we need to take the coupling losses between the connections in the model.
3. DCF: propagation loss $\alpha_{DCF} = 1$ dB km^{-1}, MFD = 6 μm, $D_{SMF} = -40$ ps/(km · nm). Again, we will need to take the coupling loss between DCF with SMF in the cavity.
4. Gain fiber: propagation loss $\alpha_{EDF} = 10$ dB m^{-1}, MFD = 8 μm, $D_{EDF} = 8$ ps/(km · nm).
5. OC 30/70: the light is assumed to output 30% and feedback to the cavity 70% at the OC.

Before we go into details, let us estimate some important parameters of the lasers:

1. Loss and gain: sometimes, it is very useful to know the cavity loss even with rough estimation, especially when optimization is critical. For example, knowing the loss gives us some basic ideas for design, such as how much gain and pump power budget. Moreover, too much gain could lead to instabilities of the laser performance. When estimating the cavity loss, it is necessary to count all the possible losses as below as examples (it could be more than that in reality):
   - Coupling loss: coupling between parts in integrated configuration. Conservatively we can simply assume coupling loss is about 0.25 dB for each intersection (the real loss could be very different depending on the fibers used in the cavity). For example, from configuration in figure 4.12 there are 10 intersections whose coupling loss is assumed about $0.5 \times 10 = 5$ dB;
   - Loss due to SA absorption from linear absorption 60% (~3.5 dB) and nonlinear value ~50% (3 dB);
   - Loss due to OC 30%: ~1.5 dB;
   - Loss due to ISO and PC: provided by manufacturers.
2. Total dispersion: It is easy to estimate dispersion of each fiber part as $L_{SMF}D_{SMF} = (1.0\text{ m}) \times 17\text{ ps/(km}\cdot\text{nm)} = 17\text{ fs nm}^{-1}$, $L_{EDF}D_{EDF} = (0.2\text{ m}) \times 8\text{ ps/(km}\cdot\text{nm)} = 1.6\text{ fs nm}^{-1}$ and $L_{DCF}D_{DCF} = (0.8\text{ m}) \times -40\text{ ps/(km}\cdot\text{nm)} = -32\text{ fs nm}^{-1}$. The total dispersion of the cavity is $(LD)_{\text{Total}} = -13.4\text{ fs nm}^{-2} < 0$.

In the following there is a MATLAB® program 4.1 for simulating the mode-lock fiber laser described in example 4.1. Readers can copy the whole program into a directory and run it by clicking the Run button in the main program MLL.m. The whole program is constructed with several functions that are close to the above descriptions. The comments in the program make it easy to understand not only the codes but also the physical and mathematical models behind the algorithms. For example, one part describes the Raman function in the GNLSE (4.30) started by a comment

```
% Raman response function approx by Blow and Wood, 1989.
```

Similarly, this part is for simulating light propagation in SA:

```
%% ————————This part for SA CNT ———————
```

And many others with explanations will be provided later after the program.

First, the main program for example 4.1.

MATLAB® Program 4.1

```
%xxxxxxxxxxxxxxxxxxxxxxxxxxxxxxxxxxxxxxxxxxxxxxxxxxxxxxxxxxxxxxxxxxxxxx
% MLL.m                                                                x
% Main Program for Fiber-based Optical Parametric Ossilation          x
% Using: RK4IP algorithm (Hunt JLT 2007)                               x
%       fucntions:                                                     x
%                 Propagation_EDF.m for propagation in EDF             x
%                 Propagation_SMF.m for propagation in SMF             x
%                                                                      x
%                        By Dan Nguyen, 06/04/21                       x
%xxxxxxxxxxxxxxxxxxxxxxxxxxxxxxxxxxxxxxxxxxxxxxxxxxxxxxxxxxxxxxxxxxxxxx
close all
clear all
clc

if ~exist('L100x120x100_10dB','dir')
    mkdir('L100x120x100_10dB');
end

c0 = 3e8;                  % (m/s)
lambdaP = 1550e-9;         % (m)% pump wavelength (3000 , 3500, 6500, 7250 nm)
wP = 2*pi*c0/lambdaP;      % Angular frequency of pump @ 1560nm
Tmax  = 60e-12;            % (s), half width of time window
Tmin  = -Tmax;             % Tmin < t < Tmax
tol = 1e-3;                % tollerence

global t N                  % vector time, globally used (even in functions)

N = 2^17;                  % number of points used to present in time domain
deltat = 2*Tmax/(N-1);     % (s), resolution of time
t = linspace(Tmin,Tmax-deltat,N);

global f                   % frequency, globally used
deltaf = 1/(2*Tmax);       % resolution of frequency
f = [0:deltaf:(N/2-1)*deltaf,-N*deltaf/2:deltaf:-deltaf]; % vertor freq
w = 2*pi*f;                % vector angular freq
wshift = -w + wP;          % plus the freq with its central freq.
```

```
                    % the minus sign is to compensate the fact that
                    % defenition of FFt in matlab and in the GNLSE are
                    % in oppositec sign
wl = 1e9*2*pi*c0./wshift;%(nm), shift to nm to present
tm = 1e12*t;        % (ps), shift to ps to present
%% ---------------This part for Raman fuction ----------------------------

% fR = 0.18;              % factor of contribution of Raman effect
 fR = 0.0;
t1 = 12.2e-15;           % (s), paramet in approx raman response function
t2 = 32e-15;             % (s), paramet in approx raman response function,
                         % 1/t2 is the bandwidth of Raman gain
t_shock = 1/wP;          % (s), paramet of self steepening & shock formation

tr = t - t(1);           % shift time vector to the root. starting point is 0

% Raman response function approx by Blow and Wood, 1989.
 hR = (t1^2+t2^2)/t1/t2^2*exp(-tr/t2).*sin(tr/t1);

 hR = hR./trapz(tr,hR);  % (normalized of integ_0^inf hR = 1)
 hR_f = fft(hR);          % raman response function in the freq domain

 %tt = 50e-15;  % contrib to effect of gain in dispers value [Agrawal OL1991]

 %% -- this part is for Er-doped fiber (EDF) -------------------------------
       n2 = 2.6e-20;          % nonlinear refractive index
     Esat = 50e-12 ;          % 50pJ saturation energy
    L_edf = 0.20;             % Length of EDF (m)
 AdB_edf  = 10.0;             % dB/m loss of fiber 97
 alfa_edf = AdB_edf/4.343;    % loss in unit 1/m
 MFD_edf  = 8.0e-6 ;          % MFD of the fiber, assume 10um (check)
 Aeff_edf = pi*(MFD_edf/2)^2;% effective core area
 gamma_edf= 2*pi*n2/lambdaP/Aeff_edf;  % 0.0013/((2)^2);

 delta_wl = 40e-9;            % bandwidth gain in nm
    Omega = (2*pi*c0/lambdaP^2)*delta_wl;
    tt    = 1/Omega

    D_edf = 10.0e6;            % Disperion of EDF in ps/(m*m)
    b2_edf = - lambdaP^2/(2*pi*c0*1e-12)*D_edf

                              % [beta1, beta2, beta3...] in ps^k/m
 beta_edf = [0 b2_edf];

                          % change unit of beta to (s^k/m)
  for ii = 1:length(beta_edf)
       beta_edf(ii) = (1e-12)^(ii)*beta_edf(ii);
  end
 %% --- this part for smf fiber -------------------------------------------

   L_smf1 = 0.3;          % 1st SMF length: 0.1(after NP)+ 2 both side ISO
   L_smf2 = 0.3;          % Lengtn of SMF before OC
   L_smf3 = 0.3;          % length of SMF after  OC
    L_dcf = 0.3;          % Length of Flex both size of WDM

   AdB_smf= 0.2;          % (dB/km) loss of silica fiber
 alfa_smf = AdB_smf/4343;% (1/m) loss
 MFD_smf  = 10.8e-6;       % MFD of the fiber, assume 10um (check)
 Aeff_smf = pi*(MFD_smf/2)^2;     % effective core area
 gamma_smf= 2*pi*n2/lambdaP/Aeff_smf;   % nonlinear coefficient (1/W/m)
```

```
    D_smf = 17e6;            % Disperion of EDF in ps/(m*m)
    b2_smf = - lambdaP^2/(2*pi*c0*1e-12)*D_smf

                          % [beta1, beta2, beta3...] in ps^k/m
beta_smf = [0 b2_smf];

                          % change unit of beta to (s^k/m)

   for ii = 1:length(beta_smf)
        beta_smf(ii) = (1e-12)^(ii)*beta_smf(ii);
   end

  %% --- this part for Dispersion compensate fiber DCF38---------------

   AdB_dcf = 1.0;          % (dB/km) loss of silica fiber
  alfa_dcf = AdB_dcf/4343;% (1/m) loss
  MFD_dcf  = 6.0e-6;       % MFD of the fiber, assume 10um (check)
  Aeff_dcf = pi*(MFD_dcf/2)^2;    % effective core area
  gamma_dcf= 2*pi*n2/lambdaP/Aeff_dcf;   % nonlinear coefficient (1/W/m)

     D_dcf = -40e6;           % Disperion of EDF in ps/(m*m)
     b2_dcf = - lambdaP^2/(2*pi*c0*1e-12)*D_dcf

                           % [beta1, beta2, beta3...] in ps^2/m
  beta_dcf = [0 b2_dcf];

                           % change unit of beta to (s^k/m)
   for ii = 1:length(beta_dcf)
        beta_dcf(ii) = (1e-12)^(ii)*beta_dcf(ii);
   end

  %% -----------This part constructs the signal 1550 nm ----------

   phase = pi*rand([1,N]);
  Unoise = 1e-4.*randn(1,N).*exp(-1i*phase);      % very small noise seeding
  Unoise_f = fft(Unoise);                         % in freq-domain (wavelength)
  It0 = abs(Unoise).^2;
  If0 = abs(Unoise_f).^2;

  Uin0 = Unoise;
  w2 = wl((wl>1500)&(wl<1600));


  %% -----------This part for SA CNT ----------
  a0 = 0.05;
  ac = 0.55;
  Ts = 750e-15;
  Es =   5e-12;

  alfa0 = SA(Uin0, a0, ac, Ts, Es);

  FB = 70/100;                  % feedback coefficient (10% in exp)
  CL1 = 96/100;                 % coupling coefficiency: loss = (1-CL)*100 (%)
  CL2 = 96/100;                 % coupling coefficiency: loss = (1-CL)*100 (%)

  GdB  = 10;                     % total gain in db
  gain = GdB/L_edf/4.343;       % gain per m (1/m)

  indc = 0;                    % loop index
  indx = 0;
```

```
loopmax = 300;                  % max # of circle trips
%% ------------------------------------------------------------------------
for loop = 1:loopmax
    indc = indc + 1

    %% propagation through EDF
    Uin_edf = Uin0;
                                        % signal get amplified thought EDF
    Uout_edf = Propagation_EDFL(Uin_edf,tol,L_edf,gain,Esat,alfa_edf,...
                                    beta_edf,gamma_edf,t_shock,fR,hR_f);%
    %% propagation through 1st section of SMF

    Uin_smf1 = sqrt(CL1)*Uout_edf;

    Uout_smf1 = Propagation_SMF(Uin_smf1,tol,L_smf1,alfa_smf,beta_smf,...
                                       gamma_smf,t_shock,fR,hR_f);   %
    %% propagation through SA
     alfa = SA(Uout_smf1, a0, ac, Ts, Es);
     Uout_sa = sqrt(1-alfa).*Uout_smf1;
     Uouf_sa = fft(Uout_sa);

     %% propagation through 2nd section of SMF

    Uin_smf2 = sqrt(CL2)*Uout_sa;

    Uout_smf2 = Propagation_SMF(Uin_smf2,tol,L_smf2,alfa_smf,beta_smf,...
                                       gamma_smf,t_shock,fR,hR_f);   %
    Uout_loop = sqrt(1-FB)*Uout_smf2;
    Uouf_loop = fft(Uout_loop);
         Iout = abs(Uout_loop).^2;
         Iouf = 10*log10(abs(Uouf_loop).^2);

    %% propagation through 3rd section of SMF

    Uin_smf3 = sqrt(FB)*Uout_smf2;

    Uout_smf3 = Propagation_SMF(Uin_smf3,tol,L_smf3,alfa_smf,beta_smf,...
                                       gamma_smf,t_shock,fR,hR_f);   %
    %% propagation through DCF

    Uin_dcf = sqrt(CL1)*Uout_smf3;

    Uout_dcf = Propagation_SMF(Uin_dcf,tol,L_dcf,alfa_dcf,beta_dcf,...
                                       gamma_dcf,t_shock,fR,hR_f);   %

    Uin0 = sqrt(CL1)*Uout_dcf;

    figure(1);
     subplot(3, 1, [1]) % present time signal
      plot(tm,It0,'r', 'LineWidth',2);
      hold on
      plot(tm,Iout,'b', 'LineWidth',1);
      xlabel ('Time (ps)','FontSize', 14);
      xlim([-60 60])
      set(gca,'XTick',-60:10:60)
      ylim([0 10])
      set(gca,'YTick',0:2:10)
      %legend('input', 'out');
      ylabel ('Power (a.u)','FontSize',12);
      grid on;
      set(gca,'FontSize',12);
      hold off
```

```
  subplot(3, 1, [2])                          % present time signal
  plot(tm,alfa0*100,'r','LineWidth',2);
  hold on
  plot(tm,alfa*100,'b','LineWidth',1);
  xlabel ('Time (ps)','FontSize', 14);
  xlim([-60 60])
  set(gca,'XTick',-60:10:60)
  ylim([54 62])
  set(gca,'YTick',54:2:62)
  %legend('\alpha(t)');
  ylabel ('Absorption (%)','FontSize',12);
  grid on;
  set(gca,'FontSize',12);
  hold off

 subplot(3, 1, [3])
 plot(wl,10*log10(If0),'LineWidth',2,'Color','r');
 hold off;
 plot(wl,Iouf,'b','LineWidth',1);
 text(1510,10, num2str(indc),'FontSize',20)
 xlabel ('Wavelength (nm)');
 ylabel ('Power (dB)');
 xlim([1500 1600])
 set(gca,'XTick',1500:10:1600)
 ylim([-80 100])
 set(gca,'YTick',-80:40:100)
 grid on;
 set(gca,'FontSize',12);
 hold off
% drawnow
   if mod(indc,2)==0,
   indx = indx +1;
   cd L100x120x100_10dB
   saveas(gcf, strcat('file', num2str(indx), '.png'));
   cd ../
   end
 end
```

The function below is for operator $\hat{D}$

```
%xxxxxxxxxxxxxxxxxxxxxxxxxxxxxxxxxxxxxxxxxxxxxxxxxxxxxxxxxxxxxxxxxxxxxx
% Function D_op: D operator
% Caculate "exp(hD){A}", in Nonlinear Schrodinger equation
%
%xxxxxxxxxxxxxxxxxxxxxxxxxxxxxxxxxxxxxxxxxxxxxxxxxxxxxxxxxxxxxxxxxxxxxx
function Aout = D_op(Ain,dz,beta,alpha,omega)

% Dispersion term in NLSE.
Dw =zeros(1,length(omega));

% Dw = sum(i*beta(i)*(-w)^i/(i!))
for jl = 1 : length(beta)
 Dw = Dw - 1i.*beta(jl)/factorial(jl).*(-omega).^(jl);
end

Aout = ifft(exp(dz*(-Dw - alpha/2)).*fft(Ain));
```

The function below is for operator $\hat{N}$

```
%xxxxxxxxxxxxxxxxxxxxxxxxxxxxxxxxxxxxxxxxxxxxxxxxxxxxxxxxxxxxxxxxxxxxxx
% Function oper_N
% Caculate "hN{A}.*A" in Nonlinear Schrodinger Equation
%
%xxxxxxxxxxxxxxxxxxxxxxxxxxxxxxxxxxxxxxxxxxxxxxxxxxxxxxxxxxxxxxxxxxxxxx
function Aout = N_op(Ain,dz,gamma,t_shock,fR,hR_w)
global t
dt = abs((t(2)-t(1)));
        % convolution term int_0^inf(hR(t)|A(z,T-t)|^2 dt
        % ie. Eq.(7) Hult 2007, or ifft(hR_w.*fft(abs(A)^2))

hR_A2 = ifft(hR_w.*fft(abs(Ain).^2))*dt;

% 1st term
N1 = 1i*gamma*((1-fR).*Ain.*abs(Ain).^2 + fR*Ain.*hR_A2);

% 2nd term.Can be wrote as gradient(F,dt) = gradient(F)/dt
 N2 = -gamma*t_shock.*gradient(((1-fR).*Ain.*abs(Ain).^2 ...
                                        + fR.*Ain.*hR_A2),dt);
Aout = dz*(N1+N2);
```

The function below is for propagation in EDF

The function below is for propagation in EDF

```
function Uout = Propagation_EDF(Uin,tol,L,g0,Es,alpha,beta,gam,ts,fR,hR)
%xxxxxxxxxxxxxxxxxxxxxxxxxxxxxxxxxxxxxxxxxxxxxxxxxxxxxxxxxxxxxxxxxxxxxxx
% This function solves the GNLE                                         x
% Using RK4IP                                                           x
% Hult JLT 2007 and Sinkin JLT 2003.                                    x
%-----------------------------------------------------------------------x
% INPUT
%
% Ei - impulses, starting field amplitude (vector)
% tol - local error limitation
% long - length of fiber
% alpha - linear loss coefficient, ie, Pz=P0*exp(-alpha*z)
% betap - dispersion polynomial coefs, [beta_1 ... beta_m], or beta(w)
% gam - nonlinearity coefficient
% ts - self steepening term (t_shock)
% fR - fraction of raman effect
% hR - the response function of Raman effect in freq domain
%------------------------------------------------------------------------
% OUTPUT

% Uout - field at the output
%-----------------------------------------------------------------------
global f t
delta_t = abs(t(2)-t(1));
omega = 2*pi*f;

Efield = Uin;

fprintf(1, '\nCaculation in process...        ');

    dz = L/10000;          % trial for 1st step
```

```
alpha0 = alpha;
 beta0 = beta;
gamma0 = gam;
  Z_prop = 0;
    ii = 0;

while Z_prop<L
    ii=ii+1;
    if Z_prop+dz>L
        dz = L-Z_prop;
    end
    Z_prop = Z_prop + dz;

    % change parameters as Z_propagation change
    tt = 40e-15;                          % T2 in Agrawal 1990 PTL

    % g = g0*exp(-sum(abs(E_z).^2)*delta_t/E_sat) (Agrawal 1991)
    % or approximation as
    g = g0./(1 + sum(abs(Efield).^2)*delta_t/Es);

    alpha = alpha0 - g;                   % loss and gain in NLSE
    beta(2) = beta0(2) + 1i*tt*tt*g;      % from eq. (2) Agrawal PTL 1991

    m = 0;                                %
    gam = gamma0*(1 + m*Z_prop);

    %---------optimization of local errors by Sinkin JLWT 2003 ------------

    Uf = rk4ip(rk4ip(Efield,omega,dz/2,beta,alpha,gam,ts,fR,hR),...
                               omega,dz/2,beta,alpha,gam,ts,fR,hR);
    Uc = rk4ip(Efield,omega,dz,beta,alpha,gam,ts,fR,hR);
    error = sqrt(sum(abs(Uf-Uc).^2))/sqrt(sum(abs(Uf).^2));

    factor = tol/error;
    if error > 2*tol          % discard the solution and recaculate
        Z_prop = Z_prop-dz;             % with a half of old stepsize
    else
       Efield = 16/15*Uf-1/15*Uc;      % improve the accuracy of the solution
    end
    dz=dz*factor^(1/5);

end

Uout = Efield;
```

The function below is for propagation in EDFL

```
function Uout = Propagation_EDFL(Uin,tol,L,g0,Es,alpha,beta,gam,ts,fR,hR)

%xxxxxxxxxxxxxxxxxxxxxxxxxxxxxxxxxxxxxxxxxxxxxxxxxxxxxxxxxxxxxxxxxxxxxxxxx
% This function solves the GNLE                                           x
% Using RK4IP                                                             x
% Hult JLT 2007 and Sinkin JLT 2003.                                      x
%                                                                         x
%-------------------------------------------------------------------------x
% INPUT
%
% Uin - input field amplitude (vector)
% tol - local error limitation
% long - length of fiber
```

```
% alpha - linear loss coefficient, ie, Pz=P0*exp(-alpha*z)
% betap - dispersion polynomial coefs, [beta_1 ... beta_m], or beta(w)
% gam - nonlinearity coefficient
% ts - self steepening term (t_shock)
% fR - fraction of raman effect
% hR - the response function of Raman effect in freq domain
%------------------------------------------------------------------------
% OUTPUT

% Uout - field at the output
%------------------------------------------------------------------------
global f t
delta_t = abs(t(2)-t(1));
omega = 2*pi*f;
c0 = 3e8;                   % (m/s)
lambdaP = 1550e-9;          % (m)% pump wavelength (3000 , 3500, 6500, 7250 nm)
wP = 2*pi*c0/lambdaP;       % Angular frequency of pump @ 1560nm
delta_wl = 20e-9;           %
OmegaP = (2*pi*c0/lambdaP^2)*delta_wl;

Efield = Uin;

fprintf(1, '\nCaculation in process...        ');


     dz = L/10000;          % the first step is chosen random
 alpha0 = alpha;
  beta0 = beta;
 gamma0 = gam;
   Z_prop = 0;
     ii = 0;

 while Z_prop<L
     ii=ii+1;
     if Z_prop+dz>L
         dz = L-Z_prop;
     end
     Z_prop = Z_prop + dz;

     % change parameters as Lrun change
     %tt = 40e-15;                        % T2 in Agrawal 1990 PTL
     tt = 1/OmegaP;

     % g = g0*exp(-sum(abs(E_z).^2)*delta_t/E_sat) (Agrawal 1991)
     % or approximation as
     g = g0./(1 + sum(abs(Efield).^2)*delta_t/Es);

     alpha = alpha0 - g;                  % loss and gain in NLSE
     beta(2) = beta0(2) + 1i*tt*tt*g;     % from eq. (2) Agrawal PTL 1991

     m = 0;                               % ???
     gam = gamma0*(1 + m*Z_prop);

     %---------optimization of local errors by Sinkin JLWT 2003 ------------

     Uf = rk4ip(rk4ip(Efield,omega,dz/2,beta,alpha,gam,ts,fR,hR),...
                                 omega,dz/2,beta,alpha,gam,ts,fR,hR);
     Uc = rk4ip(Efield,omega,dz,beta,alpha,gam,ts,fR,hR);

     error = sqrt(sum(abs(Uf-Uc).^2))/sqrt(sum(abs(Uf).^2));
```

```
    factor = tol/error;
    if error > 2*tol          % discard the solution and recaculate
         Z_prop = Z_prop-dz;            % with a half of old stepsize
    else
        Efield = 16/15*Uf-1/15*Uc;     % improve the accuracy of the solution
    end
    dz=dz*factor^(1/5);

end

Uout = Efield;
```

The function below is for propagation in SMF

```
function [Uout, numFFT] = Propagation_SMF(imp,tol,long,alpha,beta,gamma,ts,fR,hR)
% %xxxxxxxxxxxxxxxxxxxxxxxxxxxxxxxxxxxxxxxxxxxxxxxxxxxxxxxxxxxxxxxxxxxxxxxxx
% This function solves the GNLE                                            x
% Using RK4IP                                                              x
% Hult JLT 2007 and Sinkin JLT 2003.                                       x
%                                                                          x
%-------------------------------------------------------------------------x
% INPUT
%
% imp - impulses, starting field amplitude (vector)
% tolerance - local error limitation
% long - length of fiber
% alpha - linear loss coefficient, ie, Pz=P0*exp(-alpha*z)
% betap - dispersion polynomial coefs, [beta_1 ... beta_m], or beta(w)
% gamma - nonlinearity coefficient
% t_shock - self steepening term
% fR - fraction of raman effect
% hR - the response function of Raman effect in freq domain
% L_save - distances along the fiber where we save the data
%-------------------------------------------------------------------------
% OUTPUT
%
% disp - n series of temporal output signal in function of distances
% dispf - n series of spectral output signal in function of distances
% Uout - field at the output
% numFFT - number of performed FFTs
%-------------------------------------------------------------------------
global f t
omega = 2*pi*f;

E_z = imp;

fprintf(1, '\nCaculation in process...        ');
dz = long/10000;      % the first step is chosen random
alpha0 = alpha;
beta0 = beta;
gamma0 = gamma;
Z_prop = 0;
ii = 0;

while Z_prop<long
    ii=ii+1;
    if Z_prop+dz>long
        dz = long-Z_prop;
    end
```

```
    Z_prop = Z_prop + dz;

    % change parameters as Z_propagation change
    m = 0;
    alpha = alpha0*(1 + m*Z_prop);
    beta = beta0*(1 + m*Z_prop);
    gamma = gamma0*(1 + m*Z_prop);
    %---------

    Uf = rk4ip(rk4ip(E_z,omega,dz/2,beta,alpha,gamma,ts,fR,hR),...
                    omega,dz/2,beta,alpha,gamma,ts,fR,hR);

    Uc = rk4ip(E_z,omega,dz,beta,alpha,gamma,ts,fR,hR);

    error = sqrt(sum(abs(Uf-Uc).^2))/sqrt(sum(abs(Uf).^2));
    factor = tol/error;
        % discard the solution and recaculate with a half of old stepsize
    if error > 2*tol
        Z_prop = Z_prop-dz;
    else
        E_z = 16/15*Uf-1/15*Uc; % improve the accuracy of the solution
    end
    dz=dz*factor^(1/5);

    fprintf(1, '\b\b\b\b\b\b%5.2f%%', Z_prop* 100.0 /long);
end

Uout = E_z;
numFFT = 3*16*ii;
```

The function below is for RK4IP

```
%xxxxxxxxxxxxxxxxxxxxxxxxxxxxxxxxxxxxxxxxxxxxxxxxxxxxxxxxxxxxxxxxxxxxxxxxxx
% Function rk4ip
%xxxxxxxxxxxxxxxxxxxxxxxxxxxxxxxxxxxxxxxxxxxxxxxxxxxxxxxxxxxxxxxxxxxxxxxxxx
function Eout=rk4ip(A0,omega,dz,beta,alpha,gamma,ts,fR,hR)
Efield = A0;
% Aip: interaction presentation

 Aip = D_op(Efield,dz/2,beta,alpha,omega);
 k1= D_op(N_op(Efield,dz,gamma,ts,fR,hR),dz/2,beta,alpha,omega);
 k2= N_op(Aip + k1/2,dz,gamma,ts,fR,hR);
 k3= N_op(Aip + k2/2,dz,gamma,ts,fR,hR);
 k4= N_op(D_op(Aip+k3,dz/2,beta,alpha,omega),dz,gamma,ts,fR,hR);

 Eout = D_op(Aip+k1/6+k2/3+k3/3,dz/2,beta,alpha,omega)+k4/6;
```

The function below is for SA

```
function alfa = SA(Ei, a0, ac, Ts, Es)
%xxxxxxxxxxxxxxxxxxxxxxxxxxxxxxxxxxxxxxxxxxxxxxxxxxxxxxxxxxxxxxxxxxxxxxxxxx
% SA.m                                                                    x
% Caculate and display "alfa_SA(|A|)"                                     x
% dalfa(t)/dt = (a0 - alfa(t))/ts - |A(t)|^2/Esat*alfa(t)                 x
%xxxxxxxxxxxxxxxxxxxxxxxxxxxxxxxxxxxxxxxxxxxxxxxxxxxxxxxxxxxxxxxxxxxxxxxxxx

%% ---- parameters for the simulation and a part of the pulse ---------
```

```
global N t
deltat = abs(t(2)-t(1));

%% -----------This part constructs the pump @975nm ----------
Et = zeros(size(t));

alfa1  = zeros(size(t));
intga  = zeros(size(t));
intga0 = 0;
intga1 = 0;

for k = 1:N
Et(k) = Ei(k);
intga(k) = deltat*(1/Ts + abs(Et(k))^2/Es);
intga0 = intga0 + intga(k);

intga1 = intga1 + a0/Ts*deltat*exp(+intga0);
alfa1(k) = ac + exp(-intga0)*(intga1 + a0);

end

alfa = alfa1;
```

Figure 4.13 shows the simulation results of the mode-locked fiber lasers described earlier in example 4.1. The evolution of the mode-locked laser is plotted after different numbers of circulations of light inside the cavity. For instance, the results shown in each panel (a), (b), (c), (d) and (f) are when round-trip number of circulations N = 5, 30, 50, 100, 150 and 200. In each panel, the linearly-scaled light intensity (in time domain), the absorption of SA also in time domain and the logarithm-scaled spectrum of the light are shown in the top, middle and the bottom figures, respectively. In all figures, the red lines are the initial values. For the light intensity, the initial values are assumed to be spontaneous noise in the cavity. Meanwhile, the initial value of absorption is the linear absorption of the SA.

Let us first describe the evolution of the light inside the mode-locked laser cavity shown in figure 4.13. Note that, each panel in the figure shows the light intensity, absorption and spectrum at the OC after the light starting from the spontaneous noise (red line)that circulates N times in the cavity. The number of circulations is displayed by a number in the figure of light spectrum at the bottom of the panel. Let us start from the figures in panel (a) with $N = 5$. It is clear that after five circulations, the light intensity is still very weak and is not noticeably different from noise in the linear scale (top figure). As a result, the value of absorption is almost the same as linear absorption α_0 as shown in the middle figure of the panel. Only a spectrum in logarithm scale (bottom figure) can show the difference between the spectra of light that circulates in the cavity and the initial noise. Panels (b) and (c) show the light and absorption $N = 30$ and 50 circulations inside the cavity. In both panels, the light intensity increases with the number of circulations of light with gain in the cavity. However, the light intensity is still very weak and random in time domain. The results in panel (b), however, show nonlinear absorption, that the absorption is lower when light intensity high, and whenever intensity increases the absorption decreases.

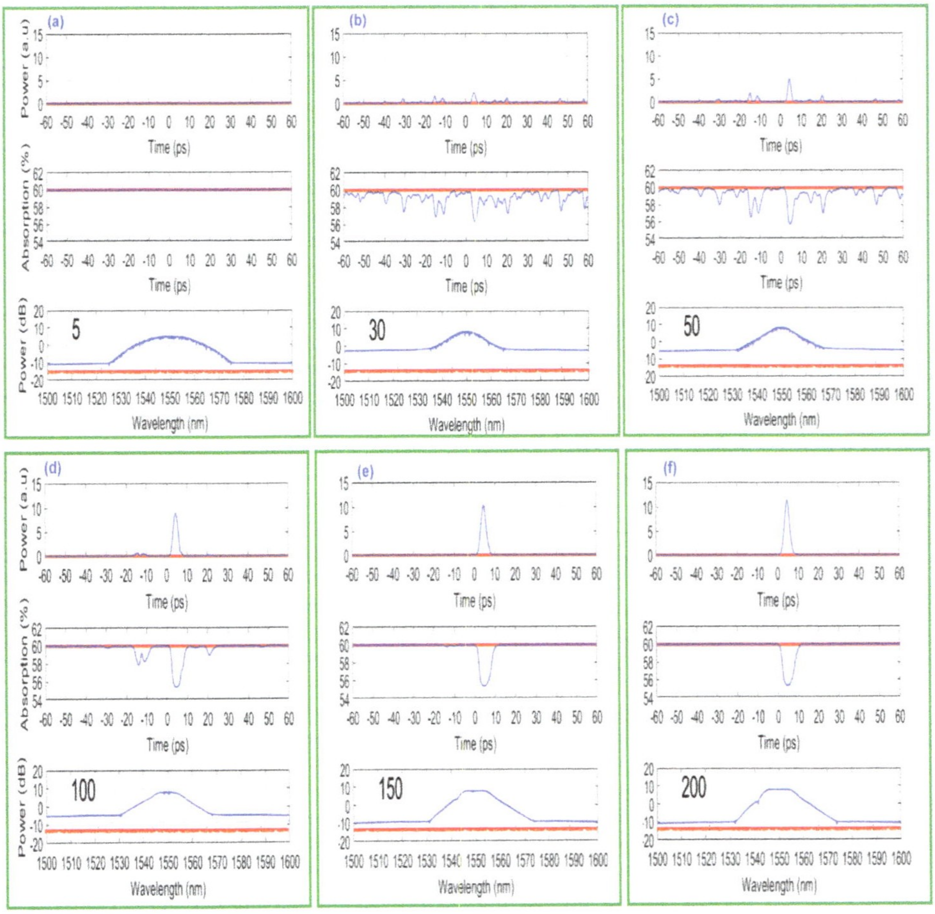

Figure 4.13. Numerical results of modeling and simulation of a mode-lock fiber laser described in example 4.1 with total dispersion $D < 0$. Total gain of the cavity is $G = 20$ dB, and total loss is ~10 dB. Panels (a)–(f) are results of light intensity (top), saturable absorption (middle) and spectra (bottom) at output coupler with circulation number $N = 5$, 30, 50 100, 150 and 200. In all figures, red lines are the starting values of light as noise in cavity in time domain (top) and frequency domain (bottom), and linear value of absorption, in these example $\alpha_0 = 60\%$. Details of the results are presented in the text.

The behavior in these figures shows the automatic synchronization between the light intensity and saturable absorption that leads to the mode-locked operation inside the cavity. Increasing the number of circulations in the cavity with gain, the light with high intensity will sense lower loss (due to lower absorption), and intensity increases further, resulting in lower loss and so on, until the absorption reaches its minimum value with modulation depth $\Delta\alpha$. On the other hand, the light with lower intensity will suffer higher loss, resulting in lower intensity and then higher loss etc. Clearly, any light part that has loss higher than gain will be suppressed right away. However, there would be different parts of light where gain is higher than loss, resulting in competition between these light parts during the circulations. Note that the energy transferred from the pump energy through the gain medium to laser signal is fixed, therefore, light parts with gain higher than loss will compete in energy during the circulation. The competition leads to the suppression of competitive parts, and only the light part that satisfies stable conditions of mode-locking include dispersion and nonlinear conditions will survive the competition. The competition can be seen clearly in the evolution from panel (c) with $N = 50$ to panel (d) with $N = 100$ and panel (e) with $N = 150$. Note that, the behavior changes very little when light propagates from 100 to 150 circulations in the cavity, and the results are almost unchanged after $N = 150$ circulations, as shown in panel (f), meaning the stable mode-locked laser is established. Note that, in this example, we assume the linear absorption is $\alpha_0 = 60\%$ and the depth modulation $\Delta\alpha = 5\%$ or minimum absorption $\alpha = 55\%$. The absorption in these two limits is clearly shown in panels (e) and (f). The results in figure 4.13 are very useful to see and understand the behavior of the mode-locked laser under conditions as described earlier. Note that, in this example, total dispersion of the cavity is negative $D < 0$.

One important point we want to stress here is that because the initial values of the light are randomly spontaneous noise, it is expected that the evolution in time of light is not exactly repeated as the same for every simulation. For example, we will not see the same changes from one circulation to the other, and the final pulse appears at the same place (in the time domain) in different simulations. However, the final results spectrum and the pulse with all characterizations (pulse duration, pulse spectrum) will be the same. The final results will be different with different gain. Readers can check the program with different gains, for example with $G = 13$, 15, 20 dB the main characterizations of the final pulses are almost the same with slight differences, but the peak power/intensity is higher with higher gain. That behavior will be changed if the total gain reaches a certain level such that the high energy light can break the mode-locking function of the SA, as shown in figure 4.14 below.

It is clear from figure 4.14 the evolution of the light with 25 dB gain is remarkably different from the lower gain in figure 4.13. First, even after a few circulations, the light intensity increases and start to affect to saturable absorber as shown in panel (a) with $N = 5$. Note that, the plot of light intensity in time domain is in linear scale and is not clear. Due to very high gain the light intensity increases significantly in different time intervals during the propagation time inside the cavity, causing lower absorption (or lower losses) in those periods, as shown in panels (b) and (c). The whole process: higher intensity → lower absorption → higher intensity … is repeated in many different time intervals, leading to multiple time pulses in a circulated time (laser repetition) and the spectrum having multiple kinks. In fact, this situation can be usually observed in experiments under some conditions that lead to unstable operation of the mode-locked laser. In general, if the gain of cavity is too high (as compared with total loss of cavity) the accumulated energy of light can be higher than saturated energy of SA, causing multiple pulses in one circulation (repetition) and multiple kinks in the spectrum. This behavior is the starting point for unstable mode-locked laser operation, since it will be repeated with a higher degree of complexity: initial pulses will be broken into a higher number of pulses. At some point, this process will lead to complete instability. We can illustrate the

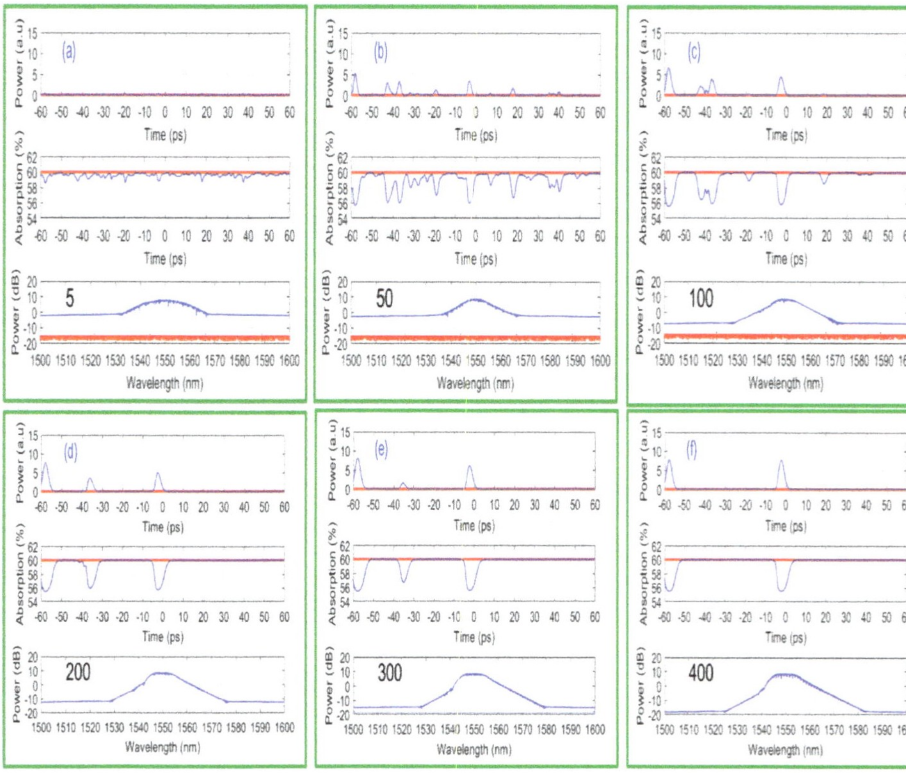

Figure 4.14. Numerical results of modeling and simulation of a mode-lock fiber laser described in example 4.1 with total dispersion $D < 0$. Total gain of the cavity is $G = 25$ dB. Panels (a–f) are results of light intensity (top), saturable absorption (middle) and spectra (bottom) at OC with circulation number $N = 5$, 50, 100, 200, 300 and 400.

situation but in different ways to show the relationship between the gain and the saturable energy of SA. In figure 4.15, we show as an example the situation in which all parameters of the cavity are fixed, only saturated energy of the SA is changed. In doing that, we can see directly how the saturable absorption plays an important role in stability of the mode-locked lasers. Note that this is a unique behavior of mode-locked lasers as compared with cw lasers. The results show in figure 4.15 that when saturated energy is reduced from 50 to 10 pJ the laser intensity increases and shortens slightly. This indicates the pulse energy is close to the saturated energy of SA as absorption reaches the minimum absorption value (in this example it is 74%). For lower saturated energy of SA, we can see double pulses and multiple kinks in the spectrum. The results are observable in experiments, as shown in figure 4.15.

Note that the unstable operation can also occur in many different situations other than the case just presented. For instance, if the PC or the isolator inside the cavity do not work properly, the interference between different light components can also lead to unstable behavior. In other cases with high intensity the high nonlinear effects can also lead to nonlinear instabilities. These are quite complicated situations and are not described in our methods.

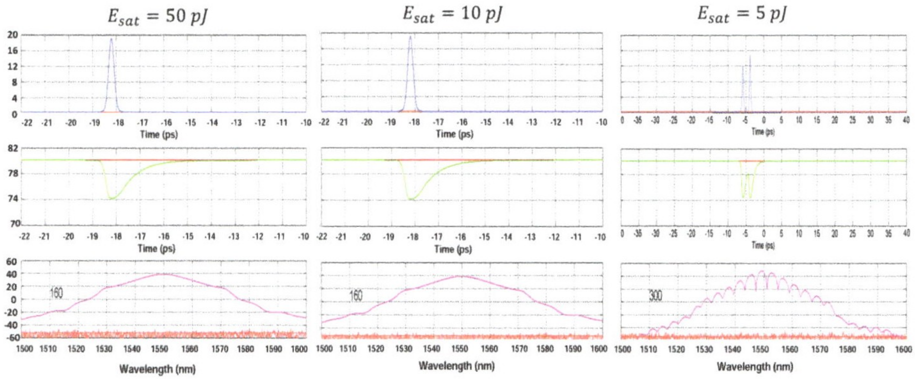

Figure 4.15. Numerical results of modeling and simulation of a mode-lock fiber laser with total dispersion $D > 0$ with different values of saturated energy of SA. All cavity parameters are fixed in all simulations, and only saturated energy of SA is changed from 50, 10 and 5 pJ. In the case $E_{sat} = 5$ pJ, the laser becomes unstable, as is observable in experiments as shown in the inset.

The results in figures 4.13 and 4.14 are with $D < 0$. Let us consider the cavity with total dispersion positive $D > 0$. To make it easy for comparison, we run same program 4.1 but make some changes to the fiber lengths in the configuration in figure 4.12. In figure 4.16, the results of simulations of cavity with $L_{smf} = 2$ m, $L_{dcf} = 0.1$ m and $L_{edf} = 0.2$ m. The total dispersion of the cavity can be estimated as $L_{SMF}D_{SMF} = (2.0\ \text{m}) \times 17\ \text{ps}/(\text{km} \cdot \text{nm}) = 34\ \text{fs nm}^{-1}$, $L_{EDF}D_{EDF} = (0.2\ \text{m}) \times 8\ \text{ps}/(\text{km} \cdot \text{nm}) = 1.6\ \text{fs nm}^{-1}$ and $L_{DCF}D_{DCF} = (0.1\ \text{m}) \times -40\ \text{ps}/(\text{km} \cdot \text{nm}) = -4\ \text{fs nm}^{-1}$. The total dispersion of the cavity is $(LD)_{\text{Total}} = 31.64\ \text{fs nm}^{-1} > 0$.

The different between results in figures 4.13, 4.16 and 4.17 for $D < 0$ and $D > 0$, and $D = 0$, respectively, is very clear. We will discuss it later in this chapter.

Explanation of program 4.1

Although we have placed many comments in program 4.1, we want to have some explanations to make it easier to understand.

Main program: The first part of the main program is listing all necessary parameters that will be used in the calculations and functions. We want to explain some points in this part as below:

1. In the section that starts from

```
%% -- this part is for Er-doped fiber (EDF) -------------------------
```

all parameters of the gain fiber as EDF are given. Note that dispersion D sometimes can be found from specifications of the fibers from the manufacturer. In the program, we use units in picosecond (ps) and meter (m). However, the function that solves the NLSE is written with dispersion as propagation constants β_n, and we only used NLSE with dispersion up to second order β_2. The gain is given as total gain (e.g., 13 dB), but in the NLSE we need to convert it to the unit gain per 1 m^{-1}.

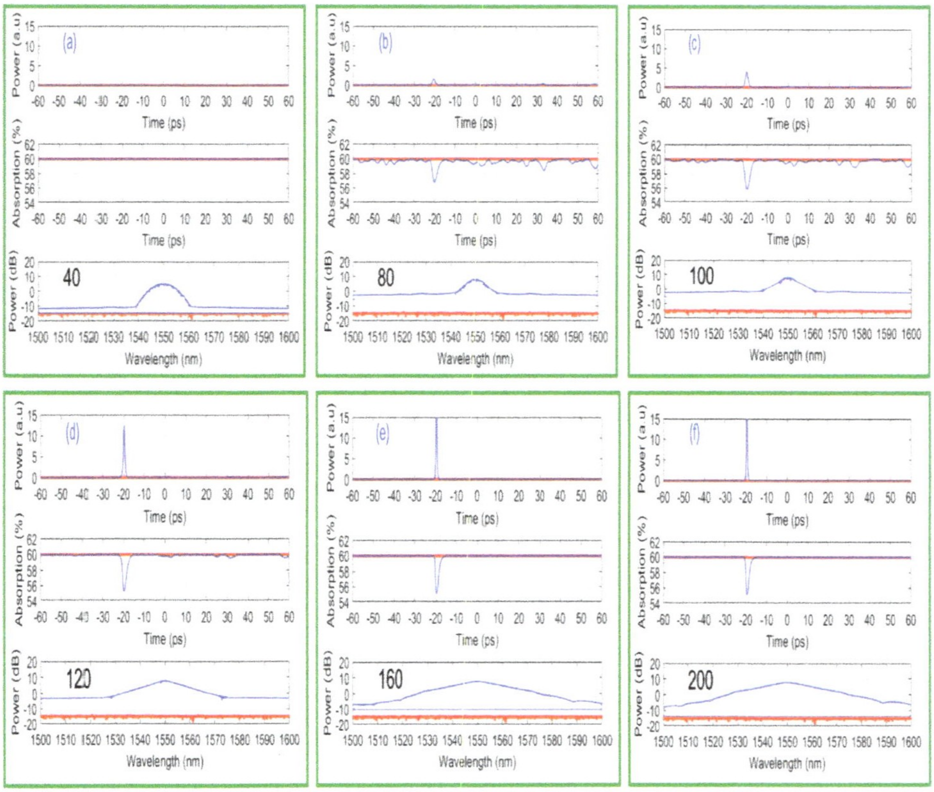

Figure 4.16. Numerical results of modeling and simulation of a mode-lock fiber laser described in example 4.1 with total dispersion $D > 0$. Total gain of the cavity is $G = 13$ dB. Panels (a–f) are results of light intensity (top), saturable absorption (middle) and spectra (bottom) at OC with circulation number $N = 40$, 80, 100, 120, 160 and 200.

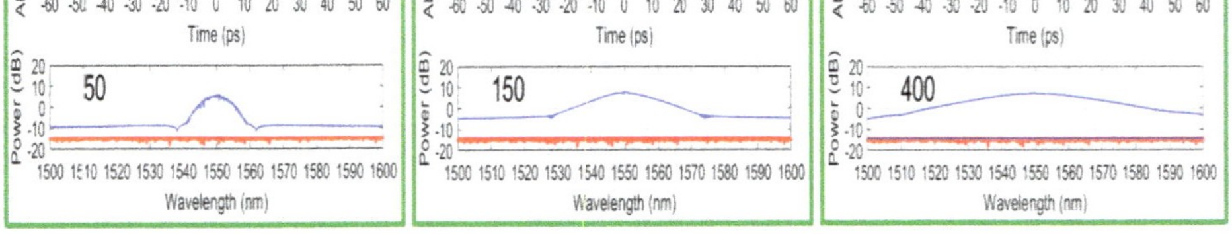

Figure 4.17. Numerical results of modeling and simulation of a mode-lock fiber laser described in example 4.1 with total dispersion $D = 0$. Total gain of the cavity is $G = 13$ dB. Panels (a–c) are results of light intensity (top), saturable absorption (middle) and spectra (bottom) at OC with circulation number $N = 50$, 150, and 400.

2. In the part for SMF that starts from

```
%% --- this part for smf fiber ------------------------------------
```

We can see there are several fiber lengths of the same SMF. In this example we just used three sections of SMF with lengths L_smf1, L_smf2 and L_smf3. The reason for that is these SMF sections are used to connect other parts in the cavity. Since we use the NLSE for one polarized component of electric field, we assume the PC works perfectly in the system, meaning we need to deal with only one polarized component of light in the whole cavity. Therefore, we do not need to describe the effect due to the PC. In many cases in interest, this is not far from the reality. The same thing is assumed for isolation ISO in the cavity. Therefore, we also ignore the counter-propagation of light inside cavity assuming that ISO works perfectly. Actually, in the real cavity more than one ISO is usually used and this condition can be acceptable. Therefore, in a simplified cavity section SMF with L_smf1 connects EDF to OC, L_smf2 with OC to DCF and finally L_smf3 connects DCF to EDF. Of course, we can add more SMF sections to describe more details of the cavity since the connections between different parts will add more coupling losses. In any case, we must account for enough loss of the whole system since it is very important in designing the real system.

3. In the part that starts from

```
%% -----------This part constructs the signal 1550 nm ----------
```

The light is started as random noise assumed to be spontaneous emission from EDF.

```
phase = pi*rand([1,N]);
Unoise = 1e-10.*randn(1,N).*exp(-1i*phase);    % very small noise seeding
Unoise_f = fft(Unoise);                          % in freq-domain
```

4. In the iteration loops, we assume the coupling losses CL1 and CL2. FB is feedback coefficient at the OC, in this case 70% of light will be feedback to the ring cavity at the OC.
5. The iteration loops:
   (i) Light starts as random noise from the input of EDF:

```
Uin_edf = Uin0;
```

   (ii) And propagates through EDF (function `Propagation_EDFL`)

```
Uout_edf = Propagation_EDFL(Uin_edf,tol,L_edf,gain,Esat,alfa_edf,
                        beta_edf,gamma_edf,t_shock,fR;hR_f);%
```

   (iii) Light after EDF couples to SMF and propagates in L_smf1 with coupling CL1.

```
Uin_smf1 = sqrt(CL1)*Uout_edf;
Uout_smf1 = Propagation_SMF(Uin_smf1,tol,L_smf1,alfa_smf,beta_smf,...
                            gamma_smf,t_shock,fR,hR_f); %
```

(iv) Light after L_smfl coupled to SA. Note that, input light to SA is output light from L_smfl `Uout_smf1` and output from SA is `Uouf_sa`.

```
%% propagation through SA
alfa = SA(Uout_smf1, a0, ac, Ts, Es);
Uout_sa = sqrt(1-alfa).*Uout_smf1;
Uouf_sa = fft(Uout_sa);
```

(v) At the OC, an amount of (1-FB) of power (or sqrt(1-FB) of amplitude) of light field output from the cavity. The other part will continue propagating in the cavity and couples to SMF of L_smf2 and then couples to DCF with L_dcf, SMF with L_smf3 and finally enters EDF to complete the circulation in the ring cavity.

Subroutines (functions)
There are several subroutines or functions: oper_D, oper_N as described in the NLSE. The propagation function `Propagation_SMF` can be used for both passive fibers DFC and SMF with different parameters, most important is different dispersions. `Propagation_EDFL` is a subroutine for numerically calculating light propagation in EDF with several parameters such as gain and dispersion, nonlinear index etc. This subroutine will call function rk4ip which is the algorithm fourth order Runge–Kutta in the interaction picture as described earlier. SA is the function for saturable absorption.

4.4 Example of modeling linear cavity mode-locked fiber lasers

In many cases, mode-locked lasers are built in linear cavity configuration in which semiconductor SA mirrors (SESAM) can be used for mode locking. A general configuration of a mode-locked fiber laser in linear cavity is presented in figure 4.18. Note that in many cases the gain medium is not an active fiber as in this configuration. However, the example we will study in this part can be used to simulate these systems with some modifications.

Example 4.2
Let us now consider the laser cavity where the gain fiber is Yb-doped fiber, and pumped by diode 975 nm. The laser operates in the 1 μm wavelength region. The other parts of the cavity are presented in the following. Figure 4.19 shows the specifications, dispersion and reflectance of a SESAM SAM-1040-40-500fs-x manufactured by BATOP Optoelectronics that will be used in example 4.2.

The CFG is chirped fiber grating that can provide a dispersion ~5 ps nm^{-1}. The filter can be assumed of Gaussian shape with bandwidth of 1 nm. The CGF

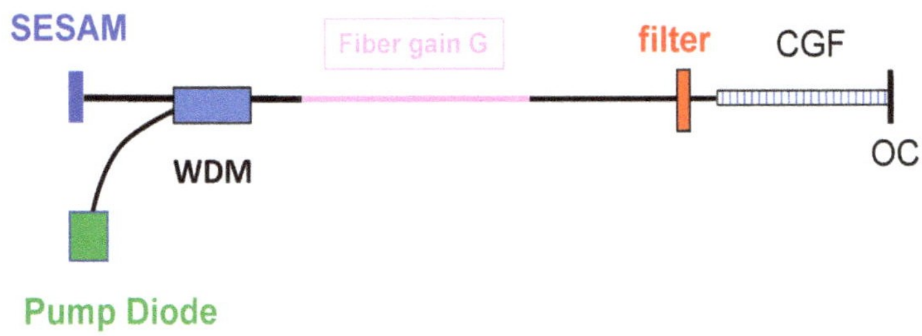

Figure 4.18. Schematic of linear cavity of mode-locked fiber laser. SESAM plays as SA for mode locking, A filter is used to control the bandwidth of the laser spectrum, CGF is a chirped grating fiber for manipulating dispersion of the cavity, and OC is output coupler. The pump diode couples to gain fiber by a WDM.

SAM™ Data Sheet SAM-1040-40-500fs-x, λ = 1040nm

Laser wavelength	λ = 1040 nm
High reflection band (R > 55%)	λ = 1010 .. 1100 nm
Absorbance	A_0 = 40 %
Modulation depth	ΔR = 24 %
Non-saturable loss	A_{ns} = 16 %
Saturation fluence	Φ_{sat} = 120 µJ/cm^2
Relaxation time constant	τ ~ 500 fs
Damage threshold	Φ = 1 mJ/cm^2
Chip area	4mm x 4mm; other dimensions on request
Chip thickness	400 µm; optional: 150 µm on request
Protection	the SAM is protected with a dielectric front layer

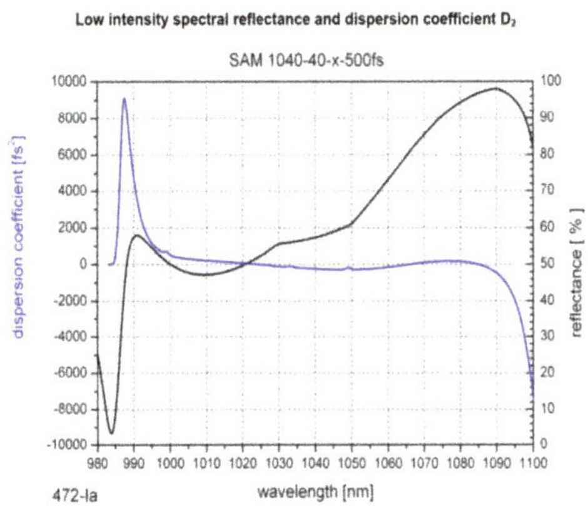

Figure 4.19. Specifications (left), reflectance spectrum and dispersion of SESAM (right) used in example 4.2.

$$IL = 2(3 + 3.5 + 0.2 + 0.2 + 0.2) + 2.2 = 16.4 \text{ dB}$$

Figure 4.20. Configuration of mode-locked laser with linear cavity and the losses in example 4.2.

transmission is 50% and reflection 50% or OC has output coefficient 50% (3 dB). The filter has loss of ~3.5 dB. The linear reflection by SESAM is 60% or ~2.2 dB. We want to stress that these specifications are taken from the real devices used in experiments of the mode-locked linear cavity fiber laser.

First, let us analyze the total loss of the cavity. We assume the parts of the cavity are connected by SMF, and the coupling loss between different parts through the SMF is about 0.2 dB denoted by X in the diagram of figure 4.20. The total loss for light propagation in one round trip of the cavity can be estimated from the output end of the CGF, and we have $IL = 16.4$ dB. The light gets amplification twice in one round trip, therefore, the threshold condition for lasing is the total gain of single-pass of the gain fiber, and is $G_0 = 8.2$ dB.

In general, the main program of program 4.2 is similar to program 4.1, but there are differences that come from configurations, linear and ring cavities, and also the different operations of SA in these cavities. For example, in ring cavity the quantity that is important from the SA is nonlinear transmission of light that gets through the SA sample. In a linear cavity with SESAM working as a mirror, the nonlinear reflection is the one that plays an important role. Therefore, the calculation of light affected by SA can be written as below.

This part is for description of SESAM. Notice that the parameters are taken from the data sheet from figure 4.19 above.

```
%% -----------This part for SA CNT ----------

a0 = 0.24;
ac = 0.16;
Ts = 500e-15;
rs = 6.6e-4;                    % spot radius in SESAm (in cm)
As = pi*rs^2;                   % spot area in cm^2
Fs = 120e-9;                    % Saturation Fluence in uJ/cm^2
Es = As*Fs                      % Saturation energy
```

This part is for calculating light affected by SESAM: the light propagates from a section of SMF, and enters mirror SESAM, therefore, the input of calculation is the output field from SMF `Uout_smf1f`. The light is then absorbed by SA (absorption coefficient `alfa`) and reflected by `Rsa = 1-alfa`. The subroutine for SA `SA (Uout_smf1f, a0, ac, Ts, Es)` is the same as in program 4.1.

```
%% Reflection through SA
alfa = SA(Uout_smf1f, a0, ac, Ts, Es);
Rsa  = 1-alfa;
Uout_sa = sqrt(Rsa).*Uout_smf1f;
Uouf_sa = fft(Uout_sa);
```

A new part in the above linear cavity is the filter which is used to control the bandwidth of the pulse spectrum. In the main program it has the simple operation

```
dw = 1.0;                       % dw: bandwidth of filter
GF = Filter(lambda0,dw);
```

The filter subroutine creates a Gaussian form in frequency domain (converted to wavelength) with center wavelength `lambda0` and bandwidth `dw` (unit nm in the program). In the iteration loop of light propagating in the cavity, the filter operation can be written as

```
%% Filter
Uout_filter = sqrt(FL)*ifft(GF.*fft(Uout_smf2b));
```

Here the FL is power loss of the filter, therefore the sqrt is for amplitude of electric field in the calculation. The filter operates upon the output field from a section SMF `Uout_smf2b`. The subroutine of the filter can be written as

```
function v = Filter(lambda0,dw)
%xxxxxxxxxxxxxxxxxxxxxxxxxxxxxxxxxxxxxxxxxxxxxxxxxxxxxxxxxxxxxxxxxxxxx
% Filter.m                                                            x
% Caculate Filter function                                            x
%xxxxxxxxxxxxxxxxxxxxxxxxxxxxxxxxxxxxxxxxxxxxxxxxxxxxxxxxxxxxxxxxxxxxx

%% ---- parameters for the simulation and a part of the pulse ---------
c0 = 3e8;                   % (m/s)
lamb0 = lambda0;            % (m)% pump wavelength
global f lambdaS            % frequency, globally used
wS = 2*pi*c0/lambdaS;       % Angular frequency of signal light
w = 2*pi*f;                 % vector angular freq
wshift = -w + wS;           % plus the freq with its central freq.
lambda = 2*pi*c0./wshift;%(nm), shift to nm to present

v = exp(-(1.3e9*(lambda - lamb0)/dw).^2);
```

Another new part of the above linear cavity that can be approximately modelled is the CGF. Modeling the whole structure of CGF is quite complicated, but in this problem the CGF function is similar to dispersion compensation component. In this example, the dispersion of the CGF in region 1040 nm is about 5 ps nm^{-1}, and the length of that grating is about 5 cm. Therefore, we can model the CGF as a dispersion compensated fiber as below.

```
%% --- this part for DSF fiber ------------------------------------------
   L_dsf = 5e-2;               % Length of (DSF) Chirped Grating Fiber
  AdB_dsf = 1.0;               % (dB/km) loss of DSF
alfa_dsf = AdB_dsf/4343;       % (1/m) loss
MFD_dsf  = 7.0e-6;             % MFD of the fiber, assume 10um (check)
Aeff_dsf = pi*(MFD_dsf/2)^2;          % effective core area
gamma_dsf= 2*pi*n2/lambdaP/Aeff_dsf;% nonlinear coefficient (1/W/m)

D_dsfxL = 5e9;                        % product D_dsf x L (ps/nm = 5e9*ps/m)
  D_dsf = D_dsfxL/L_dsf;              % Disperion of Chirp Grating in
ps/(m*m)

   b2_dsf = - lambdaP^2/(2*pi*c0*1e-12)*D_dsf
                                      % [beta1, beta2, beta3...] in ps^k/m
beta_dsf = [0 b2_dsf];

                          % change unit of beta to (s^k/m)
 for ii = 1:length(beta_dsf)
      beta_dsf(ii) = (1e-12)^(ii)*beta_dsf(ii);
 end
====
```

After these modifications from program 4.1, we can easily write the program for simulating the mode-locked fiber laser in linear cavity using SESAM. Note that in the linear cavity, for each circulation inside the cavity we have to calculate the propagation in a round trip, which means light will hit twice parts inside the cavity (filter, gain fiber) but only once the end-cavity parts such as SESAM and OC, where, in this case, the CGF plays roles of OC and DCF.

In the following we will present some simulation results and discuss some points that are useful for designing works.

Figure 4.21 shows simulation results for a cavity that has total cavity fiber length 3 m including 0.6 m of Yb-doped fiber, and total gain is $G = 8.3$ dB, just above the

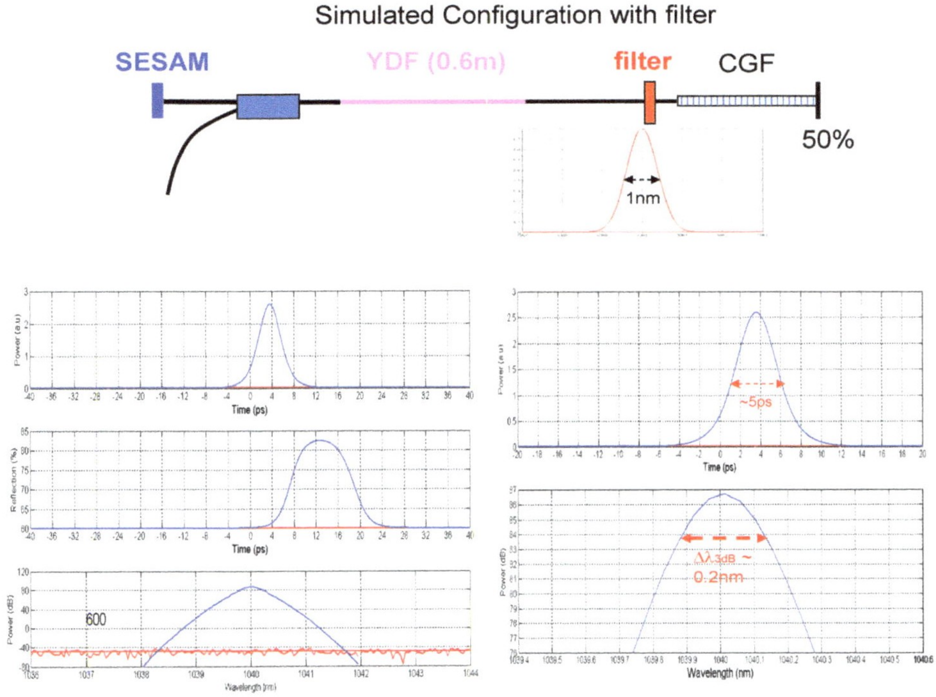

Figure 4.21. Numerical results of modeling and simulation of a mode-lock fiber laser in linear cavity using SESAM described in example 4.2. Total gain is 8.3 dB in the simulation.

threshold 8.2 dB. the filter bandwidth dw = 1 nm. Other parameters are given above for CGF and SA. Note that the SESAM has linear reflection R = 60% and modulation depth ΔR = 24%, meaning the maximum reflection is 84%. We can see the values of reflection in figure 4.21.

It is worth noting that the simulation results above are very close to the observation in experiment [24]. However, the laser is very sensitive to perturbation and can easily become unstable. Note again that the results in figure 4.21 are obtained with the total gain G = 8.3 dB, just above the threshold condition of total gain G = 8.2 dB. The results in figure 4.22 are simulated with total gain G = 9.7 dB (or 1.5 dB higher than threshold conditions.) Similar to the unstable operation in ring cavity (e.g., figure 4.15) due to relatively high energy of light inside the cavity, the SESAM is quickly saturated in a long period of time, and the nonlinear reflection has high value in that period, as shown in panel (a) with N = 44. The evolution of light inside the cavity clearly shows from panel (a) to (b), (c) and (d) that the pulse with long duration is gradually broken into two shorter pulses in each repetition. Notice how the nonlinear reflection in these panels behaves. The result of multiple pulses in each repetition of the mode-locked laser creates the multiple kinks in the output spectrum and the unstable situation of the laser, as shown in panels (d–f), similar to the observation in figure 4.15. Note that, in this example, the simulation results with total gain of 9.6 dB are still similar to the case with 8.3 dB which is

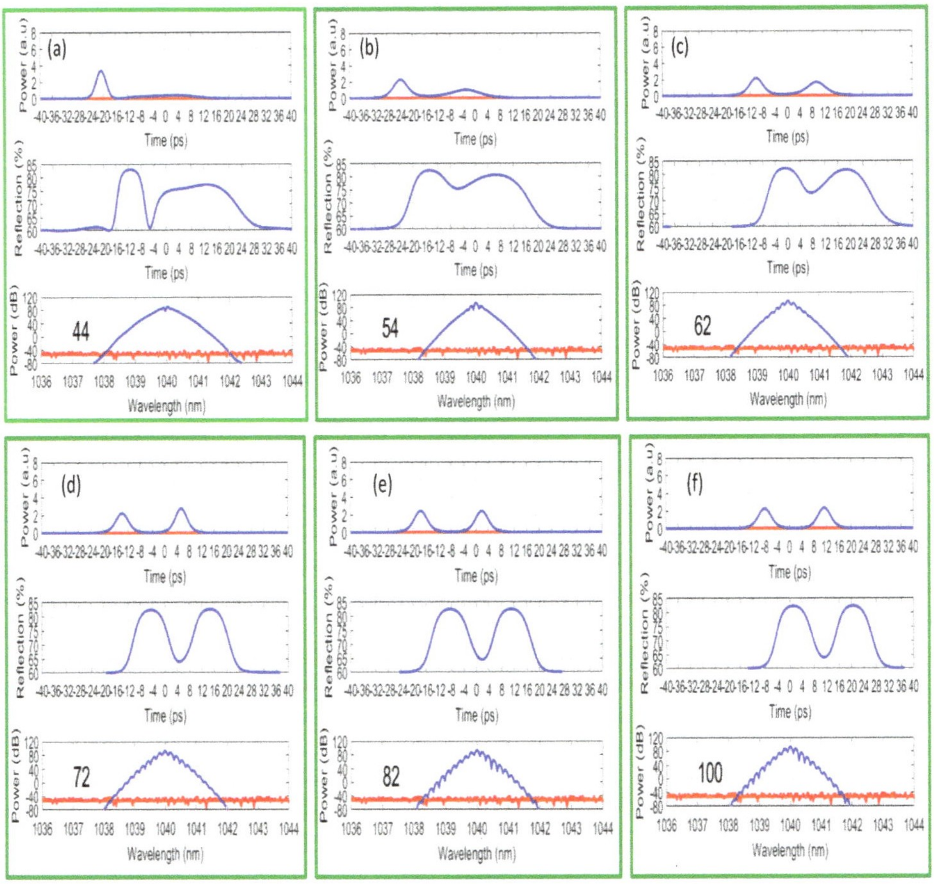

Figure 4.22. Numerical results of modeling and simulation of a mode-lock fiber laser in linear cavity using SESAM described in example 4.2. Total gain is 9.7 dB in the simulation.

stable. The results in figure 4.22 show the laser becomes unstable when the total gain is just higher than threshold by about 1.5 dB.

As discussed earlier, the reason for unstable mode-locked laser in this case is the light inside the cavity accumulating energy relatively highly as compared with saturated energy of SA. In general, replacing SESAM with higher saturated energy is a simple solution. However, in this case from simulation we can find another solution that does not require replacing SESAM by a new one.

From the results in figure 4.22, we can see that the SESAM is quickly saturated with relative low gain and the nonlinear reflection is high in a long period of time. Meanwhile, the filter with $\Delta\lambda = 1$ nm would allow a shorter pulse in the mode-locking operation. As a result, the nonlinear reflection is broken down into multiple parts that support for multiple pulses in a laser repetition. And that leads to instability due to the interference between these pulses. Now, if the light is forced to lock with longer pulse duration or narrower spectrum, then we could avoid this

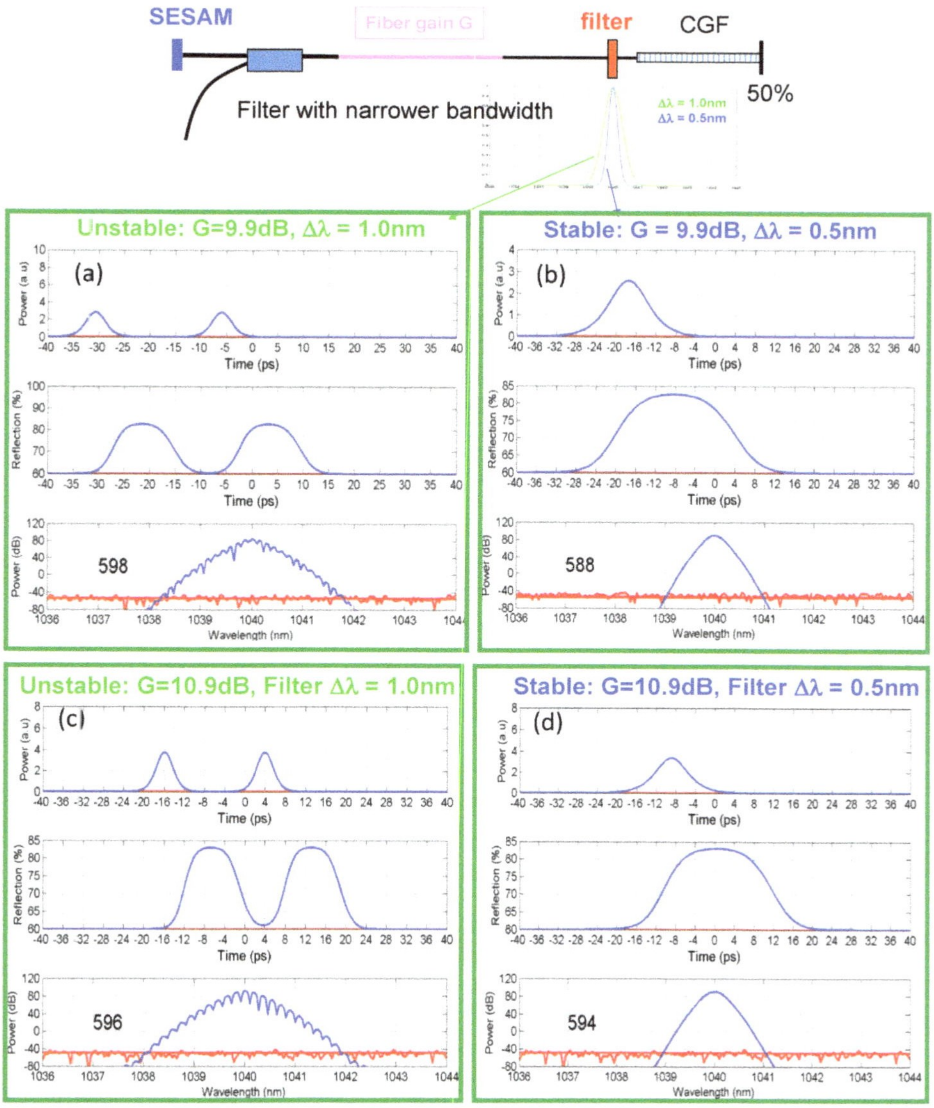

Figure 4.23. Numerical results of modeling and simulation of a mode-lock fiber laser in linear cavity using SESAM as described in example 4.2. Panels (a–c) are results of light intensity (top), reflectance by SESAM (middle) and spectra (bottom) at the OC with circulation number N = 10, 40, and 100.

phenomenon. It can be easily done by replacing a filter with narrower bandwidth. In figure 4.23 we show the simulation results using two filters with $\Delta\lambda = 1$ nm (left) and 0.5 nm (right).

Exercise 4.1

From the subroutine for saturable absorption of SA, calculate and obtain the results in figures 4.8 and 4.9.

References

[1] Keller U *et al* 1992 Solid-state low-loss intracavity saturable absorber for Nd:YLF lasers: an anti-resonant semiconductor Fabry–Perot saturable absorber *Opt. Lett.* **17** 505–7

[2] Keller U *et al* 1996 Semiconductor saturable absorber mirrors (SESAMs) for femtosecond to nanosecond pulse generation in solid-state lasers *IEEE J. Sel. Top. Quantum Electron.* **2** 435–53

[3] Keller U 2010 Ultrafast solid-state laser oscillators: a success story for the last 20 years with no end in sight *Appl. Phys.* **B100** 15–28

[4] Haus H A 1975 Theory of mode locking with a fast saturable absorber *J. Appl. Phys.* **46** 3049

[5] Haus H A 1975 Theory of mode locking with a slow saturable absorber *IEEE J. Quantum Electron.* **11** 736

[6] Haus H A 2000 Mode-locking of lasers *IEEE J. Sel. Top. Quantum Electron.* **6** 1173–85

[7] Martinez E, Fork R L and Gordon J P 1985 Theory of passively mode-locked lasers for the cast of a nonlinear complex propagation coefficient *J. Opt. Soc. Amer.* B **2** 753–60

[8] Agrawal G P 2007 *Nonlinear Fiber Optics* 4th edn (San Diego, CA: Academic)

[9] Hult J 2007 A fourth-order Runge–Kutta in the interaction picture method for simulating supercontinuum generation in optical fibers *J. Lightwave Technol.* **25** 3770–5

[10] Blow K J and Wood D 1989 Theoretical description of transient stimulated Raman scattering in optical fibers *IEEE J. Quantum Electron.* **25** 2665–73

[11] Saleh B E A and Teich M C 1991 *Fundamentals of Photonics* (New York: Wiley)

[12] Renninger W H, Wise F W, Grelu P, Shlizerman E and Kutz J N High-energy passive mode-locking of fiber lasers *Intl. J. Optics* **2012** 354156

[13] Komarov A, Leblond H and Sanchez F 2005 Multistability and hysteresis phenomena in passively mode-locked fiber lasers, *Phys. Rev.* A **71** 053809

[14] Ghatak A and Thyagarajan K 1998 *Introduction to Fiber Optics* (Cambridge: Cambridge University Press)

[15] Kodoma Y and Hasegawa A 1987 Nonlinear pulse propagation in a monomode dielectric guide, *IEEE J. Quantum Electron.* **QE-23** 510–24

[16] Dudley J M and Coen S 2002 Numerical simulations and coherence properties of super-continuum generation in photonic crystal and tapered optical fibers *IEEE J. Sel. Topics Quantum Electron.* **8** 651–9

[17] Stolen R H, Gordon J P, Tomlinson W J and Haus H A 1989 Raman response function of silica-core fibers, *J. Opt. Soc. Am. B, Opt. Phys.* **6** 1159–66

[18] Hohage T and Schmidt F 2002 On the numerical solution of nonlinear Schrödinger type equations in fiber optics *Konrad-Zuse-Zentrum für Informationstechnik* Berlin, Germany, Tech. Rep. ZIB-Report 02-04

[19] Reeves W H, Skyabin D V, Biancalana F, Knight J C, Omenetto F G, Efimov A and Taylor A J 2003 Transformation and control of ultra-short pulses in dispersion-engineered photonic crystal fibres, *Nature* **424** 511–5

[20] Hillingsøe K M, Paulsen H N, Thøgersen J, Keiding S R and Larsen J J 2003 Initial steps of supercontinuum generation in photonic crystal fibers *J. Opt. Soc. Am. B, Opt. Phys.* **20** 1887–93

[21] Siederdissen T H Z, Nielsen N C, Kuhl J and Giessen H 2006 Influence of near-resonant self-phase modulation on pulse propagation in semiconductors, *J. Opt. Soc. Am. B, Opt. Phys.* **23** 1360–70

[22] Muslu G M and Erbay H A 2005 Higher-order split-step Fourier schemes for the generalized nonlinear Schrödinger equation *Math. Comput. Simul.* **67** 581–95

[23] Caradoc-Davies B M 2000 Vortex dynamics in Bose–Einstein condensates *PhD Dissertation* Univ. Otago, Dunedin, New Zealand http://www.physics.otago.ac.nz/bec2/bmcd/phdthesis/)

[24] Inoue Y, Nguyen D T and Kieu K 2017 Stable, high efficiency, wavelength tunable fiber optic parametric oscillators *Patent US 9,557,626 B2*

Chapter 5

Chirped pulse fiber amplifiers

5.1 Background

Chirped-pulse amplification (CPA) is a technique for generating high energy, ultrashort pulse lasers. High power, ultrashort pulse lasers are usually required for numerous tasks from studying fundamental physics principles to important applications including cancer treatments, laser eye surgery, laser materials processing etc. It is worth mentioning here that the 2018 Nobel Prize in Physics was jointly awarded to Gérard Mourou and Donna Strickland for their work to develop CPA technology in the 1980s [1, 2]. CPA was the main research work of Strickland for her 1988 PhD dissertation at Rochester University [3]. CPA systems have been developed for decades using different configurations, but a typical CPA system can be described generally as in figure 5.1. The system consists of four main stages: (i) short pulse laser source (usually in picosecond or femtosecond regimes); (ii) pulse stretcher; (iii) power amplifier; and (iv) pulse compressor. In general, the energy of pulses in the first two stages is low, but it could be high to very high in the last two stages. The second stage, the pulse stretcher is used to stretch the originally ultrashort pulses into much longer duration pulses and therefore can be amplified in the third stage to much higher energy but its peak intensity can be kept under some desired thresholds. By keeping peak intensity not too high, material damage and nonlinear distortions can be minimized so that the stretched pulse can be compressed in the final stage. In the 4th stage—the compressor—the high-energy stretched pulses are compressed back as close as possible to the original pulse duration. By doing it that way, the high energy pulses can be compressed to ultrashort, and as a result pulses with high or very high peak power/intensity can be achieved. In many situations where the laser source (first stage) is either too weak and/or other components have very high losses then extra amplifiers can be added before and/or after the second stage. Similarly, extra power amplifiers can be added between the third and fourth stages. Because CPA is a matured technique with a good number of technical review papers and

doi:10.1088/978-0-7503-2272-0ch5

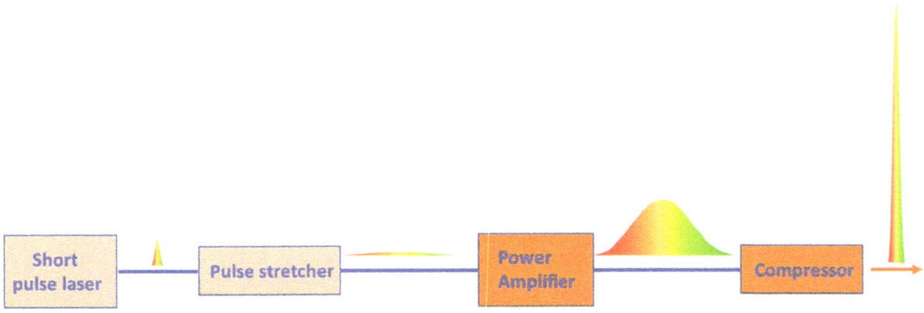

Figure 5.1. Schematic of a typical CPA system.

textbooks [1–6], readers are encouraged to study and understand both the fundamental physics and the techniques in these references.

It has been demonstrated that CPA systems that are built on bulk crystals can deliver the highest energy pulses. However, these systems are usually complex, bulky, and difficult for alignment and maintenance. As a result, it is usually difficult and expensive to use these systems in many practical applications. Therefore, developments of CPA systems that are compact and highly integrated have attracted a lot of efforts.

The fiber-based CPA systems (FCPA) were first developed in the 1990s and are highly attractive as femtosecond class laser pulse sources, owing to the excellent laser stability, reliability and thermal management enabled by the fiber platform [7–10]. The highly integrated features of the fiber-based CPA systems, whose many key parts are conveniently integrated with low loss and high stability, are their unique advantage in comparison with other systems. We have already known from earlier chapters that ultrashort lasers (first stage) can be made with mode-locked fiber lasers. Meanwhile, high power amplifiers (third stage) can be made by cladding-pumped fiber amplifiers, especially new types of fibers with large mode area (LMA) cores in which the fiber cores are much larger those that of conventional single-mode (SM) fibers. LMA fibers are specially designed to be effectively operated in the SM regime to ensure not only high power amplifiers but high beam quality. Ultrashort pulses from mode-locked fiber lasers can be stretched into much longer duration ones using chirped fiber Bragg gratings (CFBGs) or other dispersive fiber-based components. In some situations, an SM fiber with high dispersion can be used as a stretcher. In the following example that we will study in this chapter, we will present a CPA system that only uses an SM fiber that is widely used in optical communication—the SMF-28 fiber for illustrating a stretcher of pulses around 2 μm. Figure 5.2 shows a schematic of the fiber-based CPA system operated in the 2 μm-region using a Tm-doped fiber laser and fiber amplifiers.

Note that we intentionally make some clear changes in figure 5.2 by replacing several key parts in figure 5.1 by fiber-based components. Stage 1 is now a mode-locked Tm-doped fiber laser and the pulse stretcher in stage 2 is represented by a fiber, in this case SMF-28. Stage 3 is represented by a cascade of fiber amplifiers

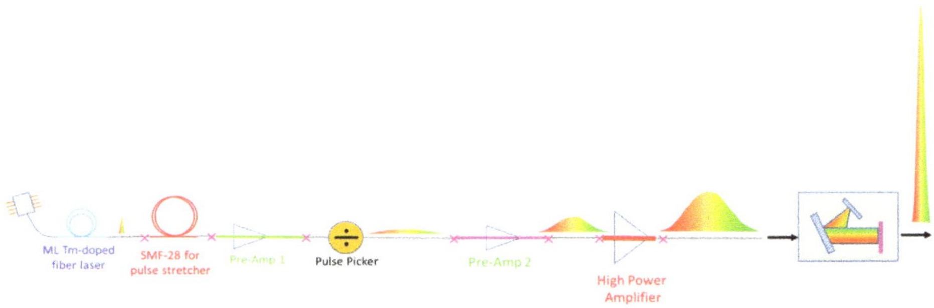

Figure 5.2. Schematic of a fiber-based CPA system operated in the 2 μm-region using a Tm-doped fiber laser and fiber amplifiers.

consisting of fiber pre-amps and high power amplifier using special LMA fibers. The fiber amplifiers in figure 5.2 are also intentionally presented in short fiber lengths, and that will be explained in more detail later. In designing the system, losses and gains of every stage including the coupling losses between stages must be carefully accounted for when optimizing the system. A well designed CPA system is to achieve as high energy as possible but to keep peak intensity under damage threshold and nonlinear distortion. Note that the repetition rate of modelocked (ML) fiber lasers is typically in the range of tens of MHz, while high energy short lasers for many applications are in the range of 100 kHz or even in kHz ranges. Therefore, usually acousto-optic modulators (AOMs) or *pulse pickers* are used to reduce the pulse repetition rate from tens of MHz to kHz range. In general, this step have very high extinction or very high loss, technically. Therefore, one or even a few pre-amps (only one shown in figure 5.2) are usually added before and/or after stage 2. The high power amplifiers (or power-amp) in fiber-based CPA systems are typically short length, high gain fibers to achieve an optimal balance of high energy output and pulse quality. The last stage—the compressor—is more complicated and usually is not made with fibers. The reason for that is the pulse energy is usually very high in this stage, and therefore peak intensity could be extremely high in any fiber platform.

In the following, we will explain the details of modeling and simulation of each part of the CPA system illustrated by a MATLAB® program. However, we will focus only on the fiber-based parts, e.g., stages 1, 2 and 3 of the system. Simulation of the compressor would require ray-tracing commercial software such as ZEMAX, and is therefore beyond the scope of this chapter. However, we will use a dispersive fiber to illustrate how the compressor works. Most importantly, the effects of high-order dispersions in compressor state can be illustrated.

5.2 Example: CPA system based on Tm-doped fiber lasers and amplifiers

In this example, we will present the modeling and simulation of a fiber-based CPA system operated in the 2 μm wavelength region. In general, such an ultrashort pulse source can be generated by ML fiber lasers using Tm-doped or Ho–Tm co-doped fibers typically operated in 1.8 to ~2 μm wavelength region, or Ho-doped fibers which can generate light with wavelengths that are ~2 to 2.1 μm or even longer. Note that in this wavelength region (called 2 μm region for short) fiber SMF-28 has very high dispersion and acceptable loss, although the fiber has been widely used in optical communication and has extremely low loss in the 1.55 μm region. In this example, we will show the reasonably low loss and high dispersion SMF can be used as a stretcher. We will use MATLAB® programs to simulate the pulse stretching amplification in the fiber-based CPA system. The effects of high orders of dispersion of the compressor will also be presented. However, the simulations of ML fiber laser and cladding-pumped fiber amplifiers for power-amp can be re-used from earlier chapters with some modifications, and will not be repeated in this chapter.

Before going into details of the MATLAB® program of modeling and simulation, let us describe briefly the physics and working mechanism of Tm-doped fiber laser and amplifiers used in the system as shown in figure 5.2.

5.2.1 Tm-doped fiber lasers and fiber amplifiers

Tm-doped fiber lasers and amplifiers have many common features as in the cases of Er-doped and Yb-doped fiber ones presented in previous chapters. However, Tm-doped fiber lasers and amplifiers have important features which would need a book to describe in detail. Here, one of our main tasks is to use Tm-doped fiber laser and amplifiers operated in the 1.8–2 μm wavelength region (2 μm region) to illustrate the works of modeling and simulations of a fiber-based CPA system. It is quite straightforward to apply the simulation method in this example to other systems of Er-doped or Yb-doped fibers and others. First, in order to understand the model and the physics of the amplification/lasing processes in Tm-doped fiber in the 2 μm region, we will describe briefly the pumping scheme and stimulated emission in the 2 μm region of Tm-doped fiber. Our aim is to make the description in the simplest way so that we can understand it in the example. Readers are encouraged to read more comprehensive description and presentation of the topics in some well known references on the topics [11–14].

Figure 5.3 shows the underlying mechanism for the highly efficient amplification process, in which a Tm^{+3}-doped fiber is pumped by laser diode of $\lambda_p \sim 790$–800 nm, and stimulated emission spans a broad region of wavelength λ_S indicated as Region I, usually called 2 μm region of Tm-doped glasses. In this pumping scheme, Tm^{+3}-ions in ground state (the H_6-state) absorb pumping photons $\sim$ 790 nm and are excited to H_4-state, from there the ions are relaxed to F_4-state by emitting photons of wavelength in Region II, usually called 1.5 μm region of Tm-doped glasses. From F_4-state the Tm^{+3}-ions can be relaxed further to H_6-state by emitting photons belonging to the broad emission region (Region I) in figure 5.4(b). The blue- and green-dashed arrows point from figures 5.4(a) and (b) to indicate the two relaxation processes corresponding to Region I and II of stimulated emission, respectively. In a low dopant concentrations regime, where Tm^{+3}-ions are spatially distanced with negligibly weak interactions, for each absorbed photon of $\lambda_p \sim 790$ nm the Tm^{+3}-ions

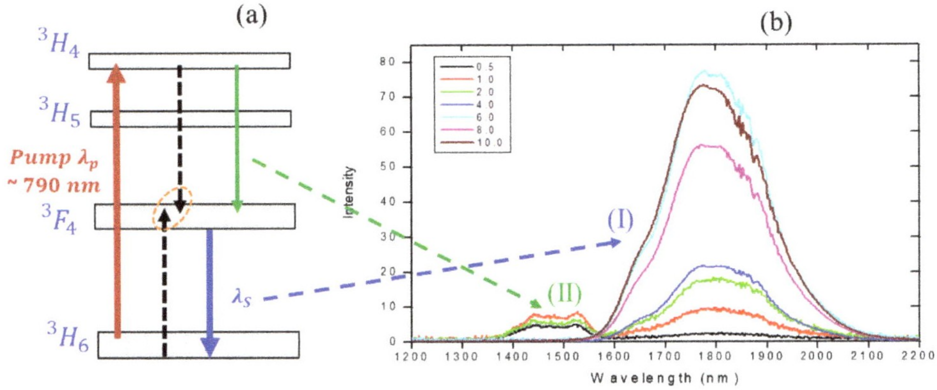

Figure 5.3. (a) Energy diagram of Tm^{+3}-ions, with pumping at ~790–800 nm and lasing/amplification at region I (2 μm region); shown in (b) stimulated emission spectrum of Tm-doped glass with different Tm_2O_3 concentrations (in weight %). Explanation is in the text.

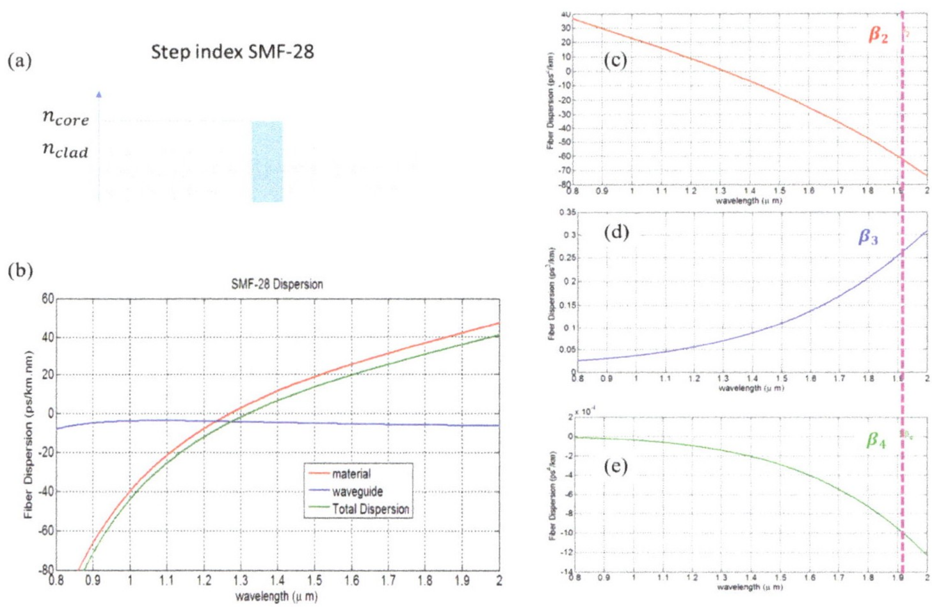

Figure 5.4. (a) Schematic of step-index SMF-28 fiber, (b) calculated dispersion of step-index SMF-28, (c–e) calculated different orders $\beta_{2,3,4}$ of the fiber. The calculations are presented in subsection 4.4.2. A red-dashed line indicates the central wavelength 1.92 μm of laser pulse in the example.

can emit one photon in Region I and one in Region II. That characteristic is clearly shown by a black curve in figure 5.4(b) for a glass doped with 1 wt% Tm concentration. Increasing the dopant concentration, the Tm^{+3}-ions become spatially close and they can transfer energy from one to others. Now, after absorbing one pumping photon, a Tm^{+3}-ion is excited into H_4-state, where it can relax to F_4-state without emitting a photon of Region I. Instead the ion (donor) transfers the energy to a nearby ion being in ground state (H_6-state). The Tm^{+3}-ion that accepted the transferred energy (acceptor) is then excited into to F_4-state, resulting in both Tm^{+3}-ions, the donor and the acceptor being in F_4-state. From there, they can emit two photons in the 2 μm region and both go down to the ground state. In other words, in high concentration regime of each absorbed pump photon (~790 nm) the Tm-doped ions can generate two photons in the 2 μm region. This cross relaxation process is presented by two black-dashed arrows in figure 5.4(a) indicating two Tm^{+3}-ions are excited to F_4-state after one pumping photon is absorbed. This processes can be used to explain the increasing emission in 2 μm region with increasing Tm-concentrations (up to a limit) as shown in figure 5.4(b). The cross relaxation process is the mechanism of high efficiency output lasing or the so-called 200% *quantum efficiency* in highly Tm-doped glass fibers [15, 16]. Note that it is possible to use another pumping scheme to generate stimulated emission in the 2 μm region. Instead of pumping at and around 790 nm, now Tm-doped fibers are pumped by laser light of wavelengths in ~1.6 to 1.7 μm (in-band pumping) so that Tm-ions in ground state (H_6-state) are excited to upper-substates of F_4-state. In this pumping scheme, a simpler 2-level system model can be used for modeling and simulation. However, pumping sources at this wavelength band are typically not high power, so this pumping scheme is usually used for low power lasers or amplifiers, but it is not suitable for high power amplifiers that are required for CPA. Meanwhile, diode lasers at and around 790 nm can be very high power, therefore, they are usually used for cladding-pumped high power Tm-doped fiber amplifiers. Furthermore, high power amplifiers in fiber-based CPA systems should employ short length, highly doped LMA core fibers to achieve an optimal balance of output pulse energy, pulse quality, and slope efficiency. The importance of these requirements will be discussed later in this chapter.

Once the pumping schemes and mechanisms of lasing and amplification in Tm-doped fibers are understood, one can apply the methods of modeling and simulation presented in chapters 2 and 3, for cladding-pumped fiber amplifiers for high-power amplifiers and ML fiber lasers for the ultrashort lasers, respectively. Therefore, we will not repeat these tasks here. Instead, we will focus on other parts of the fiber-based CPA system.

It is worth mentioning here that in precision manufacturing systems, high energy femtosecond pulses are used for cutting and drilling through various materials at faster rates without imparting a heat affected zone. Generating a high-energy femtosecond beam with fiber-based CPA systems provides the most compact and stable form factor [17, 18]. In practice, the design of the system has to take into account several important conditions, among them, restraining peak power well below the damage threshold of all components of the CPA system. Another condition for energy scaling of femtosecond fiber laser systems is to minimize the temporal pulse distortion. The distortion is caused by self-phase modulation (SPM) as the pulses propagate through the nonlinear materials, especially in the amplifiers. The magnitude of SPM is quantified by the B-integral [17, 18] defined in equation

(5.1), which is proportional to both the beam intensity, propagation length and nonlinear refractive index of the material.

$$B = \frac{2\pi}{\lambda}\int_0^L n_2 I(z)dz = \frac{2\pi}{\lambda A_{eff}}\int_0^L n_2 P(z)dz. \tag{5.1}$$

Here, λ is the laser wavelength, n_2 is the nonlinear refractive index coefficient, $I(z)$ is the pulse peak intensity along the propagation direction, $P(z)$ is the pulse peak power, L is the total fiber length, and A_{eff} is the fiber effective mode area. Typically, if the B-integral of a system is less than π radians the system is considered linear propagation, meaning the nonlinear distortion is reasonably acceptable. Above this value, nonlinear phase accumulation will impose temporally broadened pulses and/ or significant pulse pedestal, resulting in undesired beam quality.

It is clear from equation (5.1) that a simple approach for minimizing values of B-integral or nonlinear pulse distortions in fiber amplifiers is to increase A_{eff} proportionally to the desired energy of the output pulses. However, keeping $B < \pi$ in fiber amplifiers, especially high power ones is very challenging given the fact that the fiber cores are typically small, in micron-scale sizes, to operate in robust SM condition, and maintain the diffraction-limited spatial beam quality. It is well known that a fiber that satisfies condition $V = 2\pi aNA/\lambda \leqslant 2.045$ is an SM fiber. Under such conditions a large A_{eff} SM fiber can be designed by increasing core diameter and decreasing core numerical aperture (NA). However, the A_{eff} of fibers designed in that simple way have very limited sizes since fibers with too low NA tend to be very weak waveguides, and therefore are extremely sensitive to fiber bending or other perturbations. Recent progress in the field of high power fiber amplifiers has achieved LMA fibers with robust SM operation having effective mode area (A_{eff}) much larger than that of standard SM fibers. SM-operated LMF fibers have been demonstrated with A_{eff} that reaches values of hundreds or even thousands of μm^2 as compared to below 100 μm^2 of standard SM fibers. Different fiber designs having different microstructures and guiding mechanisms such as photonics crystal fibers (PCFs) [19], leakage channel fiber (LCF) [20, 21], chiral fiber [22] etc, have been developed. Increasing dopant concentrations in fiber core for higher pump absorption in short fiber lengths is another important strategy for reducing B-integral [17, 18]. Note that, however, not all host materials can be doped with high dopant concentration. The problem is complicated and belongs to material science, which is beyond the scope of this chapter. In any case, when designing a fiber-based CPA system, keep in mind that high dopant concentrations and LMA cores are very important for achieving not only high power amplifiers but also shortening the fiber length for minimizing nonlinear distortions. Readers can easily find more information of these topics in comprehensive reviews [23, 24].

5.2.2 All-fiber pulse stretcher

As stated earlier, a pulse stretcher is a dispersive device that can be used to stretch a short pulse into a much longer duration pulse. Different photonic structures or components can be used for pulse stretching such as CFBGs and dispersive fibers like dispersion compensated fiber (DCF) that have been used in examples of ML fiber lasers in chapter 4. In other systems, the chromatic dispersion of an optical component is the result of interference effects. For example, interference effects in optical resonator can generate a component dispersion, which is a huge difference with the material dispersion. Dispersion can also result from geometric effects. As can be seen in the example of a prism pair often used for dispersion compensation in early-developed ML lasers. Here, the well-known wavelength-dependent refraction at prism surfaces is the reason for wavelength-dependent path lengths, causing large chromatic dispersion of the whole system. The effects of wavelength-dependent path lengths also occur in Bragg grating pairs which are often used for dispersive pulse compression. It is worth mentioning here that the mechanisms of these above dispersive components are well understood and they have been widely used in different device applications with many available commercial productions.

In many cases, even an SM fiber with high dispersion can be used as a pulse stretcher effectively. As discussed in chapter 4, when a light pulse propagates in a dispersive material/component the pulse is stretched in time, and that is simply the working principle of a pulse stretcher. We already knew also from chapter 4 about the material dispersion, chromatic dispersion, which can be understood simply as the dependence of refractive index on frequency (or wavelength) ($n(\omega)$ or $n(\lambda)$). The material dispersion, however, is a characteristic parameter for materials, and its value is material dependent. The values of material dispersion parameters, expressed in $D(\omega)$ may not be suitable for many different applications. Meanwhile, the dispersion of a whole component, however, can be significantly different with the material dispersion. By engineering the photonics structure one can manipulate dispersion properties so that the system dispersion can be very different and controllable. A very simple component such as a SM fiber can be used to illustrate a pulse stretcher in our example of a fiber-based CPA system. As explained in chapter 4, when a light beam propagates in a waveguide like optical fiber, the total dispersion of the waveguide is summation of waveguide dispersion and material dispersion $D_{\text{Total}}(\lambda) = D_{\text{Material}}(\lambda) + D_{\text{waveguide}}(\lambda)$. In general, the effects of the dispersion can be characterized in the frequency-dependence of the propagation constant $\beta(\omega)$, which can be expanded in a Taylor series about the frequency ω_0 at which the pulse spectrum is centered:

$$\beta(\omega) = n(\omega)\frac{\omega}{c} = \beta_0 + \beta_1(\omega - \omega_0) + \frac{1}{2}\beta_2(\omega - \omega_0)^2 + \frac{1}{6}\beta_3(\omega - \omega_0)^3 + \cdots, \quad (5.2)$$

where,

$$\beta_m = \left(\frac{\partial^m \beta}{\partial \omega^m}\right)_{\omega=\omega_0}, m = 0, 1, 2 \ldots. \quad (5.3)$$

It should be stressed again that for the fiber-based dispersive components, the most widely used ones in practice are DCFs and CBGFs. Again, these two fibers were briefly mentioned and used in examples of ML fiber lasers in chapter 4.

In this chapter, we will study the problems of modeling and simulation of fiber-based CPA systems operated in the 2 μm wavelength region using a step-index fiber—the SMF-28 fiber for pulse stretcher. Note that SMF-28 is the mostly-used fiber loss in 1.5-μm wavelength band of the optical communication due to its extremely low loss in that region, ~0.2 dB km^{-1}. The use of SMF-28 fiber in the following example is two folds. First, it is convenient for us to make use of the dispersion of waveguides that we have learned earlier in chapter 4. All parameters of the fiber are well-known, and the dispersion of the fiber can be calculated without any problem. Second, we can focus on the modeling and simulation of pulse stretching, and especially on the effects of high-order dispersion in the pulse compressing instead of spending a large effort on calculating dispersion of a different stretcher. It is worth stressing that this example is based on the developments of a real fiber-based CPA system operated in the 2 μm region, in which such a component is not very well developed and and available at the time. The example, therefore, illustrates a flexible way in designing a system utilizing available components.

Let us consider a simple step-index fiber SMF-28 shown in figure 5.4(a). Since this is a simple fiber structure, the calculations of dispersion of the fiber are quite simple, and were presented in subsection 4.2.2 in chapter 4. The calculated waveguide dispersion and total dispersion of the step-index SMF using equation (4.11).

$$D_{\text{Waveguide}} = -\frac{n_{\text{clad}}\Delta n}{\lambda c} \times \frac{Vd^2(bV)}{dV^2}. \tag{5.4}$$

Here $V = 2\pi a\sqrt{n_{\text{core}}^2 - n_{\text{clad}}^2}/\lambda$ is V-number, a is core radius of the waveguide (see more details in chapter 4).

Calculated material and waveguide dispersions of the step-index SMF-28 are shown in figure 5.4(b), and different high orders of dispersion parameters β_m are shown in figures 5.4(c)–(e).

We want to stress again that SMF-28 is one of the best fibers with very low loss (α = 0.2 dB km^{-1}) at the 1.5 μm region in optical communication. In the 2 μm-region, the fiber loss ~7.5 dB km^{-1} is significantly higher than in the 1.5 μm region. This is a trade-off between positive and negative effects that we must consider carefully in the system design. As mentioned earlier, many photonics components including dispersive devices in the 2 μm region were in development stage at the time. Therefore, on one hand there were not many choices for a pulse stretcher in a 2 μm fiber-based CPA system. However, on the other hand the choice of simple fiber for stretcher was based on good considerations, including modeling and simulation. It turned out the choice was very good.

In the following part, we will present the main results of modeling and simulation of the pulse stretcher, and then the pulse propagation in fiber amplifiers of the CPA system. By using a 'virtual' fiber with dispersion that can be changed to arbitrary value as a compressor we can consider the impacts of high order dispersion of the compressor on the final results of a CPA system. We will first explain in simple terms the operation of the system, and MATLAB® programs will be presented after that. Again, the examples of fiber amplifiers and ML fiber lasers have been presented in chapters 3 and 4, respectively, and will not be repeated here. The modeling and simulation methods and the MATLAB® programs for these systems can be applied for the problems of Tm-doped fibers with some modifications. A simple description of the model and rate equation of Tm-doped systems with pumping scheme and lasing/amplification processes will be provided for basic understanding. After understanding these problems in earlier chapters and the model of Tm-doped fibers, readers can develop MATLAB® programs to simulate the Tm-doped fiber lasers and amplifiers.

Let us now consider the problem of simulating a fiber-based CPA system, in which the source is an ultrashort laser with pulse of 500 fs with center wavelength of 1.92 μm as shown in figure 5.5(a). The laser pulse is then stretched in time by propagating over the long distance of an SM fiber, in the example 1 km of SMF-28 fiber, figure 5.5(b). First, let us assume the fiber loss is 3 dB km^{-1} or ~50% power loss so that we can easily check the accuracy of simulation results. Figure 5.5(c) illustrates the relative spectrum of the pulse propagating in 1 km of fiber. It is clear that from figure 5.5(c) the peak power of the pulse after propagating in 1 km of fiber is ~50% of the input, agreeing with the assumption of loss 3 dB km^{-1} (50% power loss). This is a simple but useful way, especially for beginners who want to test the accuracy of the modeling results. Notice that the plot of normalized spectrum in figure 5.6(e) shows the pulse spectrum is preserved when the light

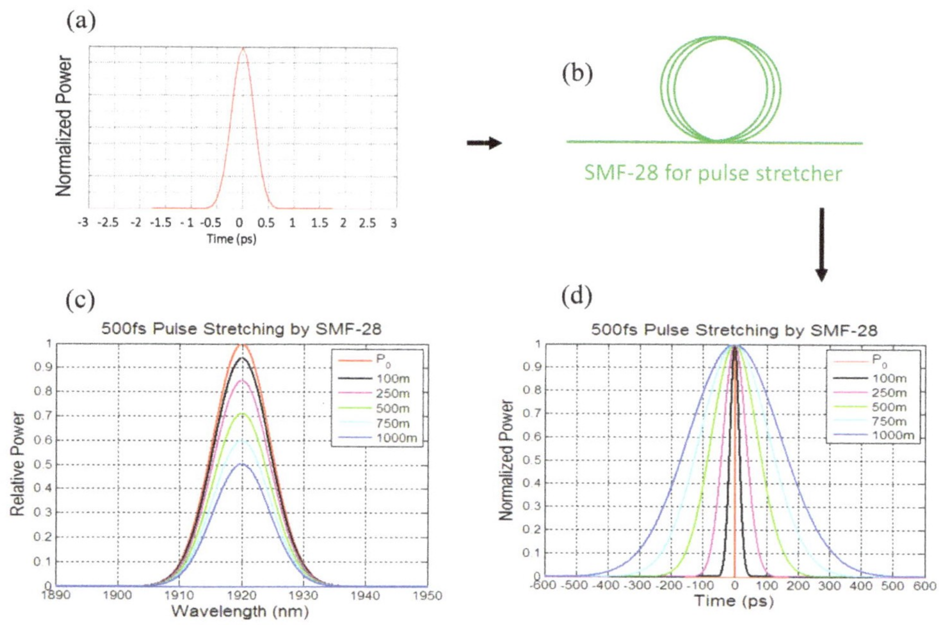

Figure 5.5. Stretching of a pulse in fiber SMF-28. (a) Input pulse of 500 fs with center wavelength 1.92 μm, (b) Step-index SMF-28 as a pulse stretcher. Input pulse propagates in the SMF-28 and is stretched in time due to fiber dispersion. (c) Relative-power pulse spectrum and pulse shapes, respectively, at different propagating distances. (d) Normalized-power pulse spectrum and pulse shapes, respectively, at different propagating distances.

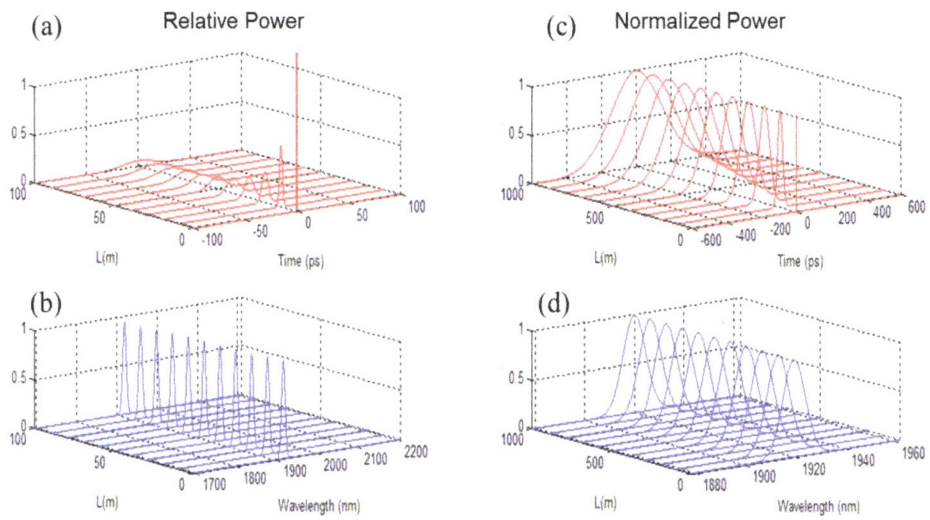

Figure 5.6. Stretching of a pulse of 500 fs with center wavelength 1.92 μm in SM fiber with assumed loss of 3 dB km^{-1}. (a) and (b) Relative-power pulse spectrum and pulse shapes, respectively, propagating in first 100 m fiber. (c) and (d) Normalized-power pulse spectrum and pulse shapes, respectively, at different propagating distances, up to 1000 m.

propagates in different lengths of the fiber. That is understandable, given the source in the stretcher is very weak so that all nonlinear effects that cause distortion are negligible. Meanwhile, the plots in the time domain, figures 5.5(d) and (f) show relative and normalized-power pulse in different lengths of the fiber. Significantly, the pulse is stretched into much longer duration from 500 fs of input pulse (red). Figure 5.5(d) shows the peak power of the stretched pulse decreases significantly in just 100 m. In the simulation the fiber loss is assumed 3 dB km^{-1}, therefore, the reduction of peak power is due to stretching effect. Figure 5.5(f) shows more clearly the stretched effects where the input pulse with 500 fs duration (red curve) is stretched to 500 ps (blue curve) in 1 km of fiber. The results in figure 5.5 illustrate the concept of stretching an ultrashort pulse to a much longer time duration pulse with much lower peak power. Due to very low peak power, the stretched pulse can be amplified to much higher energy, without risk of increasing the peak power (intensity) so much that it may cause damage to the system. Ultrashort pulse with high peak intensity can also cause distortions to the laser beam itself due to nonlinear effects that make the laser beam less useful or even completely useless in many applications. Meanwhile, during the stretching the pulse spectrum must be preserved as close as possible to the original one. Any distortion of spectrum in earlier stages such as stretcher and pre-amps must be avoided since the distortions will be amplified significantly in the high power amplifiers of the CPA system.

As can be seen from the above example, the length of stretching fiber could be very long, therefore, in order to keep B-integral values small the power of input pulse and its peak power should be low or very low. The example shows by using 1 km of SMF-28 a pulse of 500 fs and center wavelength at 1.92 mm can be stretched to 500 ps or a factor of 1000 times longer with the peak power proportionally reduced from the original unstretched pulse. In principle, this stretched pulse can be amplified by a similar factor more than an unstretched pulse to reach the same level of peak power that is considered not useful.

In order to make the results more clear, relative-power pulses propagating in the first 100 m are shown in time domain (figure 5.6(a)) and in wavelength domain (figure 5.6(b)). We can see the reduction of peak power due to stretching is significant in just 100 m in figure 5.6(a), given the propagation loss in simulation is assumed to be 0.3 dB for 100 m (or 3 dB km^{-1}). The reduction due to propagation loss 0.3 dB for the 100 m propagation is reflected very well in the plot of relative power spectrum in figure 5.6(b). Figures 5.6(c) and (d) are the normalized power plots in time- and wavelength domains for stretching in 1000 m.

Figures 5.5 and 5.6 are just for illustration of how a dispersive component/device, in this case a long SM fiber can be used as pulse stretcher. A 500-fs laser pulse with center-wavelength at 1.92 μm can be stretched into a pulse of 500 ps duration just by propagating in 1 km of SMF-28 fiber.

Let us now dig deeper into the architecture of a real fiber-based CPA system. First, as mentioned earlier, the propagation loss of SMF-28 at 1.92 mm is ~7.5 dB km^{-1}, which is quite high compared to the simulations in figures 5.5 and 5.6. Second, typically the repetition rate of ML fiber lasers is in the range of tens of MHz, while high energy short lasers for many applications are in the range of 100 kHz or even in kHz ranges. Therefore, usually AOMs or pulse pickers are used to reduce the pulse repetition rate from tens of MHz to kHz range [17, 18]. In general, this step could be very high extinction or very high loss. It is important to stress here that the ultrashort ML lasers for the high quality source of a CPA system typically are in the low power range, typically from few to tens of milliwatt (mW). Therefore, in order to compensate the losses due to stretcher and pulse pickers, a few amplifiers (pre-Amps) are usually added before and/or after pulse pickers aiming to boost average power of the pulse before it enters stage 3—the high power amplified stage. It is important to note that the high power amplifier plays a very important role in a CPA system since reaching very high energy is one of the main goals of the CPA system. However, it must be designed carefully so that values of the B-integral, which directly relates to nonlinear distortion, must be kept reasonably low for many applications.

In the following, a MATLAB® program is presented to simulate the results in figures 5.5 and 5.6 above.

MATLAB® Program 5.1

The following program simulates the fiber propagation in a long distance of SMF-28 fiber for stretching the pulse. The program will call functions that have been detailed in chapter 4 such as rk4ip.m for fourth order Runge–Kutta in interaction presentation, and Propagation_SMF.m for light propagation in SMF fiber. Note that the loss and dispersion of the fiber in this example is for 1.92 μm, that is, higher than that at 1.55 μm in optical communication.

```
%xxxxxxxxxxxxxxxxxxxxxxxxxxxxxxxxxxxxxxxxxxxxxxxxxxxxxxxxxxxxxxxxxxxxxxxxx
% Strecher.m Program 5.1                                                 x
% Main Program for pulse strecher in CPA design                          x
% Using: RK4IP algorithm (Hunt JLT 2007)                                 x
%       fucntions:                                                       x
%                  Propagation_EDF.m for propagation in EDF              x
%                  Propagation_SMF.m for propagation in SMF              x
%                     Dan Nguyen 06/05/2021                              x
%xxxxxxxxxxxxxxxxxxxxxxxxxxxxxxxxxxxxxxxxxxxxxxxxxxxxxxxxxxxxxxxxxxxxxxxxx
close all
clear all
clc

if ~exist('Stretcher','dir')
    mkdir('Stretcher');

end

%% ---- parameters for the simulation and a part of the pulse ---------

c0 = 3e8;                  % (m/s)
lambdaP = 1920e-9;         % (m)% pump wavelength (3000 , 3500, 6500, 7250 nm)
wP = 2*pi*c0/lambdaP;      % Angular frequency of pump @ 1560nm
TFWHM = 500e-15;           % (s) full width at half maximum of pump pulse
Tmax  = 100e-12;            % (s), half width of time window
Tmin  = -Tmax;             % Tmin < t < Tmax
tol = 1e-3;                % tollerence


global t                   % vector time, globally used (even in functions)

N = 2^13;                  % number of points used to present in time domain
deltat = 2*Tmax/(N-1);     % (s), resolution of time
t = linspace(Tmin,Tmax-deltat,N);


global f                   % frequency, globally used
deltaf = 1/(2*Tmax);       % resolution of frequency
f = [0:deltaf:(N/2-1)*deltaf,-N*deltaf/2:deltaf:-deltaf]; % vertor freq
w = 2*pi*f;                % vector angular freq
wshift = -w + wP;          % plus the freq with its central freq.
                           % the minus sign is to compensate the fact that
                           % defenition of FFt in matlab and in the GNLSE are
                           % in opposited sign
wl = 1e9*2*pi*c0./wshift;%(nm), shift to nm to present
tm = 1e12*t;               % (ps), shift to ps to present
limit_t = [-30 30];    % (ps)
limit_f = [1890 1950];   % (nm)

%% -----------This part constructs the pump @975nm ----------


C = 0;                     % chirp of the pulse
form_pulse = 'gaus'        % 'gaus'
if form_pulse == 'sech'
    T0 = TFWHM/(2*log(1+sqrt(2))); % if pulses are secant-hyperbol (soliton)
    U0 = sech(t/T0);       %
else
    T0 = TFWHM/(2*sqrt(log(2)));      % if pulses are gaussian (bellshape)
    U0 = exp(-0.5*(t/T0).^2*(1+i*C));
end
```

```
% ------------- from mode-locked laser before EDF ------------------------

R = 42e6;                  % repetation rate 42 MHz
Tau = 1/R;
P = 1*1e-4                 % output of seed laser is about 1mW
P0 = 0.93*P*Tau/TFWHM      % (W), peak power of pulses

Uint = sqrt(P0)*U0;        % (W^0.5), Input pulse E-field, in time domain
Uinf = fft(Uint);         % Input spectrum, freq domain

Iint = abs(Uint).^2;
Iinf = abs(Uinf).^2;

%---------------

t_shock = 1/wP;            % (s), paramet of self steepening & shock formation
fR = 0.18;                 % factor of contribution of Raman effect
t1 = 12.2e-15;             % (s), paramet in approx raman response function
t2 = 32e-15;               % (s), paramet in approx raman response function,
                           % 1/t2 is the bandwidth of Raman gain

tr = t - t(1);             % shift time vector to the root. starting point is 0

% Raman response function approx by Blow and Wood, 1989.
hR = (t1^2+t2^2)/t1/t2^2*exp(-tr/t2).*sin(tr/t1);

hR = hR./trapz(tr,hR); % (??? why divided, not affect to results ???)
hR_f = fft(hR);            % raman response function in the freq domain
tt = 50e-15;               % effect of gain in dispers value [Agrawal OL1991]

% --- This part for streching fiber ---------------------------------------
-
%      Assume fiber is SMF-28
%------------------------------------------------------------------------

    n2 = 2.6e-20;             % nonlinear refractive index
   MFD = 10e-6 ;              % MFD of the fiber, assume 10um (check)
  L_st = 20;                  % Length in m of stretching fiber
AdB_st = 0.1;                   % (dB/km) loss of strech fiber
alfa_st= AdB_st/4343;         % (1/m) loss
Aeff = pi*(MFD/2)^2;          % effective core area
gamma_st = 2*pi*n2/lambdaP/Aeff  % 0.0013/((2)^2);

                           % [beta1, beta2, beta3...] in ps^k/m
beta_st = [0 -6.2 2.6 -10.40].*[0 1e-2 1e-4 1e-7];

                           % change unit of beta to (s^k/m)
 for ii = 1:length(beta_st)

     beta_st(ii) = (1e-12)^(ii)*beta_st(ii);

 end
 AdB_cp = 0.1;                % dB/km loss of silica fiber
 alfa_cp=AdB_cp/4343;        % loss in unit 1/m
 MFD = 10e-6 ;               % MFD of the fiber, assume 10um (check)
 Aeff = pi*(MFD/2)^2;        % effective core area
 n2 = 2.6e-20;               % nonlinear refractive index
 gamma_cp = 2*pi*n2/lambdaP/Aeff  % 0.0013/((2)^2);
 L_cp = 20;              % (50m) length of EDF (m)
```

```
                              % [beta1, beta2, beta3...] in ps^k/m
beta_cp = [0 6.2 -2.6 10.40].*[0 1e-2 1e-4 1e-7];

                              % change unit of beta to (s^k/m)
 for ii = 1:length(beta_cp)
      beta_cp(ii) = (1e-12)^(ii)*beta_cp(ii);
 end


 %-------Progagation in strecher fiber -------------------
 %       signal 1920nm propagates throu SMF w/o gain

Ltotal = L_st + L_cp;
Nz = 20;               % number of display
dL = Ltotal/Nz;


  Lz = 0;
indc = 0;


for nz=1:Nz+1
    indc = indc + 1;
  Lz(nz) =(nz-1)*dL;
  if Lz < L_st
     Uout1 = Propagation_SMF(Uint,tol,Lz(nz),alfa_st,...
                          beta_st,gamma_st,t_shock,fR,hR_f);%
     Uint = Uout1;
   else
     Lz2(nz) = (nz-1)*dL - L_st;
     Uout1 = Propagation_SMF(Uint,tol,Lz2(nz),alfa_cp,...
                             beta_cp,gamma_cp,t_shock,fR,hR_f);%
   end


Uouf1 = fft(Uout1);
Iout1(:,indc)  = abs(Uout1./max(Uout1)).^2;              % intensity of
output
Iouf1(:,indc)  = abs(Uouf1).^2;              % intensity in freq domain


%-----This part presents the DATA ----------------------------------


  figure(1);                        % present time signal
  subplot(2, 1, [1])
    plot3(tm,Lz(nz)*ones(size(tm)), Iout1(:,nz),'b');

   hold on;
   axis([min(tm) max(tm) 0 Ltotal 0 1])
   xlabel('Time (ps)');
   ylabel('L(m)');
   zlabel('Relative power');
   grid on;

   subplot(2, 1, [2])
   plot3(wl,Lz(nz)*ones(size(tm)), Iouf1(:,indc)/max(Iinf),'b');
   hold on;
   axis([min(wl) max(wl) 0 Ltotal 0 1])
   xlabel('Wavelength (nm)');
```

```
    ylabel('L(m)');
    zlabel('Relative power');
    grid on;

cd Stretcher
    saveas(gcf, strcat('pulse', num2str(indc), '.jpg'));
    cd ../

   drawnow
end
```

Figure 5.7 below illustrates the simulation of the first two stages in which the real loss 7.5 dB km^{-1} of SMF-28 at 1.92 μm is used, and a pre-amp #1 with 25 dB gain is added into the system. The simulation results of normalized power and relative power are plotted in time domain. Note that we use the same colors for the pulses and the fibers in which the pulses propagate. For example, in the diagram of figure 5.7 the fiber SMF-28 (stretcher) and the fiber amplifier (Pre-Amp #1) are in blue and green, respectively. Correspondingly, in plots of simulation results the propagating pulses in SMF-28 and in fiber amplifier are colored blue and green, respectively (with the input pulse in red). The pink crosses are for couplings, usually fusion splices coupling, between different fiber parts, with coupling losses of 0.5 dB assumed in simulations.

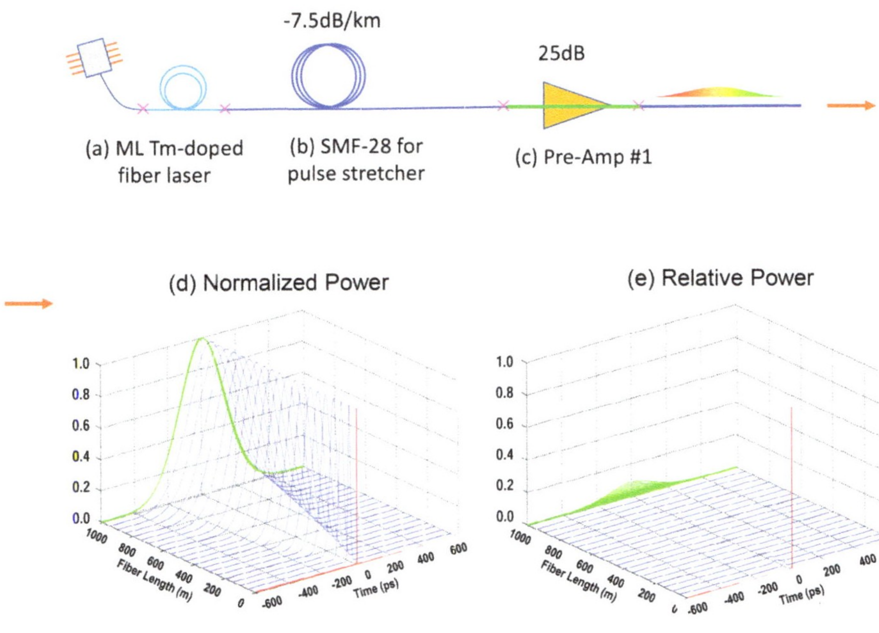

Figure 5.7. Schematic of the first stages of a CPA system with a pre-amp added after the stretcher. (a) ML Tm-doped fiber laser, 500-fs duration, and center-wavelength 1.92 μm, (b) a stretcher by SMF-28 of 1000 m with 7.5 dB propagation loss, (c) pre-amp #1 with 25 dB gain, (d) normalized-power spectrum and (e) relative-power pulse propagation in these fibers. The colors of light pulses propagating in (d) and (e) are blue and green, respectively corresponding to the same color components (b) and (c). The pink crosses stand for the couplings between different fiber parts.

In plots (d) and (e), the input pulse (red) propagates in the stretcher (blue) and fiber amplifier (green). The normalized power spectrum plot shows the pulse is stretched in 1 km of SMF-28 (blue) and then in 10 m of fiber amplifier (green). Meanwhile, in the plot of relative power, the stretched pulse has peak intensity that drops significantly in just 20 m. The peak intensity is very low in the entire 1000 m of SMF-28 (blue). The pulse gets amplified in pre-amp #1 and its energy and peak intensity increase significantly in 10 m of the fiber amplifier. The pulse duration changes very little in 10 m of the fiber amplifier (green). The results allow us to see behavior of the pulse in each part of the system with all losses and gains accounted for. Although the above simulations are quite simple, the results are useful for optimization, especially when the total budget of pump power is under some constraint conditions.

As mentioned earlier, ML fiber lasers usually have repetition rate in the range of tens of MHz which is too high for many applications. Therefore, in practice one or a few pulse pickers are used to reduce the pulse repetition rate from tens of MHz to kHz range, resulting in high loss. Figure 5.8 shows as an example, in which a pulse picker with 15 dB loss that reduces repetition from 60 MHz to 100 kHz is added. To compensate the loss due to the pulse picker, and more importantly to boot the power of the pulse before it enters the high power amplifier.

Note that because the energy and peak power/intensity are low in the pre-amplifiers, especially in pre-amp #1, the fiber length of pre-amp #1 can be flexible as *B*-integral values are typically low in the stages. For example, let us assume input

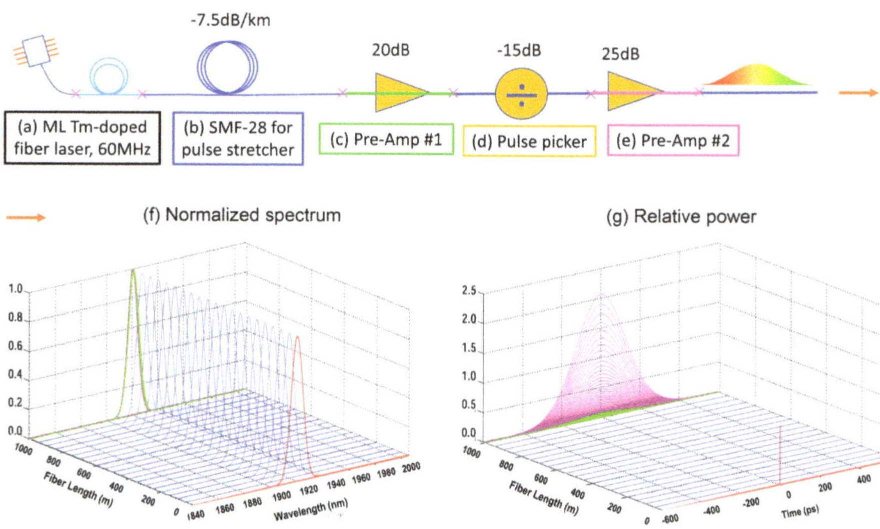

Figure 5.8. Schematic of the first stages of a CPA system similar to figure 5.7 but adding a pulse picker (d) and a new fiber amplifier pre-amp #2 (e). (f) Normalized power spectrum and (g) relative power pulse propagation in time domain. The input pulse is in red and the colors of light pulses propagating in (b) stretcher (blue), (c) pre-amp #1 (green) and (e) pre-amp #2 (pink) correspond to the same color components (b), (c) and (e). The pink crosses stand for the couplings between different fiber parts.

power 1 mW (0 dBm) and each coupling loss in diagram of figure 5.8 is 0.5 dB, then we can easily estimate the output power from pre-amp #1 is about 11.5 dBm or slightly higher than 10 mW. That's why in the example, the fiber length pre-amp #1 uses is quite long, $L = 10$ m. The situation becomes more delicate in pre-amp #2, where the average power reaches 20 dBm or 100 mW. At this average power level, peak power of femtosecond pulses becomes quite high, but the stretched pulse now becomes 500 ps, although, the design of pre-amp #2 must be careful to account for the effects of fiber length and peak power into the B-integral. A fiber with higher dopant concentration should be used for this fiber amplifier. The simulation in the example shown in figure 5.8 is with fiber length of pre-amp #2 isbeing 1 m versus 10 m of pre-amp #1.

The results can be obtained from running the following MATLAB® program.

MATLAB® Program 5.2

```
%xxxxxxxxxxxxxxxxxxxxxxxxxxxxxxxxxxxxxxxxxxxxxxxxxxxxxxxxxxxxxxxxxxxxxxxxxxx
% SA01 Program                                                             x
% Program for pulse Streching, Amplification in CPA System                x
%      Stretching fiber   : Lst, beta_st, alfa_st                          x
%      Amplification fiber: Lam, beta_am, g_am, alfa_am                    x
% Using: RK4IP algorithm (Hunt JLT 2007)                                   x
%        fucntions:                                                        x
%                   Propagation_EDF.m for propagation in EDF               x
%                   Propagation_SMF.m for propagation in SMF               x
%                                                                          x
%xxxxxxxxxxxxxxxxxxxxxxxxxxxxxxxxxxxxxxxxxxxxxxxxxxxxxxxxxxxxxxxxxxxxxxxxxxx
close all
clc

if ~exist('SA01','dir')
    mkdir('SA01');
end

%if ~exist('SA01b','dir')
%    mkdir('SA01b');
%end

%if ~exist('SA01c','dir')
%    mkdir('SA01c');
%end
%% ---- parameters for the simulation and a part of the pulse ---------

c0 = 3e8;                    % (m/s)
lambdaP = 1920e-9;           % (m)% pump wavelength (3000 , 3500, 6500, 7250 nm)
wP = 2*pi*c0/lambdaP;        % Angular frequency of pump @ 1560nm
TFWHM = 500e-15;             % (s) full width at half maximum of pump pulse
Tmax  = 600e-12;              % (s), half width of time window
Tmin  = -Tmax;               % Tmin < t < Tmax
tol = 1e-3;                  % tollerence

global t                     % vector time, globally used (even in functions)

N = 2^14;                    % number of points used to present in time domain
deltat = 2*Tmax/(N-1);       % (s), resolution of time
t = linspace(Tmin,Tmax-deltat,N);
```

```
global f                   % frequency, globally used
deltaf = 1/(2*Tmax);       % resolution of frequency
f = [0:deltaf:(N/2-1)*deltaf,-N*deltaf/2:deltaf:-deltaf]; % vertor freq
w = 2*pi*f;                % vector angular freq
wshift = -w + wP;          % plus the freq with its central freq.
                           % the minus sign is to compensate the fact that
                           % defenition of FFt in matlab and in the GNLSE are
                           % in opposited sign
wl = 1e9*2*pi*c0./wshift;%(nm), shift to nm to present
tm = 1e12*t;               % (ps), shift to ps to present
limit_t = [-50 50];   % (ps)
limit_f = [1890 1950];  % (nm)

%% ----------This part constructs the pump @975nm ----------

C = 0;                     % chirp of the pulse
form_pulse = 'gaus'        % 'gaus'
if form_pulse == 'sech'
    T0 = TFWHM/(2*log(1+sqrt(2))); % if pulses are secant-hyperbol
(soliton)
    U0 = sech(t/T0);      %
else
    T0 = TFWHM/(2*sqrt(log(2)));     % if pulses are gaussian (bellshape)
    U0 = exp(-0.5*(t/T0).^2*(1+i*C));
end

% ------------ from mode-locked laser before EDF ------------------------

R = 42e6;                  % repetation rate 42 MHz
Tau = 1/R;
Pin = 1*1e-3                 % output of seed laser is about 1mW

P0 = 0.93*Pin*Tau/TFWHM   % (W), peak power of pulses

Uint = sqrt(P0)*U0;       % (W^0.5), Input pulse E-field, in time domain
Uinf = fft(Uint);         % Input spectrum, freq domain

Iint = abs(Uint).^2;
Iinf = abs(Uinf).^2;
check0  = sum(Iint);         % Input power = integral of (Uint)^2

%---------------

t_shock = 1/wP;           % (s), paramet of self steepening & shock formation
fR = 0.18;                % factor of contribution of Raman effect
t1 = 12.2e-15;            % (s), paramet in approx raman response function
t2 = 32e-15;              % (s), paramet in approx raman response function,
                          % 1/t2 is the bandwidth of Raman gain

tr = t - t(1);           % shift time vector to the root. starting point is 0

% Raman response function approx by Blow and Wood, 1989.
hR = (t1^2+t2^2)/t1/t2^2*exp(-tr/t2).*sin(tr/t1);

hR = hR./trapz(tr,hR); % (??? why divided, not affect to results ???)
hR_f = fft(hR);            % raman response function in the freq domain
tt = 50e-15;               % effect of gain in dispers value [Agrawal OL1991]
```

```
% --- This part for streching fiber ----------------------------------------
---
%      Assume fiber is SMF-28
%---------------------------------------------------------------------------
Lst = 1000;                 % Length of Strecher  (m)
La1 = 10;                   % Length of Amplifier (m)

    n2 = 2.6e-20;            % nonlinear refractive index
  MFD1 = 10e-6 ;             % MFD of the fiber, assume 10um (check)
AdB_st = 7.5;                % (dB/km) loss of strech fiber
alfa_st= AdB_st/4343;        % (1/m) loss
Aeff1 = pi*(MFD1/2)^2;       % effective core area
gamma_st = 2*pi*n2/lambdaP/Aeff1  % 0.0013/((2)^2);

                          % [beta1, beta2, beta3...] in ps^k/m
beta_st = [0 -6.2 2.6 -10.40].*[0 1e-2 1e-4 1e-7];

                          % change unit of beta to (s^k/m)
 for ii = 1:length(beta_st)
      beta_st(ii) = (1e-12)^(ii)*beta_st(ii);
 end

%% -- this part is for gain fiber
AdB_am1 = 7.5;               % dB/km loss of silica fiber
alpha_am1=AdB_am1/4343;      % loss in unit 1/m
MFD2a = 10e-6 ;            % MFD of the fiber, assume 10um (check)
Aeff2a = pi*(MFD2a/2)^2;   % effective core area

 n2a1 = 2.6e-20;             % nonlinear refractive index
 gamma_am = 2*pi*n2a1/lambdaP/Aeff2a  % 0.0013/((2)^2);

                          % [beta1, beta2, beta3...] in ps^k/m
 beta_am1 = [0 -6.2 2.6 -10.40].*[0 1e-2 1e-4 1e-7];


                          % change unit of beta to (s^k/m)
  for ii = 1:length(beta_am1)
       beta_am1(ii) = (1e-12)^(ii)*beta_am1(ii);
  end


  %-------Progagation in strecher fiber -------------------
  %       signal 1920nm propagates throu SMF w/o gain


        CL1 = 0.5;
  TotalLoss = (AdB_st*Lst/1000 + AdB_am1*La1/1000)
  TotalGain = TotalLoss + CL1 + 17
        GdB = TotalGain/La1       % gain dB/m
          g1 = GdB/4.343          % linear gain in per m
  Esat = 20e-9 ;                  % 20nJ saturation energy


 Nst = 20;
 dLst = Lst/Nst;
 Nam = 40;
 dLam = La1/Nam;
 Nz   = Nst + Nam;
 Ltot = Lst + La1;
```

```
 Lz = 0;

figure(1)
plot3(tm,0*ones(size(tm)),Iint/max(Iint),'r','LineWidth',2);
hold on;

for nz=1:Nz+1
   if nz < Nst+1                  % stretcher
        Lz(nz) = nz*dLst;
        Uout1 = Propagation_SMF(Uint,tol,Lz(nz),alfa_st,...
                          beta_st,gamma_st,t_shock,fR,hR_f);%
    Uouf1 = fft(Uout1);
    Iout1(:,nz)  = abs(Uout1).^2/max(Iint);      % intensity of output
    Inormt1(:,nz) = abs(Uout1./max(Uout1)).^2;    % intensity of output
    Iouf1(:,nz)   = abs(Uouf1).^2/max(Iinf);     % intensity in freq domain
    Inormf1(:,nz)= abs(Uouf1./max(Uouf1)).^2;     % intensity in freq domain

     plot3(tm,Lz(nz)*ones(size(tm)),Iout1(:,nz),'b','LineWidth',0.5);
    % plot3(tm,Lz(nz)*ones(size(tm)),Inormt1(:,nz),'b','LineWidth',0.5);
    hold on;
      if nz==Nst
        check1 = sum(abs(Uout1).^2);
        P1 = check1/check0*Pin
      end
   else nz<=Nst+Nam+1            % amplification


  L2(nz) = (nz-Nst-1)*dLam;
  Lz(nz) = Lst + L2(nz);
  Uin2  = sqrt(1)*Uout1;
  Uout2 = Propagation_EDF(Uout1,tol,L2(nz),g1,Esat,alpha_am1,...
                                   beta_am1,gamma_am,t_shock,fR,hR_f);%
  Uouf2 = fft(Uout2);
  Iout2(:,nz)   = abs(Uout2).^2/max(Iint);
  Inormt2(:,nz) = abs(Uout2./max(Uout2)).^2;
  Iouf2(:,nz)   = abs(Uouf2).^2/max(Iinf);
  Inormf2(:,nz)   = abs(Uouf2./max(Uouf2)).^2;

   plot3(tm,Lz(nz)*ones(size(tm)),Iout2(:,nz),'g','LineWidth',0.5);
  % plot3(tm,Lz(nz)*ones(size(tm)),Inormt2(:,nz),'g','LineWidth',0.5);

  if nz==Nz
      check2 = sum(abs(Uout2).^2);
      P2 = check2/check0*Pin
  end

end

%-----This part presents the DATA -----------------------------------
   axis([min(tm) max(tm) 0 Ltot 0 1])
   xlabel('Time (ps)', 'FontSize',16);
   ylabel('L(m)', 'FontSize',16);
   zlabel('Relative Power', 'FontSize',16);
   %zlabel('Normalized Power', 'FontSize',16);
   grid on;

  cd SA01
   saveas(gcf, strcat('SAC01R','.bmp'));
  cd ../
```

```
   drawnow
end

figure(2);                              % present time signal
    plot(tm,Iint/max(Iint),'LineWidth',2,'Color','r');
    %semilogy(tm,Iint/max(Iint),'LineWidth',2,'Color','k');
    hold on;
    plot(tm,Iout1(:,Nst),'LineWidth',2,'Color','b');
    plot(tm,Iout2(:,Nst+Nam),'LineWidth',2,'Color','g');
    %semilogy(tm,Iout1(:,Nst),'LineWidth',2,'Color','b');
    %semilogy(tm,Iout2(:,Nst+Nam),'LineWidth',2,'Color','g');
    hold off;
    legend('P_0', 'Stretcher', 'Amp I');
    xlabel ('Time (ps)','FontSize',14);
    ylabel ('Relative Power','FontSize',14);
    grid on;
    set(gca,'FontSize',12);
    axis([-600 600 0 Inf]);
    set(gca,'xtick',[-600 -500 -400 -300 -200 -100 0 100 200 300 400 500
600])


     title('(1km SMF) & (10m Amp #1)','FontSize',14)
   %cd SA01b
   %  saveas(gcf, strcat('SA01Rb','.bmp'));
   %cd ../



     figure(3);                              % present time signal
     plot(wl,Iinf/max(Iinf),'LineWidth',2,'Color','r');
     %semilogy(wl,Iinf/max(Iinf),'LineWidth',2,'Color','k');
     hold on;
     plot(wl,Iouf1(:,Nst),'LineWidth',2,'Color','b');
     plot(wl,Iouf2(:,Nst+Nam),'LineWidth',2,'Color','g');
     %semilogy(wl,Iouf1(:,Nst),'LineWidth',2,'Color','b');
     %semilogy(wl,Iouf2(:,Nst+Nam),'LineWidth',2,'Color','g');
     hold off;
     legend('P_0', 'Stretcher', 'Amp #1');
     xlabel ('Wavelength (nm)','FontSize',14);
     ylabel ('Relative Power','FontSize',14);
     grid on;
     set(gca,'FontSize',12);
     axis([1890 1950 0 Inf]);
     set(gca,'xtick',[1890 1900 1910 1920 1930 1940 1950])
     title('(1km SMF) & (10m Amp #1)','FontSize',14)
    drawnow

 ==========
```

5.2.3 Third stage: cladding-pump Tm-doped fiber amplifiers in 2 μm-region

Let us now discuss about the third stage, the stage of the high power amplifier of the CPA system. In this example we want to obtain a pulse with average power of 10 W and *B*-integral value is about 2–3. We consider a cladding-pumped Tm-doped fiber amplifier with fiber core doped with 4 wt% Tm concentration, core diameter $d = 26$ μm, $NA = 0.04$, which is close to the limit in practice. As described earlier, the fiber is pumped by high power diodes with wavelength $\lambda_p = 790$ nm, and amplified signal is in the 2 μm region.

The modeling method for cladding-pumped fiber amplifier has been described in chapter 3 for Yb-doped and Er–Yb co-doped fiber amplifiers. Readers can apply the modeling method and MATLAB® programs for these systems to Tm-doped fibers by replacing the rate equations for Tm-doped system. As stated earlier, for high power Tm-doped fiber amplifiers in the 2 μm region, a cladding-pumped scheme with pumping wavelength ~790–800 nm has been widely used due to available high power pump diodes and very high efficiency (so-called 200% quantum efficiency). The pumping and amplification processes are presented in figure 5.9, and the rate equations can be derived from the energy diagram as

$$\begin{aligned}
\frac{dN_1}{dt} &= W_{01}N_0 - \left(W_{10} + \frac{1}{\tau_1}\right)N_1 + \beta_{21}\frac{N_2}{\tau_2} + \beta_{31}\frac{N_3}{\tau_3} + k_{31,\,01}N_3N_0 \\
&\quad - k_{10,\,13}N_1^2 + k_{10,\,12}N_1N_0 - k_{10,\,12}N_1^2, \\
\frac{dN_2}{dt} &= \beta_{32}\frac{N_3}{\tau_3} - \frac{N_2}{\tau_2} - k_{31,\,01}N_3N_0 + k_{10,\,12}N_1^2, \\
\frac{dN_3}{dt} &= R_{03}N_1 - \frac{N_3}{\tau_3} - k_{31,\,01}N_3N_0 - k_{10,\,13}N_1^2, \\
N_{\mathrm{Total}} &= N_0 + N_1 + N_2 + N_3.
\end{aligned} \tag{5.5}$$

Where τ_1 is the lifetime of the upper level $^3H_4$, and $1/\tau_3$, $1/\tau_2$ account for the multi-phonon nonradiative relaxations $^3F_4 \rightarrow (^3H_5, {}^3H_4, {}^3H_6)$, and $^3H_5 \rightarrow (^3H_4, {}^3H_6)$, respectively. R_{03} is pumping rate; W_{01}, and W_{10} are signal absorption and stimulated

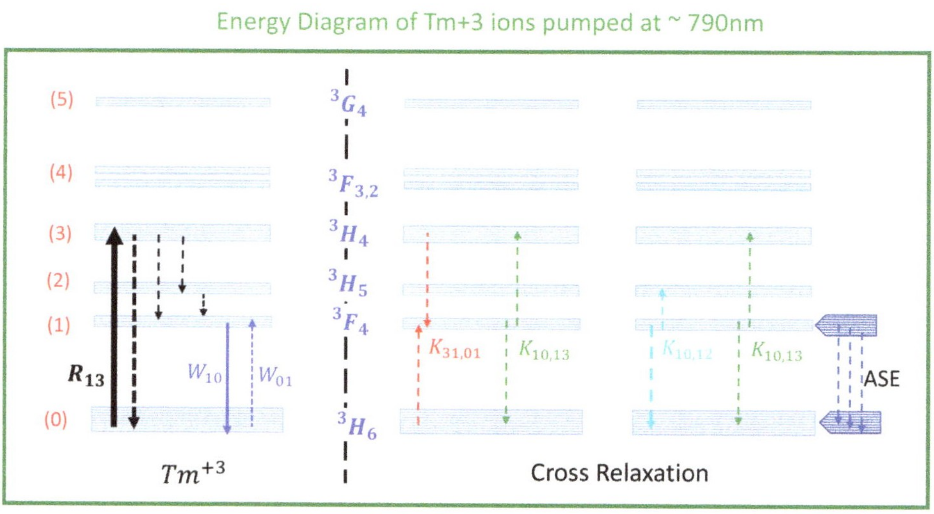

Figure 5.9. Energy diagram of Tm-ion pumped at 790 nm. Left: pump photons with pumping rate R_{03}(solid back arrow) excite Tm-ions from ground state (0) or $^3H_6$ state to excited state (3) or $^3H_4$-state. From there the ions relax to lower states with branch ratios β_{3i}, $i = 0,1,2$. The stimulated emission and absorption rates W_{10}, W_{01} of signal in the 2 μm region are presented by blue down and up arrows, respectively. Right: cross relaxation processes in the system.

emission transition rates of Tm-ions, respectively (see more details in chapter 3). The cross sections, branch ratios β_{ij} and cross-relaxation coefficients $k_{ij,mn}$ are material dependent and can be found in [1–6]. Moreover, the coefficients $k_{ij,mn}$ also depend on the Tm concentrations.

We show in figure 5.10 an example with simulation results of a high power Tm-doped fiber amplifier with core diameter 26 μm and NA = 0.04, 4 wt% Tm concentration. Note that the simulation is for cw regime and average power P_{ave} is calculated and plotted in the figure. The average power is then used to calculate the peak power P_0 for different pulse durations τ (200, 400 and 800 ps) and repetition rate f =100 KHz (T = 1/f) as

$$P_0 = P_{av}\left(\frac{T}{\tau}\right). \tag{5.6}$$

Then, the values of B-integral can be calculated along the fiber together with average power (blue curve) of the signal as shown in the upper plot in figure 5.10. The pumping residue (green) and absorption (pink) powers are shown in the lower plot of

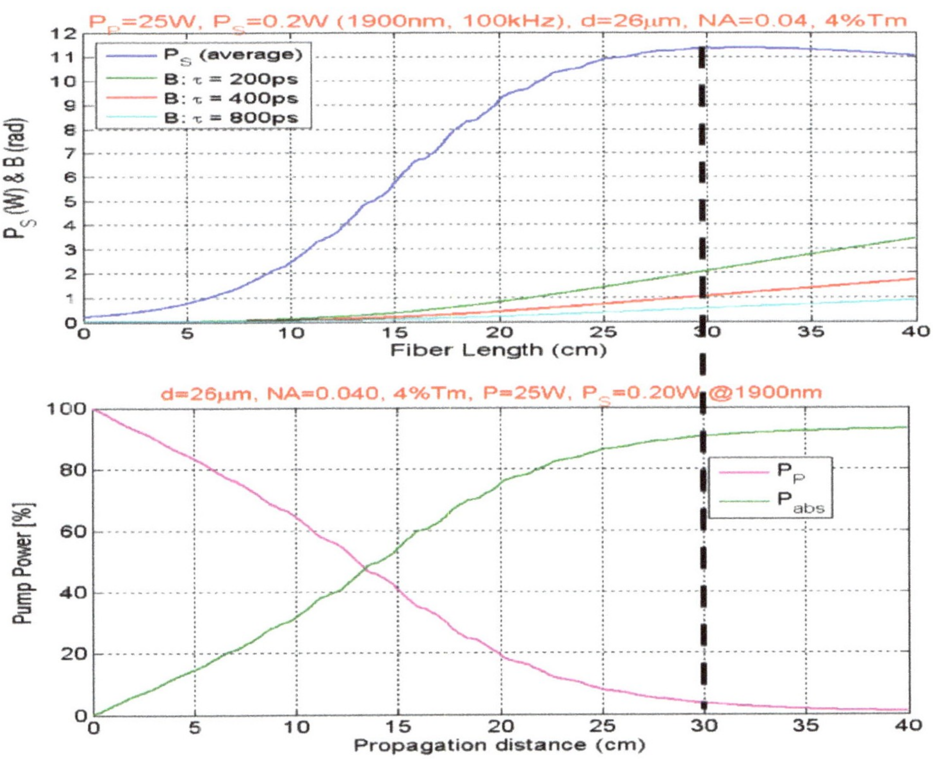

Figure 5.10. Upper: average power of signal (blue) and B-integral calculated with different pulse duration. Lower: pump residue (green) and absorption (pink) along the fiber. The vertical dashed line show the values of signal average power, B-integral values and pump power at the same distance, where signal has peak power.

figure 5.10. The simulation results in figure 5.10 are simple but very useful for optimization of the CPA system. The results allow us to optimize the fiber length (with available pump power) with optimum signal power and B-integral values. It should be pointed out that for high power fiber amplifiers, besides the key parameters of the fiber design such as dopant concentration and core size, the pumping scheme is very important. Furthermore, it is important to have input power relatively high so that in a short length of fiber the signal can reach the desired power, so that the B-integral values can be kept acceptably low.

5.2.4 Effects of high order dispersion in pulse compression

Let us now consider the problem of the pulse compressor in a CPA system. As stated earlier, the energy of the pulse after a high power amplifier is typically very high, therefore, the peak power of the compressed pulse could be very high or even extremely high. Guiding such pulses with high peak power in optical fibers whose core sizes are in micron-scale is a very bad idea: extremely high intensity could easily cause damage to the fiber, or at least nonlinear effects could make the distorted pulse useless. Therefore, it is clear that optical fiber is not suitable for a pulse compressor in a CPA system. Instead, we can use other dispersive devices such as grating pairs that can be used to manipulate the dispersion of the device based on controlling the wavelength-dependent path lengths. Simulation of the grating-based compressor can be done using ray-tracing methods, conveniently with commercial software such as ZEMAX, and therefore is beyond the scope of this chapter. However, a dispersive fiber can still be used to study the working principle of a general compressor. Moreover, it helps to understand the effects of high orders of dispersion in the compressor to the final results of a CPA system.

In the following example, we will study the effects of high-order dispersions of the whole CPA system to the final pulse. In particular, we assume the dispersion of the stretcher and other fiber parts such as fiber amplifiers are known and presented in figures 5.4–5.8. We will show how the differences between the compressor dispersion and those of other parts that would impact the final results of the CPA system. The results would be very useful since if the dispersion of the fiber stretcher and fiber amplifiers are known, the simulation results can be used to analyze and optimize the compressor. To do that, let us now consider an example in which a pulse compressor is a 'virtual' fiber, whose different orders of dispersion β_n can be numerically changed to any values without changing other physical parameters. Of course this fiber is *unreal* because the fiber dispersion is determined by the fiber geometry and refractive index profile. However, this fiber can be used for the purpose of considering the effects of compressor dispersion. Furthermore, the nonlinear effects are completely ignored in this fiber or the nonlinear term is assumed negligibly small and is turned off in simulations. Because the compressed pulses in a CPA system typically have extremely high peak power, therefore, the simulations are just for illustrating the dispersion effect in the final results. Note that the final results in a real device would always include both nonlinear and dispersion effects. However, the use of 'virtual' fiber is still worth considering as it demonstrates the impacts of different terms of high order dispersion to the final pulse after compressing. Furthermore, in practice the results can be used for calibrating the system with very low power when the nonlinear effects are negligible. Once we understand the dispersion impacts we can optimize the compressor or stretcher and other parts.

As described earlier in chapter 4, in general, dispersion of a dispersive device can be characterized by the frequency dependence of the propagation constant $\beta(\omega)$, which can be expanded in a Taylor series (5.3) and (5.4). The dispersion terms are included in the generalized nonlinear Schrödinger equation (GNLSE) to describe the pulse propagation.

$$\frac{\partial A(z, T)}{\partial z} = (\hat{D} + \hat{N})A(z, T), \tag{5.7}$$

whereas the dispersion $\hat{D}$ and nonlinear $\hat{N}$ operators are given by

$$\hat{D} = -\frac{\alpha}{2} - \left(\sum_{n\geqslant 2}\beta_n \frac{i^{n-1}}{n!}\frac{\partial^n}{\partial T^n}\right), \tag{5.8}$$

$$\hat{N} = i\gamma\left(1 + \frac{1}{\omega_0}\frac{\partial}{\partial T}\right) \times \{(1 - f_R)\}\, |A|^2 + f_R \int_0^\infty h_R(\tau)|A(z, T - \tau)|^2 d\tau. \tag{5.9}$$

The detailed derivation of GNLSE and the meaning of the terms in equations (5.7–5.9) can be found in chapter 4. Note that, in practice only the first few terms of series (5.7), up to the fourth order dispersion (FOD) should be taken into account in simulations since values of higher order terms are negligibly small.

In the example that a compressor is simulated as a virtual fiber as stated earlier utilizing the MATLAB® program for simulating pulse propagation in SMF, we can change the values of high order dispersion in the virtual fiber to investigate the effects of compressor dispersion to the final results. The assumption of virtual fiber can be easily satisfied in the simulations by turning off the nonlinear term in equation (5.9) with setting nonlinear coefficient $\gamma = 0$. The nonlinear effect can also be considered negligibly small if very weak input pulse is used in solving the GNLSE equation (5.7). Either way, the nonlinear effects can be considered numerically negligible. Applying the assumptions, we are now able to run a MATLAB® program (see below) to simulate a simplified CPA system compromised of two main parts: stretcher (may include fiber amplifiers) and compressor, which is a virtual fiber with the high orders dispersion parameters changeable.

In figure 5.11 we show the simulation results with effect of the difference of the third order of dispersion (TOD) between the stretcher and the compressor. The simulations are performed with the assumption that the second order of dispersion (SOD) and FOD terms of the compressor and the stretcher of the simplified CPA system are completely compensated, or $L_{Comp}\beta_{2,4}(Comp) = -\, L_{Stret}\beta_{2,4}(Stret)$.

In figure 5.11 the stretcher (including fiber amplifiers) is represented by a green fiber (a) and a general compressor (b), which is assumed as a virtual fiber described above.

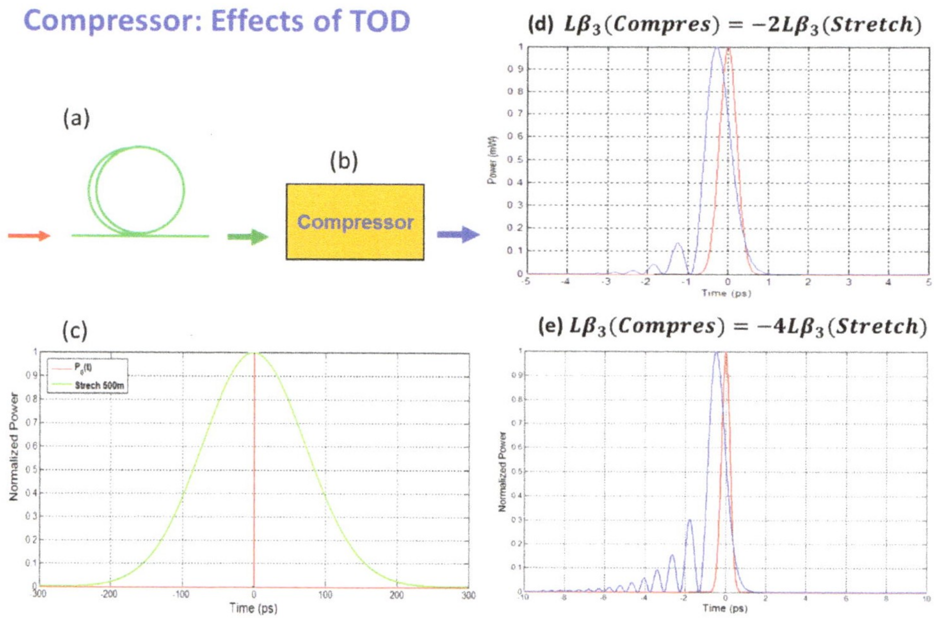

Figure 5.11. (a) SMF fiber (green) represents a stretcher (including fiber amplifier), (b) a simplified compressor is simulated as a 'virtual' dispersive fiber, in which dispersion can be changed numerically without modifying physical parameters of the fiber. (c) Original pulse (red) and stretched pulse (green) after propagating 500 m of SMF fiber. (d) TOD effect with $L\beta_3(Comp) = -2L\beta_3(Stret)$. Original pulse (red) and compressed pulse (blue). (e) TOD effect with $L\beta_3(Comp) = -4L\beta_3(Stret)$ (see more details in the text).

The input pulse is 500 fs with center-wavelength at 1.92 μm (red) and stretched pulse in 500 m of SMF-28 (green) as plotted in (c). The product of the length and SOD and FOD of the compressor $L\beta_{2,4}(Comp)$ and the stretcher $L\beta_{2,4}(Stret)$ are assumed to be equal but opposite sign $L_{Comp}\beta_2(Comp) = -L_{Stret}\beta_2(Stret)$. However, for simplifying notations $L\beta_2(Comp) = -L\beta_2(Stret)$ are used in the text and figures. The simulations are performed with the assumption that only TOD of the two parts are different. By doing that we can see the effect of differences in TOD of the compressor and stretcher $L\beta_3(Comp) = -2L\beta_3(Stret)$ and $L\beta_3(Comp) = -4L\beta_3(Stret)$ as in figures (d) and (e), respectively. These figures show clearly the TOD effect in the final result (blue) as compared with the original pulse (red). In general, the larger the TOD difference between compressor and stretcher, the higher the number of waves in the final pulse.

Similarly, figure 5.12 shows simulation results with only effect of FOD. In this case, both SOD and TOD terms of stretcher and compressor are completely compensated, and set to be equal values but opposite signs. Only the difference of FOD between compressor and stretcher is considered. For example, figure 5.12(d) and (e) compare the final (blue) and original (red) pulses in which $L\beta_4(Comp) = -100L\beta_4(Stret)$ and $L\beta_4(Comp) = -1000L\beta_4(Stret)$, respectively.

By turning on and off different terms of dispersion series in the GNLSE or changing the values of $L\beta_n(Comp)$ we can easily see how such changes or difference of dispersion impact the final results. Given that we already knew the values of the

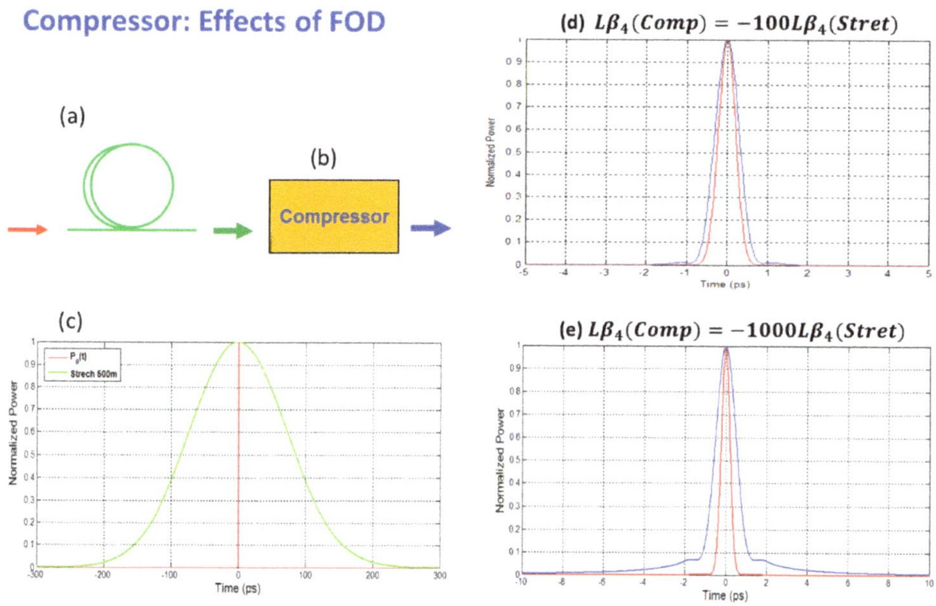

Figure 5.12. (a) SMF fiber (green) represents a stretcher (including fiber amplifier), (b) a simplified compressor is simulated as a 'virtual' dispersive fiber, in which dispersion can be changed numerically without modifying physical parameters of the fiber. (c) Original pulse (red) and stretched pulse (green) after propagating 500 m of SMF fiber. (d) FOD effect with $L\beta_4(Comp) = -100L\beta_3(Stret)$. Original pulse (red) and compressed pulse (blue). (e) FOD effect with $L\beta_4(Comp) = -1000L\beta_4(Stret)$.

stretcher and amplifier stages, the results are useful to understand the behavior of the pulse under different situations of compressor dispersion. In practice, this helps to analyze and optimize the system although the real compressor is much more complicated than the virtual fiber used in this example.

So far we have only considered the effect of each dispersion term in the GNLSE. The results give us a general sense about the behavior of the final pulse that correlates with each order of dispersion. Now let us consider a more complicated situation in which both effects of TOD and FOD occur at once. Figure 5.13(a) and (b) are simulated results of the simplified system described above with fixed TOD difference $L\beta_3(Comp) = -4L\beta_3(Stret)$ in both cases, but different FOD, (a) $L\beta_4(Comp) = -L\beta_4(Stret)$ and (b) $L\beta_4(Comp) = -100L\beta_4(Stret)$. Note that in the program the value of TOD β_3 is about three orders of magnitude larger than FOD β_4. Therefore, it should not be a surprise to see that the results in both figures are very similar even if the difference of FOD increases by two orders of magnitude. In other words, in this case the FOD effect is still much weaker than TOD, and that is a typical situation but does not always happen in practice. Meanwhile, the results in figure 5.13(c) and (d) show when the FOD effect dominates the distortion of the final pulse. In both cases, the FOD difference is three orders of magnitude $L\beta_4(Comp) = -1000L\beta_4(Stret)$ which become comparable with the TOD magnitude. If the TOD is absent then the FOD causes

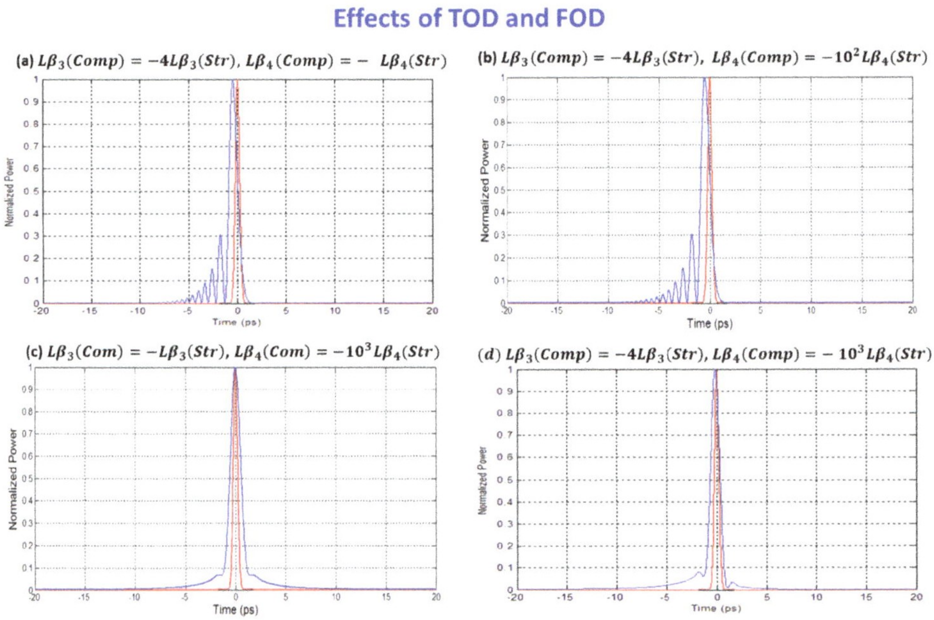

Figure 5.13. Simulated results with TOD difference $L\beta_3(Comp) = -4L\beta_3(Stret)$, but different FOD (a) with $L\beta_4(Comp) = -L\beta_4(Stret)$ and (b) $L\beta_4(Comp) = -100L\beta_4(Stret)$. Simulated results with FOD difference $L\beta_4(Comp) = -1000L\beta_4(Stret)$, but different TOD (c) $L\beta_3(Comp) = -L\beta_3(Stret)$ and (d) $L\beta_3(Comp) = -4L\beta_3(Stret)$.

symmetry distortion in the final pulse (c). Meanwhile, the effects of both TOD and FOD are shown in (d).

MATLAB® Program 5.3

In the following MATLAB® program the reader can make any change of different orders of the dispersion of the stretcher and/or compressor.

```
%xxxxxxxxxxxxxxxxxxxxxxxxxxxxxxxxxxxxxxxxxxxxxxxxxxxxxxxxxxxxxxxxxxxxxxx
% SnC02 Program                                                         x
% Program for pulse Streching & compressing in CPA System               x
%     Stretching fiber    : Lst, beta_st, alfa_st                       x
%     Amplification fiber: Lam, beta_am, g_am, alfa_am                  x
% Using: RK4IP algorithm (Hunt JLT 2007)                                x
%        fucntions:                                                     x
%                    Propagation_EDF.m for propagation in EDF           x
%                    Propagation_SMF.m for propagation in SMF           x
%                                                                       x
%xxxxxxxxxxxxxxxxxxxxxxxxxxxxxxxxxxxxxxxxxxxxxxxxxxxxxxxxxxxxxxxxxxxxxxx
close all
clear all
clc

if ~exist('SnC02tf','dir')
    mkdir('SnC02tf');
end
```

```
%% ---- parameters for the simulation and a part of the pulse ---------

c0 = 3e8;                   % (m/s)
lambdaP = 1920e-9;          % (m)% pump wavelength (3000 , 3500, 6500, 7250 nm)
wP = 2*pi*c0/lambdaP;       % Angular frequency of pump @ 1560nm
TFWHM = 500e-15;            % (s) full width at half maximum of pump pulse
Tmax  = 60e-12;             % (s), half width of time window
Tmin  = -Tmax;              % Tmin < t < Tmax
tol = 1e-3;                 % tollerence

global t                    % vector time, globally used (even in functions)

N = 2^13;                   % number of points used to present in time domain
deltat = 2*Tmax/(N-1);      % (s), resolution of time
t = linspace(Tmin,Tmax-deltat,N);


global f                    % frequency, globally used
deltaf = 1/(2*Tmax);        % resolution of frequency
f = [0:deltaf:(N/2-1)*deltaf,-N*deltaf/2:deltaf:-deltaf]; % vertor freq
w = 2*pi*f;                 % vector angular freq
wshift = -w + wP;           % plus the freq with its central freq.
                            % the minus sign is to compensate the fact that
                            % defenition of FFt in matlab and in the GNLSE are
                            % in opposited sign
wl = 1e9*2*pi*c0./wshift;%(nm), shift to nm to present
tm = 1e12*t;                % (ps), shift to ps to present

limit_t = [-50 50];    % (ps)
limit_f = [1890 1950];  % (nm)

%% -----------This part constructs the pump @975nm ----------


C = 0;                      % chirp of the pulse
form_pulse = 'gaus'         % 'gaus'
if form_pulse == 'sech'
    T0 = TFWHM/(2*log(1+sqrt(2))); % if pulses are secant-hyperbol
(soliton)
    U0 = sech(t/T0);        %
else
    T0 = TFWHM/(2*sqrt(log(2)));     % if pulses are gaussian (bellshape)
    U0 = exp(-0.5*(t/T0).^2*(1+i*C));
end

% ------------- from mode-locked laser before EDF ------------------------


R = 42e6;                   % repetation rate 42 MHz
Tau = 1/R;
P = 1*1e-4                  % output of seed laser is about 1mW
P0 = 0.93*P*Tau/TFWHM       % (W), peak power of pulses

Uint = sqrt(P0)*U0;         % (W^0.5), Input pulse E-field, in time domain
Uinf = fft(Uint);          % Input spectrum, freq domain


Iint = abs(Uint).^2;
Iinf = abs(Uinf).^2;

%---------------
```

```
t_shock = 1/wP;          % (s), paramet of self-steepening & shock formation
fR = 0.18;               % factor of contribution of Raman effect
t1 = 12.2e-15;           % (s), paramet in approx raman response function
t2 = 32e-15;             % (s), paramet in approx raman response function,
                         % 1/t2 is the bandwidth of Raman gain
tr = t - t(1);           % shift time vector to the root. starting point is 0

% Raman response function approx by Blow and Wood, 1989.
hR = (t1^2+t2^2)/t1/t2^2*exp(-tr/t2).*sin(tr/t1);

hR = hR./trapz(tr,hR); % (??? why divided, not affect to results ???)
hR_f = fft(hR);          % raman response function in the freq domain
tt = 50e-15;             % effect of gain in dispers value [Agrawal OL1991]

% --- This part for streching fiber ----------------------------------
%     Assume fiber is SMF-28
%-------------------------------------------------------------------
    n2 = 2.6e-20;            % nonlinear refractive index
   MFD = 10e-6 ;             % MFD of the fiber, assume 10um (check)
  Lst = 200;                 % Length in m of stretching fiber
 AdB_st = 0.1;                 % (dB/km) loss of strech fiber
 alfa_st= AdB_st/4343;       % (1/m) loss
 Aeff = pi*(MFD/2)^2;        % effective core area
gamma_st = 2*pi*n2/lambdaP/Aeff  % 0.0013/((2)^2);

                         % [beta1, beta2, beta3...] in ps^k/m
beta_st = [0 -6.2 2.6 -10.40].*[0 1e-2 1e-4 1e-7];

                         % change unit of beta to (s^k/m)
 for ii = 1:length(beta_st)
      beta_st(ii) = (1e-12)^(ii)*beta_st(ii);
 end

AdB_cp = 0.1;            % dB/km loss of silica fiber
alfa_cp=AdB_cp/4343;     % loss in unit 1/m
MFD = 10e-6 ;            % MFD of the fiber, assume 10um (check)
Aeff = pi*(MFD/2)^2;     % effective core area
n2 = 2.6e-20;            % nonlinear refractive index
gamma_cp = 2*pi*n2/lambdaP/Aeff  % 0.0013/((2)^2);
Lcp = 40;                % (50m) length of EDF (m)

                         % [beta1, beta2, beta3...] in ps^k/m
beta_cp = 5*[0 6.2 -5*2.6 1000*10.40].*[0 1e-2 1e-4 1e-7];

                         % change unit of beta to (s^k/m)
 for ii = 1:length(beta_cp)
      beta_cp(ii) = (1e-12)^(ii)*beta_cp(ii);
 end

%-------Progagation in strecher fiber -------------------
%       signal 1920nm propagates throu SMF w/o gain

Nz1 = 20;
dL1 = Lst/Nz1;
Nz2 = 4;
dL2 = Lcp/Nz2;
Ltotal = Lst + Lcp;
Nz  = Nz1 + Nz2;
```

```
  Lz = 0;
 indc = 0;
 figure(1)
 plot3(tm,0*ones(size(tm)),Iint/max(Iint),'b','LineWidth',0.5);
 hold on;

 for nz=1:Nz+1
    if nz < Nz1+1
         Lz(nz) = nz*dL1;
         Uout1 = Propagation_SMF(Uint,tol,Lz(nz),alfa_st,...
                          beta_st,gamma_st,t_shock,fR,hR_f);%
     Uouf1 = fft(Uout1);
     Iout1(:,nz)  = abs(Uout1).^2/max(Iint);      % intensity of output
     Inorm1(:,nz) = abs(Uout1./max(Uout1)).^2;   % intensity of output
     Iouf1(:,nz)  = abs(Uouf1).^2/max(Iinf);      % intensity in freq domain
     Inormf1(:,nz)= abs(Uouf1./max(Uouf1)).^2;     % intensity in freq domain

      plot3(tm,Lz(nz)*ones(size(tm)),Iout1(:,nz),'b','LineWidth',0.5);
     % plot3(tm,Lz(nz)*ones(size(tm)),Inorm1(:,nz),'b');

    hold on;

   else
     L2(nz) = (nz-Nz1-1)*dL2;
     Lz(nz) = Lst + L2(nz);
     Uout2 = Propagation_SMF(Uout1,tol,L2(nz),alfa_cp,...
                           beta_cp,gamma_cp,t_shock,fR,hR_f);%
    Uouf2 = fft(Uout2);
    Iout2(:,nz)  = abs(Uout2).^2/max(Iint);    % intensity of output
    Inorm2(:,nz) = abs(Uout2./max(Uout2)).^2;  % intensity of output
    Iouf2(:,nz)  = abs(Uouf2).^2/max(Iinf);    % intensity in freq domain
    Inormf2(:,nz)  = abs(Uouf2./max(Uouf2)).^2;    % intensity in freq
domain
    plot3(tm,Lz(nz)*ones(size(tm)),Iout2(:,nz),'r','LineWidth',0.5);
    % plot3(tm,Lz(nz)*ones(size(tm)),Inorm2(:,nz),'r');
  end
%-----This part presents the DATA ----------------------------------
    axis([min(tm) max(tm) 0 Ltotal 0 1])
    xlabel('Time (ps)', 'FontSize',14);
    ylabel('L(m)', 'FontSize',14);
    zlabel('Relative Power', 'FontSize',14);
    %zlabel('Normalized Power', 'FontSize',14);
    grid on;
    cd SnC02tf
    saveas(gcf, strcat('SnC02', num2str(indc), '.jpg'));
    cd ../
   drawnow
end
==============
```

Exercise 5.1

Modify the MATLAB® program above and obtain simulation results shown in figure 5.14 for a simplified system comprised of two fibers, one for stretching and the other fiber for compressing.

Exercise 5.2

Modify the above MATLAB® programs to simulate an all-fiber system comprised of a stretcher, a fiber amplifier and a fiber compressor. The results are

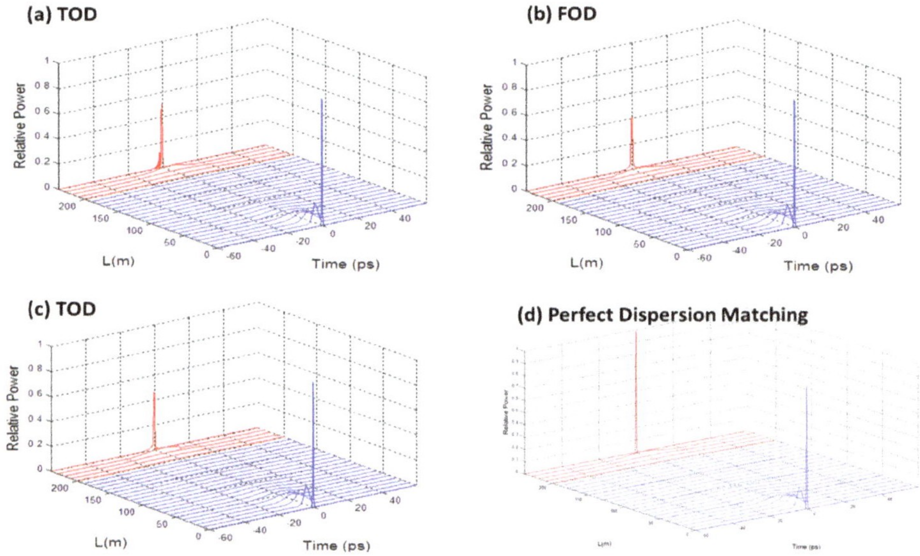

Figure 5.14. Simulated results of a simplified system comprised of two fibers: 200 m of SMF-28 fiber for stretching, and 50 m of a virtual fiber for compressing. (a) TOD effect, (b) FOD effect, (c) both TOD and FOD effects, and (d) perfect matching dispersion between stretching and compressor.

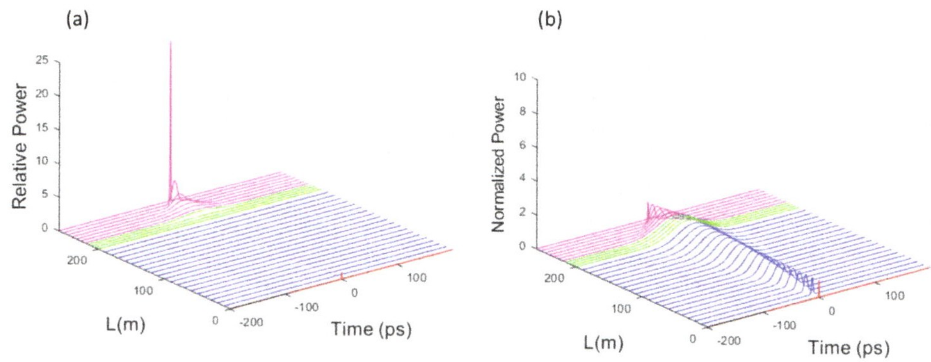

Figure 5.15. Simulated results of a simplified system comprised of 200 m of SMF-28 fiber for stretching, and 10 m of a gain fiber (green) and 100 m of virtual fiber (pink) with perfect matching dispersion.

plotted in figure 5.15 with input (red), stretched pulse (blue), amplified (green) and compressor (pink).

Exercise 5.3

When considering the dispersion effects in the compressor using the concept of virtual fiber, it is important to turn off the nonlinear term in program or using very weak power input pulse. Otherwise the nonlinear in the compressing fiber would generate nonlinear distortions as shown below. Modify the program and reproduce

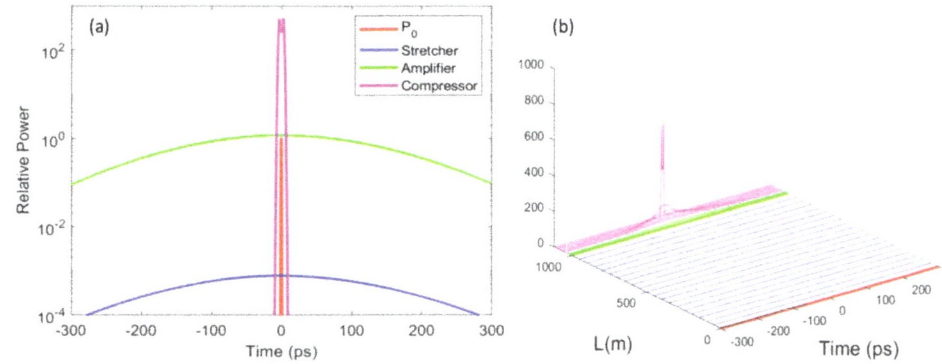

Figure 5.16. Simulated results of a simplified CPA system but the nonlinear effects in the compressor are not turned off or the input pulse is not very weak.

the simulation results for a fiber-based CPA system including stretcher (1 km SMF-28, blue), amplifier (10 m, green) and a compressor (pink), which is 100 m of a fiber of matching dispersion but do not turn off the nonlinear term in the GNLSE (figure 5.16).

References

[1] Strickland D and Mourou G 1985 Compression of amplified chirped optical pulses *Opt. Comm.* **56** 219–21

[2] Maine P, Strickland D, Bado P, Pessot M and Mourou G 1988 Generation of ultrahigh peak power pulses by chirped pulse amplification *IEEE J. Quantum Electron.* **24** 398–403

[3] Strickland D 1988 Development of an ultra-bright laser and an application to multi-photon ionization *PhD Dissertation* University of Rochester

[4] Fermann M E, Galvanauskas A and Sucha G 2003 *Ultrafast Lasers: Technology and Applications* (New York: Marcel Dekker)

[5] Keppler S *et al* 2016 The generation of amplified spontaneous emission in high-power CPA laser systems *Laser Photonics Rev.* **10** 264

[6] Paschotta R 2017 Chirped-pulse amplification *RP Photonics Encyclopedia*

[7] Taverner D, Galvanauskas A, Harter D, Richardson D J and Dong L 1996 Generation of high energy pulses using a large mode area erbium doped fibre amplifier *Technological Digest of Conference on Lasers and Electro-Optics*, paper CFD5. OSA Publishing

[8] Galvanauskas A, Sartania Z and Bischoff M 2001 Millijoule femtosecond all-fiber system Presented at Conference on Lasers and Electro-Optics, Baltimore, Paper CMA1.

[9] Price J H V 2003 The development of high power, pulsed fiber laser systems and their applications *PhD Thesis* Optoelectronics Research Centre: University of Southampton

[10] He F, Price J H V, Vu K T, Malinowski A, Sahu J K and Richardson D J 2006 Optimisation of cascaded Yb fiber amplifier chains using numerical modeling, *Opt. Express* **14** 12846–58

[11] Miniscalo W 1993 Optical and electronics properties of rare earth ions in glasses *Rare Earth Doped Fiber Lasers and Amplifiers* ed M Digonnet (Optical Engineering, vol 37) (New York: Marcel Dekker)

[12] Florez A *et al* 1999 Application of standard and modified Judd-Ofelt theories to thulium doped glass *J. Non-Crys. Solids* **247** 215

[13] Jackson S D and King T A 1999 Theoretical modeling of Tm-Doped silica fiber lasers *J. Lightwave Technol.* **17** 948
[14] Jackson S D and King T A 1998 Efficient gain-switched operation of a Tm-doped silica fiber laser *IEEE J. Quantum Electron.* **34** 779
[15] Wu J, Yao Z, Zong J and Jiang S 2007 Highly efficient high-power thulium-doped germanate glass fiber laser *Opt. Lett.* **32** 638–40
[16] Wu J, Jiang S, Luo T, Geng J, Peyghambarian N and Barnes N P 2006 Efficient thulium-doped 2-μm germanate fiber laser *IEEE Photon. Technol. Lett.* **18** 334–6
[17] Peng X *et al* 2013 High efficiency, monolithic fiber chirped pulse amplification system for high energy femtosecond pulse generation *Opt. Express* **21** 25440–51
[18] Peng X *et al* 2014 Monolithic fiber chirped pulse amplification system for millijoule femtosecond pulse generation at 1.55 μm *Opt. Express* **22** 2459–64
[19] Knight J C *et al* 1998 Large mode area photonic crystal fibre *Electron. Lett.* **34** 1347
[20] Dong L *et al* 2009 All-glass large-core leakage channel fibers *IEEE Sel. Top. Quantum Electron.* **15** 47
[21] Dong L *et al* 2007 Leakage channel optical fibers with large effective area *J. Opt. Soc. Am.* B **24** 1689
[22] Lefrancois S *et al* 2013 High-energy similariton fiber laser using chirally-coupled core fiber *Opt. Lett.* **38** 43
[23] Brodericka N G R *et al* 1999 Large mode area fibers for high power applications *Opt. Fiber Technol.* **5** 185–96
[24] Stutzk F *et al* 2014 Designing advanced very-large-mode-area fibers for power scaling of fiber-laser systems *Optica* **1** 233–42

www.ingramcontent.com/pod-product-compliance
Ingram Content Group UK Ltd.
Pitfield, Milton Keynes, MK11 3LW, UK
UKHW051857100926
8556IPUK00005B/666